Azure Architecture Explained

A comprehensive guide to building effective cloud solutions

David Rendón

Brett Hargreaves

BIRMINGHAM—MUMBAI

Azure Architecture Explained

Group Product Manager: Preet Ahuja
Publishing Product Manager: Suwarna Rajput
Senior Editor: Sayali Pingale
Technical Editor: Nithik Cheruvakodan
Copy Editor: Safis Editing
Project Manager: Sean Lobo
Proofreader: Safis Editing
Indexer: Sejal Dsilva
Production Designer: Ponraj Dhandapani
Senior Marketing Coordinator: Linda Pearlson
Marketing Coordinator: Rohan Dobhal

First published: September 2023

Production reference: 1230823

Published by Packt Publishing Ltd.
Grosvenor House
11 St Paul's Square
Birmingham
B3 1RB, UK

ISBN 978-1-83763-481-1

www.packtpub.com

I would like to express my deepest gratitude to my beloved family, whose unwavering support and unconditional love have made this journey possible. To my remarkable son; you are the shining star that fills my world with joy and purpose. Your laughter and innocent curiosity inspire me to be the best version of myself. And to my extraordinary wife; you are the anchor that keeps our family grounded, and the light that guides us through every storm. This book is a testament to the immeasurable impact you have had on my life, and I dedicate its words to each and every one of you.

– David Rendón

To my amazing wife and our three wonderful children. Their understanding of my working extra hours to write books makes the process so much easier – even if they still don't know what I actually do for a living!

I'd also like to thank my incredible colleagues at IR77 Ltd. Thank you for your hard work, dedication, and friendship. I'm so lucky to work with such a talented team.

Finally, this book is for all of you who are passionate about learning and technology. I hope it inspires you to continue exploring the world of technical innovation.

– Brett Hargreaves

Foreword

At Microsoft, we work hard to create tools that enable businesses to build amazing solutions using the latest technologies. Azure in particular is a platform that has changed the world completely by providing the newest and most advanced services available, from AI, mass data storage, and data science to web apps, reporting, and analytics – in fact, just about every technology you can think of (and a few more!).

However, with tools as powerful and flexible as Azure, to get the most out of them, you need to know how to use them properly.

Security is, of course, a key aspect of any solution, and mistakes can easily be made if the services provided aren't configured correctly. Similarly, costs can quickly get out of hand if you don't manage them or make the best use of the right component for the job.

This is why *Azure Architecture Explained* is vital and does an amazing job of explaining some of the fundamental processes and designs to make sure your Azure solutions are secure, performant, resilient, and, of course, cost-effective.

David and Brett will walk you through how to implement security through identity and access management best practices, as well as implementing and using services such as Microsoft Defender and Entra.

You'll learn how to choose and architect the right solutions for your needs, choose the right storage and compute mechanisms, as well as implement networking and governance processes to further secure and control your applications.

The book finishes with a great section on using infrastructure-as-code – using Bicep, Azure Pipelines, and CI/CD practices to deploy your systems quickly and easily in a controlled and repeatable manner.

Azure Architecture Explained has just the right mix of conceptual architecture and practical hands-on explanations to help you quickly and confidently start building state-of-the-art solutions using Microsoft's best practices.

Sarah Kong

Microsoft Technical Trainer at Microsoft | Learning Room Expert | Coach | Blogger | Podcaster | Public Speaker

Contributors

About the authors

David Rendón, Microsoft MVP and Microsoft Certified Trainer, is a highly regarded expert in the Azure cloud platform. With over 15 years of experience as an IT professional, he has been deeply committed to Microsoft technologies, especially Azure, since 2010.

With a proven track record of leading and driving strategic success, David has over seven years of management experience, technical leadership, and collaboration skills.

David delivers private technical training classes worldwide, covering EMEA, South America, and the US, and he is a frequent speaker at renowned IT events such as Microsoft Ignite, Global Azure, and local user group gatherings in the US, Europe, and Latin America. Stay connected with David on LinkedIn at `/daverndn`.

Brett Hargreaves is a principal Azure consultant for Iridium Consulting, who has worked with some of the world's biggest companies, helping them design and build cutting-edge solutions. With a career spanning infrastructure, development, consulting, and architecture, he's been involved in projects covering the entire solution stack using Microsoft technologies.

He loves passing on his knowledge to others through books, blogging, and his online training courses.

About the reviewers

Vaibhav Gujral is a director at Capgemini, where he drives cloud innovation and excellence for BFSI clients across geographies. As a trusted technology advisor, he helps clients shape their cloud transformation solutions by understanding their business drivers, priorities, and challenges. He specializes in cloud strategy and governance, cloud security, cloud architecture, application architecture, microservices architecture, and FinOps and DevOps practices.

He has over 17 years of IT experience and is a 4x Microsoft Azure MVP, a prestigious award given to exceptional technical community leaders who share their expertise and passion for Azure. He runs the Omaha Azure User Group and is a regular speaker at conferences.

I'd like to thank my family and friends who understand the time and commitment it takes to make community contributions such as reviewing books. I would also like to thank Packt for offering me the opportunity to review this book.

Bram van den Klinkenberg has worked in IT for over 20 years. He grew from a helpdesk employee to a Windows server administrator to a change manager, and moved back to the technical side of IT as a DevOps engineer. In the past eight years, his focus has been on Azure and he has specialized in automation, Kubernetes, container technology, and CI/CD.

As a consultant, he has fulfilled roles as a senior cloud engineer, senior platform engineer, senior DevOps engineer, and cloud architect.

I would like to thank Packt and the writers for asking me to review their book. It has been a new and good experience that has also taught me new things and given me new insights.

Kasun Rajapakse, a DevOps engineer at Robeco Nederland, is a cloud enthusiast from Sri Lanka with over 9 years of experience. Specializing in Kubernetes, he shares his expertise through technical blogs at `https://kasunrajapakse.me`. As a speaker, Microsoft Certified Trainer, and MVP, Kasun actively contributes to the community. With certifications from Microsoft, AWS, and Google Cloud, he is dedicated to advancing cloud technologies. In addition to his blogging endeavors, Kasun actively contributes to the technology community through his engagements as a speaker at user groups and conferences.

Table of Contents

3

Using Microsoft Sentinel to Mitigate Lateral Movement Paths 57

Part 2 – Architecting Compute and Network Solutions

4

Understanding Azure Data Solutions 79

8

Understanding Networking in Azure 193

9

Securing Access to Your Applications 251

Part 3 – Making the Most of Infrastructure-as-Code for Azure

10

11

12

Preface

In today's rapidly evolving technological landscape, the community requires comprehensive guidance to fully explore the advanced features and use cases of Azure. This book provides you with a clear path to designing optimal cloud-based solutions in Azure. By delving into the platform's intricacies, you will acquire the knowledge and skills to overcome obstacles and leverage Azure effectively.

The book establishes a strong foundation, covering vital topics such as compute, security, governance, and infrastructure-as-code. Through practical examples and step-by-step instructions, the book empowers you to build custom solutions in Azure, ensuring a hands-on and immersive learning experience.

By the time you reach the final pages of this book, you will have acquired the knowledge and expertise needed to navigate the world of cloud computing with confidence. Operating a cloud computing environment has become indispensable for businesses of all sizes, and Azure is at the forefront of this revolution. Discover strategies, best practices, and the art of leveraging the Microsoft cloud platform for innovation and organizational success. This book equips you with the tools to harness the full potential of Azure and stay ahead in today's competitive digital landscape.

Who this book is for

The book is targeted toward Azure architects who develop cloud-based computing services or focus on deploying and managing applications and services in Microsoft Azure. They are responsible for various IT operations, including budgeting, business continuity, governance, identity, networking, security, and automation. It's for people with experience in operating systems, virtualization, cloud infrastructure, storage structures, and networking and who want to learn how to implement best practices in the Azure cloud.

What this book covers

Chapter 1, Identity Foundations with Azure Active Directory and Microsoft Entra, covers key topics in IAM, including authentication, authorization, collaboration, and the significance of digital identities.

Chapter 2, Managing Access to Resources Using Azure Active Directory, provides an overview of Azure Active Directory and its capabilities for IAM, covering key components such as Azure Active Directory Connect, Azure Active Directory Application Proxy, Conditional Access, and Privileged Identity Management.

Chapter 3, Using Microsoft Sentinel to Mitigate Lateral Movement Paths, explores how Microsoft Sentinel detects and investigates security threats, compromised identities, and malicious actions. It emphasizes the importance of mitigating lateral movement, using Sentinel to prevent attackers from spreading within a network and accessing sensitive information.

Chapter 4, Understanding Azure Data Solutions, explores data storage options in Azure, including considerations for structured, semi-structured, and unstructured data. It covers Azure Storage accounts and SQL options and highlights Cosmos DB as a powerful NoSQL database solution for global solutions.

Chapter 5, Migrating to the Cloud, covers the migration of on-premises workloads to Azure, discussing strategies such as lift and shift, refactor, rearchitect, or rebuild. It explores options for moving compute to Azure, including scale sets and web apps for minimal code changes. Additionally, it addresses migrating SQL databases to Azure, considering questions, the potential issues, and utilizing the DMA tool for analysis and migration.

Chapter 6, End-to-End Observability in Your Cloud and Hybrid Environments, emphasizes the significance of a unified monitoring strategy across various environments, including Azure, on-premises, and other cloud providers.

Chapter 7, Working with Containers in Azure, provides insights into Azure containers, including their usage compared to Azure virtual machines, the features and use cases of Azure Container Instances, and the implementation of Azure container groups. It also explores the features and benefits of Azure Container Registry and the automation capabilities provided by ACR Tasks. Furthermore, it covers Azure Container Apps, its components, and how it enables running microservices on a serverless platform.

Chapter 8, Understanding Networking in Azure, emphasizes implementing controls to prevent unauthorized access and attacks. Designing a secure network is crucial in Azure, and this chapter explores the network security options, tailored to meet organizational security needs.

Chapter 9, Securing Access to Your Applications, emphasizes the importance of considering application architecture to secure access and explores tools such as VNet integration, SQL firewalls, Azure Firewall, Application Gateway, Front Door, Azure Key Vault, and managed identities to achieve this.

Chapter 10, Governance in Azure – Components and Services, addresses how Azure governance is crucial for the effective management of cloud infrastructure, compliance, security, cost optimization, scalability, and consistency. This chapter covers key components such as management groups, policies, blueprints, resource graphs, and cost management, highlighting the need for continuous improvement.

Chapter 11, Building Solutions in Azure Using the Bicep Language, discusses how Azure Bicep offers numerous benefits for organizations using Azure cloud services, simplifying resource provisioning through infrastructure-as-code templates. This enables consistent and repeatable deployments, reduces errors, and facilitates version control.

Chapter 12, Using Azure Pipelines to Build Your Infrastructure in Azure, helps you understand how Azure Pipelines automates software development pipelines, minimizing errors and enabling development teams to concentrate on producing high-quality software. This chapter also covers Azure DevOps setup, repository configuration with Azure Repos, the creation of build and release pipelines, and verifying resource creation in the Azure environment.

Chapter 13, Continuous Integration and Deployment in Azure DevOps, discusses how incorporating CI/CD with Azure Pipelines enhances software delivery with improved quality, speed, and efficiency. This comprehensive platform automates the software delivery process, allowing teams to detect and resolve issues early, resulting in fewer bugs and stable releases.

Chapter 14, Tips from the Field, provides an overview of top best practices for organizations, including Azure governance, monitoring, access management, network security, and container deployment.

To get the most out of this book

Software/hardware covered in the book	Operating system requirements
An Azure subscription	None
Azure PowerShell	Windows, Linux, or macOS

If you are using the digital version of this book, we advise you to type the code yourself or access the code from the book's GitHub repository (a link is available in the next section). Doing so will help you avoid any potential errors related to the copying and pasting of code.

Download the example code files

You can download the example code files for this book from GitHub at `https://github.com/PacktPublishing/Azure-Architecture-Explained`. If there's an update to the code, it will be updated in the GitHub repository.

We also have other code bundles from our rich catalog of books and videos available at `https://github.com/PacktPublishing/`. Check them out!

Conventions used

There are a number of text conventions used throughout this book.

`Code in text`: Indicates code words in text, database table names, folder names, filenames, file extensions, pathnames, dummy URLs, user input, and Twitter handles. Here is an example: When an Azure AD tenant is created, it comes with a default `*.on.microsoft.com` domain. A custom domain name such as `springtoys.com` can be added to the Azure AD tenant to make usernames more familiar to the users.

A block of code is set as follows:

```
{
  "Logging": {
    "LogLevel": {
      "Default": "Information",
      "Microsoft.AspNetCore": "Warning"
    }
```

Bold: Indicates a new term, an important word, or words that you see on screen. For instance, words in menus or dialog boxes appear in **bold**. Here is an example: As the modern IT landscape continues to evolve, so does the importance of effective **identity and access management** (**IAM**) solutions.

> **Tips or important notes**
> Appear like this.

Get in touch

Feedback from our readers is always welcome.

General feedback: If you have questions about any aspect of this book, email us at `customercare@packtpub.com` and mention the book title in the subject of your message.

Errata: Although we have taken every care to ensure the accuracy of our content, mistakes do happen. If you have found a mistake in this book, we would be grateful if you would report this to us. Please visit `www.packtpub.com/support/errata` and fill in the form.

Piracy: If you come across any illegal copies of our works in any form on the internet, we would be grateful if you would provide us with the location address or website name. Please contact us at `copyright@packt.com` with a link to the material.

If you are interested in becoming an author: If there is a topic that you have expertise in and you are interested in either writing or contributing to a book, please visit `authors.packtpub.com`.

Share Your Thoughts

Once you've read *Azure Architecture Explained*, we'd love to hear your thoughts! Scan the QR code below to go straight to the Amazon review page for this book and share your feedback.

`https://packt.link/r/1837634815`

Your review is important to us and the tech community and will help us make sure we're delivering excellent quality content.

Download a free PDF copy of this book

Thanks for purchasing this book!

Do you like to read on the go but are unable to carry your print books everywhere?

Is your eBook purchase not compatible with the device of your choice?

Don't worry, now with every Packt book you get a DRM-free PDF version of that book at no cost.

Read anywhere, any place, on any device. Search, copy, and paste code from your favorite technical books directly into your application.

The perks don't stop there, you can get exclusive access to discounts, newsletters, and great free content in your inbox daily

Follow these simple steps to get the benefits:

1. Scan the QR code or visit the link below

   `https://packt.link/free-ebook/9781837634811`

2. Submit your proof of purchase
3. That's it! We'll send your free PDF and other benefits to your email directly

Part 1 – Effective and Efficient Security Management and Operations in Azure

This section addresses how organizations can confidently leverage the power of the cloud while safeguarding their assets and maintaining a strong security posture.

This part has the following chapters:

- *Chapter 1, Identity Foundations with Azure Active Directory and Microsoft Entra*
- *Chapter 2, Managing Access to Resources Using Azure Active Directory*
- *Chapter 3, Using Microsoft Sentinel to Mitigate Lateral Movement Paths*

1

Identity Foundations with Azure Active Directory and Microsoft Entra

In today's rapidly changing digital landscape, businesses need to embrace cloud technology to remain competitive. Microsoft Azure provides a powerful suite of cloud services, enabling organizations to achieve scalability, agility, and cost-effectiveness. However, adopting Azure can be a daunting task, with a wide range of tools and services to navigate.

This book aims to simplify the process by providing a comprehensive guide to the most essential Azure topics, including managing access to resources, mitigating security threats with Microsoft Sentinel, understanding data solutions, and migrating to the cloud. With a focus on practical applications and real-world scenarios, this book also covers end-to-end observability, working with containers, networking, security principals, governance, building solutions with the Bicep language, and using Azure Pipelines for continuous integration and deployment. The book also includes tips from the field, sharing best practices and common pitfalls to avoid. By the end of this book, readers will have a solid foundation in Azure technologies and be well equipped to implement cloud solutions that drive their organization's success.

As the modern IT landscape continues to evolve, so does the importance of effective **identity and access management** (**IAM**) solutions. Authentication and authorization, engaging and collaborating with employees, partners, and customers, and the significance of digital identities are just a few critical concepts that must be considered by organizations to maintain secure and efficient operations.

Azure **Active Directory** (**AD**), a cloud-based identity management service, is an integral component of Microsoft Entra. Microsoft Entra, a powerful identity-driven security tool, offers a comprehensive perspective on IAM in diverse environments. This chapter will delve into the importance of IAM in contemporary organizations, emphasizing the pivotal role of solutions such as Azure AD and Microsoft Entra in bolstering security measures.

In this chapter, we'll cover the following main topics:

- Protecting users' identities and securing the value chain – the importance of IAM in decentralized organizations
- Authentication and authorization in Azure
- Engaging and collaborating with employees, partners, and customers
- The significance of digital identities in the modern IT landscape
- Securing cloud-based workloads with Microsoft Entra's identity-based access control

Let's get started!

Protecting users' identities and securing the value chain – the importance of IAM in decentralized organizations

Over the last decade, organizations have been decentralizing and outsourcing non-core functions to suppliers, factories, warehouses, transporters, and other stakeholders in the value chain, making it more complex and vulnerable. This is most notable in global manufacturing and retail, where decentralization is crucial to introduce efficiency, lower costs, and decrease supply chain disruption risks.

These companies are pursuing multiple strategies to maximize the value of the various functions across multiple external businesses. Each resource access can grant bridges to several security domains, making it a potential entry point for unauthorized users. This can lead to malicious intent or accidental information access by unknowing users.

As digital transformation continues to change how we interact with businesses and other users, the risk of identity data being exposed in breaches has increased, causing damage to people's social, professional, and financial lives. What are your beliefs about protecting users' identities?

In our opinion, every individual has the right to own and control their identity securely, with elements of their digital identity stored in a way that preserves privacy.

Organizations must have a comprehensive cybersecurity strategy to protect the value chain from security risks. A robust strategy involves a multi-layered approach that includes network segmentation, data encryption, secure access controls, and continuous monitoring to identify potential security breaches.

It's also crucial to implement policies for data access and management across the value chain to control who has access to sensitive information and how it's used. As organizations continue to decentralize and outsource non-core functions to suppliers, it's essential to establish trust between partners and have transparency in data management to ensure data security and privacy.

Therefore, data protection and access control are essential for organizations to maintain the confidentiality, integrity, and availability of their digital assets. IAM is a critical component of modern cybersecurity, encompassing a range of technologies and processes that enable organizations to control user access to applications, systems, and data.

IAM is crucial to maintaining the security of an enterprise's digital assets, including confidential data, applications, and systems. By implementing IAM, organizations can ensure that only authorized individuals can access sensitive information, reducing the risk of data breaches and cyberattacks. IAM also provides an efficient way to manage user accounts, credentials, and permissions, making adding or removing users as necessary easier.

IAM is a crucial technology framework that enables organizations to ensure that their resources are only accessed by authorized individuals. The framework includes two main functions: authentication and authorization. In the next section, we will discuss how IAM solutions can help organizations reduce security risks and protect their sensitive data from unauthorized access and data breaches.

Authentication and authorization in Azure

IAM is a technology framework that helps organizations ensure that the right people have access to the right resources. IAM includes two main functions: **authentication** and **authorization**.

Authentication is the process of verifying the identity of a user. It ensures that a user is who they claim to be before they can access an organization's resources. For example, when you log in to your email account, you must enter your username and password. This form of authentication helps the email provider ensure that you are the legitimate user of the account.

Authorization, conversely, is the process of determining what resources a user is allowed to access after their identity has been verified. For instance, once you have logged in to your email account, the email provider uses authorization to determine what you can do with your account. For example, you may have permission to read emails, compose emails, and send emails, but you may not have permission to delete emails. Authorization helps ensure that users only have access to the resources they are authorized to use.

Another vital component related to the preceding two concepts is **multifactor authentication** (**MFA**). Think of MFA as a security process that requires users to provide two or more credentials to access a system or application. These credentials can include something the user knows (such as a password), something the user has (such as a smart card or mobile phone), or something the user is (such as a fingerprint or facial recognition). By requiring multiple authentication factors, MFA makes it more difficult for unauthorized individuals to access sensitive information or systems, even if they do obtain one of the user's credentials.

For example, a bank may require MFA when a user tries to access their online banking account. After entering their username and password, the user is prompted to enter a unique code generated by a mobile app or sent via text to their phone. This code is a second factor of authentication that proves the user's identity beyond their login credentials. By requiring this extra step, the bank ensures that only the authorized user can access their account, even if someone else has obtained their login information.

With IAM, organizations can streamline their access management processes, reducing the burden on IT staff and improving overall efficiency. Additionally, IAM can help organizations comply with regulatory requirements, such as the **Health Insurance Portability and Accountability Act** (**HIPAA**) or **General Data Protection Regulation** (**GDPR**), by providing auditable access controls and ensuring user access aligns with policy requirements.

Effective IAM solutions help organizations enforce security policies and comply with regulations by ensuring users can access only the resources they need to do their jobs.

IAM solutions also provide audit trails and visibility into user activity, making identifying and mitigating security incidents and compliance violations easier. By implementing robust IAM strategies, organizations can reduce security risks and protect their sensitive data from unauthorized access and data breaches.

Engaging and collaborating with employees, partners, and customers

Collaboration and communication are critical components of a successful organization, and they can be challenging to achieve without the proper infrastructure in place. The IT team of an organization may struggle to provide secure access for external users, leaving employees isolated and limited to email communications, which can lead to inefficiencies in managing marketing campaigns and hinder the exchange of ideas between team members. However, with the proper infrastructure that supports IAM, organizations can improve productivity, reduce costs, and increase work distribution while fostering a culture of teamwork and sharing. Improved visibility and consistency in managing project-related information can help teams track tasks and commitments, respond to external demands, and build better relationships with partners and external contributors.

Organizations need to prioritize collaboration capabilities and invest in the right tools and technologies to realize these benefits. This can include everything from shared workspaces and project management platforms to video conferencing and secure access controls. By providing employees with the tools they need to work together effectively, businesses can create a more dynamic and responsive organization better equipped to compete in a rapidly changing marketplace.

The significance of digital identities in the modern IT landscape

In today's digital age, digital identities are essential for accessing IT-related services. An identity strategy goes beyond just provisioning and adding or removing access but determines how an organization manages accounts, standards for validation, and what a user or service can access.

Reporting on activities that affect identity life cycles is also an essential component of an identity strategy. A well-formed identity infrastructure is based on guidelines, principles, and architectural designs that provide organizations with interoperability and flexibility to adapt to ever-changing business goals and challenges.

An effective identity infrastructure should be based on integration and manageability standards while being user-friendly and secure. In order to simplify the end user experience, the infrastructure should provide easy-to-use and intuitive methods for managing and accessing digital identities. With a well-designed and implemented identity infrastructure, organizations can reduce the risk of unauthorized access to their IT resources and improve their overall security posture. Additionally, a standardized identity infrastructure can facilitate collaboration between organizations and make it easier for users to access resources across multiple organizations.

Also, with the growing trend of organizations seeking to invest in cloud services to achieve modernization, cost control, and new capabilities, IAM capabilities have become the central pillar for cloud-based scenarios. Azure AD has become a comprehensive solution that addresses these requirements for both on-premises and cloud applications. The following section provides insights into common scenarios and demonstrates how Azure AD can help with planning and preparing organizations to use cloud services effectively.

Modernizing your IAM with Microsoft Azure AD

Microsoft's Azure AD is a cloud-based IAM service designed to help organizations manage access to resources across different cloud environments. With Azure AD, organizations can control access to cloud applications, both Microsoft and non-Microsoft, through a single identity management solution. This enables employees to access the tools and information they need from any device, anywhere in the world, with increased security and efficiency.

The following figure highlights the structure of Azure AD.

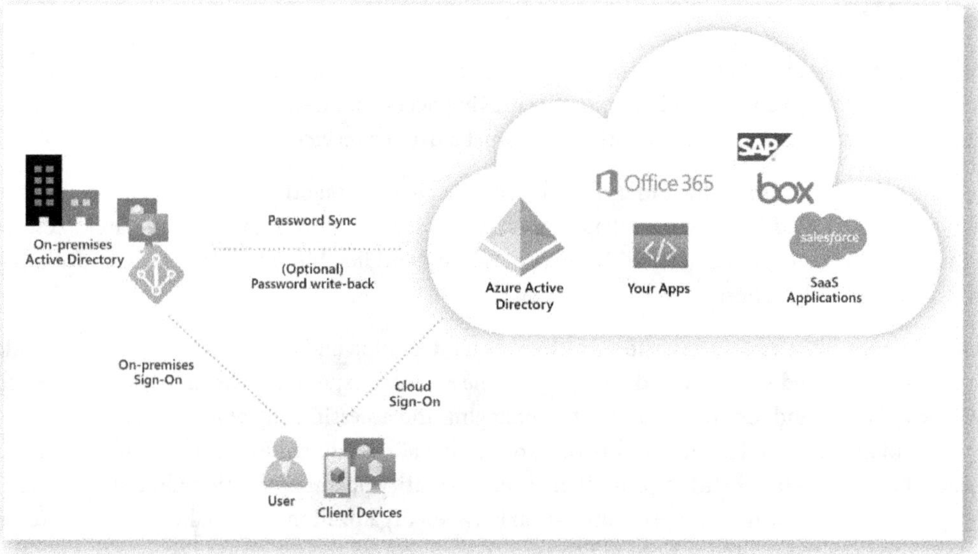
On-premises Active Directory
Password Sync
(Optional) Password write-back
On-premises Sign-On
Cloud Sign-On
User
Client Devices
Azure Active Directory
Office 365
Your Apps
SAP
box
salesforce
SaaS Applications

Figure 1.1 – Azure AD

Azure AD provides several benefits for organizations looking to modernize their IT infrastructure. It offers seamless integration with other Azure services and enables IT administrators to manage user identities and security policies and access resources from a central location. Additionally, it provides MFA and Conditional Access policies to help protect against identity-based attacks.

Organizations can also use Azure AD to manage access to third-party applications, including **Software as a Service** (**SaaS**) applications, such as Salesforce, Box, and Dropbox, providing a consistent and secure user experience across different cloud environments.

However, IAM tasks can significantly burden IT departments, taking up valuable time that could be spent on higher-value work. A crucial piece of an IAM solution is its life cycle management capabilities.

Life cycle management

Provisioning new users can be tedious, requiring administration and configuration across multiple systems. Users may have difficulty obtaining the necessary access to perform their jobs, causing delays and inefficiencies.

For example, the IT team of SpringToys, an online retail organization, may have to access and configure multiple identity utilities and repositories to onboard a new user for online services, making the process even more complicated. With an ad hoc manual method, achieving stringent levels of control and compliance with necessary regulatory standards can be challenging. Each time an employee needs to access an IT service, IT staff must manually handle the request and perform administrative tasks to enable access, creating inefficiencies and delays that impact productivity. By implementing a robust IAM solution, organizations can reduce the burden on IT staff, streamline IAM processes, and improve security and compliance posture.

Effective management of the identity life cycle can bring numerous benefits to organizations, including reducing the time and cost of integrating new users and improving security by controlling access to resources centrally.

By maximizing the investments in existing on-premises identities, organizations can extend them to the cloud, reducing the time for new users to access corporate resources and streamlining the provisioning process. Consistent application of security policies enhances the security posture and reduces exposure to outdated credentials. It also minimizes business interruptions and reduces the time and cost required to enable applications to be accessible from the internet.

Additionally, the increased capacity of IT to develop core application features and the ability to delegate specific administration tasks can lead to increased flexibility and auditing capabilities, enhancing the overall efficiency and effectiveness of IAM solutions.

Leveraging the Microsoft Cloud Adoption Framework

If your organization is on its journey of adopting Azure IAM, consider leveraging the Microsoft **Cloud Adoption Framework** (**CAF**) for Azure (`https://bit.ly/azurecaf`), a guide that helps organizations create and implement strategies for cloud adoption in their business.

It provides a set of best practices, guidance, and tools for different stages of cloud adoption, from initial planning to implementation and optimization. The framework is designed to help organizations develop a comprehensive cloud adoption plan, create a governance structure, and identify the right tools and services for their specific business needs.

The CAF comprises multiple stages: *strategy*, *plan*, *ready*, *migrate*, *innovate*, *secure*, *manage*, and *govern*. Each stage includes a set of recommended practices, tools, and templates that help organizations to assess their readiness, build a cloud adoption plan, migrate applications and data to the cloud, and optimize cloud resources.

The following figure highlights the CAF stages:

Define strategy
Understand motivations
Business outcomes
Business justification
Prioritize project
Plan
Digital estate
Initial organization alignment
Skills readiness plan
Cloud adoption plan
Ready
Operating model
Landing zone concepts
Design area guidance
Implementation Options
Adopt
Migrate
First workload migration
Expanded scenarios
Best practice validation
Process improvements
Innovate
Innovation guide
Expanded scenarios
Best practice validation
Process improvements
Secure
Risk insights
Business resilience
Asset protection
Manage
Business commitments
Operations baseline
Operations maturity
Govern
Methodology
Benchmark initial best practice
Governance maturity

Figure 1.2 – Microsoft CAF for Azure

The framework is flexible and can be customized to fit an organization's specific needs. It is designed to work with different cloud services and technologies, including Microsoft Azure, **Amazon Web Services** (**AWS**), and Google Cloud.

Also, the CAF includes a specific IAM design area that focuses on providing guidance and best practices for designing secure and scalable IAM solutions in the Azure cloud platform. This includes managing identities, implementing authentication and authorization mechanisms, and establishing proper governance and compliance policies. By following the Azure IAM design principles, organizations can ensure their cloud environments are secure and compliant and effectively manage access to their cloud resources.

Utilize this framework to expedite your cloud adoption process. The accompanying resources can assist you in every stage of adoption. These resources, including tools, templates, and assessments, can be applied across multiple phases: `https://bit.ly/azure-caf-tools`.

Azure AD terminology, explained

Azure AD is a system used to manage access to Microsoft cloud services. It involves several terms that are important to understand. Identity is something that can be authenticated, such as a user with a username and password or an application with a secret key or certificate. An account is an identity that has data associated with it.

Azure AD supports two distinct types of security principals: user principals, which represent user accounts, and service principals, which represent applications and services. A user principal encompasses a username and password, while a service principal (also referred to as an application object/registration) can possess a secret, key, or certificate.

An Azure AD account is an identity created through Azure AD or another Microsoft cloud service, such as Microsoft 365. The account administrator manages billing and all subscriptions, while the service administrator manages all Azure resources.

The owner role helps manage Azure resources and is built on a newer authorization system, called Azure **role-based access control** (**RBAC**). The Azure AD Global Administrator is automatically assigned to the person who created the Azure AD tenant and can assign administrator roles to users.

An Azure tenant is a trusted instance of Azure AD created when an organization signs up for a Microsoft cloud service subscription. A custom domain name can be added to Azure AD to make usernames more familiar to users.

When an Azure AD tenant is created, it comes with a default `*.on.microsoft.com` domain. A custom domain name such as `springtoys.com` can be added to the Azure AD tenant to make usernames more familiar to the users.

For example, imagine SpringToys wanting to use Microsoft Azure to store and manage its data. They would need to create an Azure subscription, which would automatically generate an Azure AD directory for them. They would then create Azure AD accounts for each employee who needs access to the company's data stored in Azure.

Each employee's Azure AD account would be associated with their Microsoft 365 account, which they use to log in to their work computer and access company resources. The company could also add a custom domain name to Azure AD so that employees can use email addresses with their company's domain name to log in to their Azure AD account, such as `john@springtoys.com`. The company would also need to assign roles to each employee's Azure AD account, such as the owner role or service administrator role, to manage access to Azure resources. In broad terms, Azure roles govern permissions for overseeing Azure resources, whereas Azure AD roles govern permissions for managing Azure AD resources.

The following table summarizes the Azure AD terminology:

Concept	Description
Identity	An object that can be authenticated
Account	An identity that has data associated with it
Azure AD account	An identity created through Azure AD or another Microsoft cloud service
Azure AD tenant/directory	A dedicated and trusted instance of Azure AD, a tenant is automatically created when your organization signs up for a Microsoft cloud service subscription

Azure AD is a crucial aspect of cloud security that enables organizations to control access to their resources and data in the cloud.

Securing applications with the Microsoft identity platform

Managing the information of multiple usernames and passwords across various applications can become challenging, time-consuming, and vulnerable to errors. However, this problem can be addressed using a centralized identity provider. Azure AD is one such identity provider that can handle authentication and authorization for various applications. It provides several benefits, including conditional access policies, MFA, and **single sign-on (SSO)**. SSO is a significant advantage as it enables users to sign in once and automatically access all the applications that share the same centralized directory.

More broadly speaking, the Microsoft identity platform simplifies authentication and authorization for application developers. It offers identity as a service and supports various industry-standard protocols and open source libraries for different platforms. Developers can use this platform to build applications that sign in to all Microsoft identities, get tokens to call Microsoft Graph, and access other APIs. Simply put, by utilizing the Microsoft identity platform, developers can reduce the complexity of managing user identities and focus on building their applications' features and functionality.

Microsoft's identity platform can help organizations streamline identity management and improve security. Organizations can take advantage of features such as conditional access policies and MFA by delegating authentication and authorization responsibilities to a centralized provider such as Azure AD. Furthermore, developers can benefit from the platform's ease of use, supporting various industry-standard protocols and open source libraries, making it easier to build and integrate applications.

By integrating your app with Azure AD, you can ensure that your app is secure in the enterprise by implementing Zero Trust principles.

As a developer, integrating your app with Azure AD provides a wide range of benefits that help you secure your app in the enterprise. One of the significant benefits of using Azure AD is the ability to authenticate and authorize applications and users. Azure AD provides a range of authentication methods, including SSO, which can be implemented using federation or password-based authentication. This simplifies the user experience by reducing the need for users to remember multiple passwords.

Another benefit of using Azure AD is the ability to implement RBAC, which enables you to restrict access to your app's features based on a user's role within the organization. You can also use OAuth authorization services to authenticate and authorize third-party apps that access your app's resources.

The Microsoft identity platform supports multiple protocols for authentication and authorization. It is crucial to understand the differences between these protocols to choose the best option for your application.

One example is the comparison between OAuth 2.0 and SAML. OAuth 2.0 is commonly used for authorization, while SAML is frequently used for authentication. The OAuth 2.0 protocol allows users to grant access to their resources to a third-party application without giving the application their login credentials. On the other hand, SAML provides a way for a user to authenticate to multiple applications using a single set of credentials. An example of SAML being used in the Microsoft identity platform is with **Active Directory Federation Services** (**AD FS**) federated to Azure AD.

Another example is the comparison between **OpenID Connect** (**OIDC**) and SAML. OIDC is commonly used for cloud-based applications, such as mobile apps, websites, and web APIs. It allows for authentication and SSO using a JSON web token. SAML, on the other hand, is commonly used in enterprise applications that use identity providers such as AD FS federated to Azure AD. Both protocols support SSO, but SAML is commonly used in enterprise applications.

The following table summarizes the protocols and descriptions and their typical usage scenarios:

Protocol	Description	Use Cases
OAuth	OAuth is used for authorization, granting permissions to manage Azure resources	When managing permissions to access and perform operations on Azure resources
OIDC	OIDC builds on top of OAuth 2.0 and is used for authentication, verifying the identity of users	When authenticating users and obtaining information about their identity
SAML	SAML is used for authentication and is commonly used with identity providers, such as AD FS, to enable SSO in enterprise applications	When integrating with enterprise applications and identity providers, particularly with AD FS federated to Azure AD

Understanding these protocols and their differences can help you choose the best option for your application and ensure secure and efficient authentication and authorization.

As more companies transition their workloads to the cloud, they face the challenge of ensuring the security of their resources in these new environments. In order to effectively manage access to cloud-based workloads, organizations must establish definitive user identities and control access to data, while also ensuring authorized operations are performed. This is where Microsoft Entra comes in – which provides a set of multiple components that provide identity-based access control, permissions management, and identity governance to help organizations securely manage their cloud-based workloads.

Securing cloud-based workloads with Microsoft Entra's identity-based access control

When transitioning workloads to the cloud, companies must consider the security implications of moving their resources. They need to define authorized users, restrict access to data, and ensure that employees and vendors only perform authorized operations. To centrally control access to cloud-based workloads, companies must establish a definitive identity for each user used for every service. This identity-based access control ensures that users have the necessary permissions to perform their jobs while restricting unauthorized access to resources.

Microsoft Entra comprises a set of multiple components, including the following:

- Azure AD
- Microsoft Entra Permissions Management
- Microsoft Entra Verified ID
- Microsoft Entra workload identities
- Microsoft Entra Identity Governance
- Microsoft Entra admin center

Let's look at them in detail.

Azure AD

To simplify the process of securing cloud-based resources, Azure AD, a cloud-based IAM service that is part of Microsoft Entra, offers features such as SSO and MFA, which helps protect both users and data. By learning the basics of creating, configuring, and managing users and groups of users, organizations can effectively control access to their cloud-based resources. Additionally, by managing licenses through Azure AD, organizations can ensure that their employees and vendors have access to the necessary tools to perform their jobs while maintaining a secure environment.

Azure AD provides three ways to define users, which are helpful for different scenarios. The first way is cloud identities, which only exist in Azure AD. These can include administrator accounts and users managed directly in Azure AD. Cloud identities are deleted when removed from the primary directory, making them an excellent option for managing temporary access to Azure resources. The following figure represents the cloud identity.

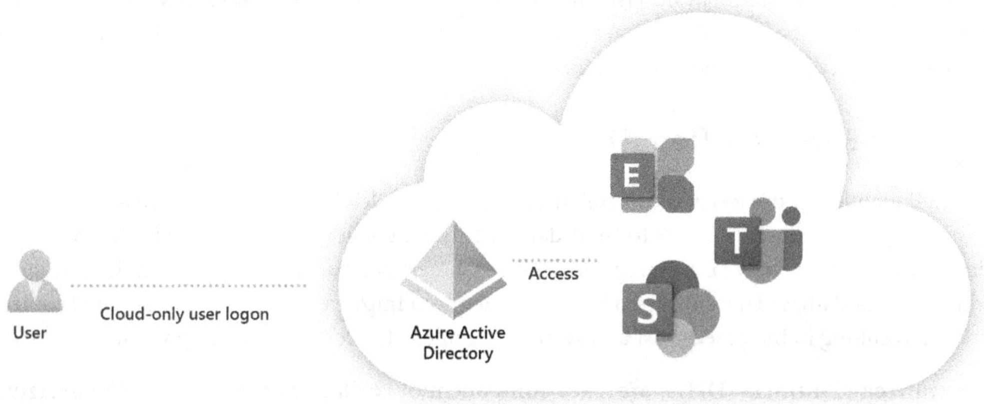
User
Cloud-only user logon
Azure Active Directory
Access
E
T
S

Figure 1.3 – Cloud identity

The second way is directory-synchronized identities, which exist in an on-premises AD. These users are brought into Azure through a synchronization activity with Azure AD Connect, making them useful for organizations with existing on-premises infrastructure.

You can leverage directory synchronization with **Pass-through Authentication** (**PTA**) or SSO with AD FS.

Finally, there are guest users that might exist outside of Azure or can be on a different Azure AD tenant. These can be accounts from other cloud providers or Microsoft accounts, such as an Xbox Live account. Guest users are invited to access Azure resources. They can be removed once their access is no longer necessary, making them an excellent option for external vendors or contractors who require temporary access.

Managing permissions is a critical aspect of Zero Trust security and is increasingly challenging for organizations adopting a multi-cloud strategy. With the proliferation of cloud services and identities, high-risk cloud permissions are exploding, creating a larger attack surface for organizations. IT security teams are pressured to ensure access to their expanding cloud estate is secure and compliant. However, the inconsistency of cloud providers' native access management models makes it even more complex for security and identity teams to manage permissions and enforce least privilege access policies across their entire environment.

Microsoft Entra Permissions Management

Organizations need a **cloud infrastructure entitlement management** (**CIEM**) solution such as Microsoft Entra Permissions Management to enable comprehensive visibility into permissions assigned to all identities across multi-cloud infrastructures such as Microsoft Azure, AWS, and **Google Cloud Platform** (**GCP**). Microsoft Entra Permissions Management can detect and right-size unused and excessive permissions while continuously monitoring permissions to maintain a least privilege access policy. By implementing a CIEM solution such as Permissions Management, organizations can improve their cloud security posture and better manage access to their cloud-based resources.

Microsoft Entra Verified ID

The digital identity we use today is controlled by other parties, leading to potential privacy concerns. Users give apps and devices access to their data, making it challenging to track who has access to which information. Securely exchanging data with consumers and partners is difficult in the enterprise world. A standards-based decentralized identity system can improve user and organizational control over data, resulting in increased trust and security for apps, devices, and service providers.

Decentralized identifiers (**DIDs**) are a key component of **verifiable credentials** (**VCs**) in Azure AD. DIDs are unique identifiers created in a decentralized system and are not controlled by a central authority. DIDs can be used to represent individuals, organizations, devices, and other entities in a secure and privacy-preserving way. They can also be used to prove ownership of digital assets, such as domain names or social media handles.

Azure AD supports using DIDs and VCs to enable secure and trusted digital identities. This allows organizations to reduce the reliance on traditional usernames and passwords and instead use more secure and privacy-preserving methods for identity verification. The article also highlights the benefits of using DIDs and VCs, including increased security, privacy, and interoperability. It provides resources for developers and organizations to use DIDs and VCs in Azure AD.

Microsoft Entra workload identities

In the world of cloud computing, a workload identity is essential for authenticating and accessing other resources and services securely and efficiently. Workload identities can take different forms, such as a user account that an application uses to access a database or a service role attached to an instance with limited access to a specific resource. Regardless of its form, a workload identity ensures that the software entity can securely access the resources it needs while also helping to prevent unauthorized access and data breaches.

In Azure AD, a workload identity is a way for a software program, such as an application or service, to identify and authenticate itself when accessing other services and resources. There are three types of workload identities in Azure AD: applications, which are like templates that define how a program can access resources; service principals, which are like local copies of applications that are specific to a particular tenant; and managed identities, which are a special type of service principal that don't require a developer to manage passwords or credentials.

Here are a few examples of how you can leverage workload identities:

- You can use a managed identity to access resources protected by Azure AD without the need to manage credentials or keys to authenticate your identity

- You can use workload identity federation to access Azure AD-protected resources without needing to manage secrets or credentials for workloads running in supported scenarios such as GitHub Actions, Kubernetes, or compute platforms outside Azure
- You can use *access reviews* for service principals to review and audit the access of service principals and applications assigned to privileged directory roles in Azure AD
- You can leverage Conditional Access policies for workload identities to control access to resources based on certain conditions or policies and use continuous access evaluation to monitor and evaluate access to resources in real time
- You can use Identity Protection to detect and respond to identity-related risks and threats for your workload identities and apply security policies to protect your identities from cyberattacks

As organizations embrace digital transformation, the need for the secure and efficient management of access to resources becomes increasingly important. Microsoft Entra Identity Governance is a tool designed to address this need, enabling companies to balance productivity and security by ensuring the right people have access to the right resources. Identity Governance uses a foundation of identity life cycle management to keep track of who has access to what resources and ensure that access is updated as needed.

Microsoft Entra Identity Governance

Microsoft Entra Identity Governance is a tool that helps organizations balance the need to keep their data secure and ensure employees can get their work done efficiently. It helps by ensuring the right people have access to the right things, and the company can keep an eye on who is accessing what. This helps reduce the risk of someone getting access to something they shouldn't have and helps the company ensure employees can still do their jobs.

Identity Governance helps organizations to manage access to their resources in a way that balances productivity and security. It is designed to answer questions such as "Who should have access to which resources?" and "How can we ensure that access is appropriate and secure?" To do this, Identity Governance relies on a foundation of identity life cycle management, which involves keeping track of who has access to what resources and making sure that access is updated as needed. This process helps organizations ensure that their resources are protected while enabling their employees to get the access they need to do their jobs.

Sometimes, organizations need to work with people outside of their own company. Azure AD B2B collaboration is a feature that allows companies to safely share their apps and services with other people, such as guests and partners from different organizations. This way, organizations can maintain control over their own data while still allowing others to use their resources. Microsoft Entra entitlement management will enable organizations to decide which users from other organizations can request access and become guests in their directory. It will also remove these guests when they no longer need access.

Microsoft Entra admin center

Microsoft launched the Entra admin center for its Microsoft 365 and Azure AD customers. And you can log in to the portal using your Microsoft 365 account. The Entra admin center provides customers with better security, governance, and compliance features for their organization.

The portal is accessible through the following URL: `https://entra.microsoft.com`.

As you can see, Microsoft Entra helps organizations to make sure the right people have access to the right things. It does this by verifying who someone is and allowing them to access the apps and resources needed to do their job. Microsoft Entra works across different environments, such as cloud and on-premises systems. It also makes it easier for people to access what they need by using smart tools to make quick decisions about who should have access to what.

Summary

This chapter covered several important topics related to IAM in the modern IT landscape. We discussed authentication and authorization, which are crucial components of any IAM solution. Then, we moved on to explore the importance of engaging and collaborating with employees, partners, and customers, as well as the role that digital identities play in this process.

We provided an overview of Azure AD, a cloud-based IAM service that enables organizations to control access to cloud applications. Finally, we discussed how Microsoft Entra's identity-based access control can help organizations secure their cloud-based workloads by establishing definitive identities for each user and controlling access to resources.

In the next chapter, we will discuss the core IAM capabilities that can be utilized with Azure AD.

2

Managing Access to Resources Using Azure Active Directory

This chapter aims to give you a brief overview of the **Identity and Access Management** (**IAM**) capabilities you can leverage using Azure Active Directory, also known as Microsoft Entra ID, including step-by-step configurations to properly secure identities and access to resources in your organization.

This chapter will address key components related to IAM, such as Azure Active Directory, and the most relevant and recently released features and components, such as Azure Active Directory Connect, Azure Active Directory Application Proxy, Azure Active Directory Conditional Access, and Privileged Identity Management.

In this chapter, we'll cover the following main topics:

- Understanding the need for IAM
- Understanding Azure Active Directory (now Microsoft Entra ID)
- Understanding the capabilities of Microsoft Entra ID
- Hybrid identity – integrating your on-premises directories (Azure AD Connect sync and cloud sync)

Let's get started!

Understanding the need for IAM

Having controls and mechanisms to successfully manage access to resources and applications, while managing control over the identities of the people who are a part of our organization, has been a major business need for the past few years.

Undoubtedly, we all have seen a significant change in the field since 2020, where organizations have sped up the process to broaden their digital transformation initiatives. And, more often than not, identity management and identity protection are overlooked as components, with the only purpose to provide user credentials and manage access to SaaS applications such as Office 365 and a few familiar others.

During the COVID-19 pandemic, we were all impacted to some level in carrying out our daily work, and organizations needed to accommodate remote management and remote working for staff. A clear strategy to enable a hybrid work approach became critical to organizations of different sizes and verticals.

Organizations that fully understood the need for, and successfully enabled, remote work were able to enable access for their entire staff to services and applications, establishing the right security stack to effectively handle the right resources, infrastructure, and operations management.

Therefore, when driving a clear strategy to effectively manage access to resources in an organization and enable the protection of identities, it is of paramount importance that we fully understand the foundations of identity management.

Companies face the challenge of providing access to all the aforementioned types of users, which can include employees, business partners, and customers. How can we achieve the proper management level for all these user types while maintaining the right balance between privacy and security?

We all realize the need for organizations to manage employees' credentials and access to resources inside and outside the company, usually referred to as **user type**. The *user type* relates to the kind of account that a specific user holds and may include remote users, remote groups, local users, database users, and maintenance users, among others.

Here's where an IAM solution comes into play. Before diving into specific technologies, understanding the core needs of your business is of utmost importance to leverage the right identity management solution.

Here are a few questions for you to consider when adopting a solution:

- Will the solution help you integrate with your business partners' solutions and customers?
- Is it scalable and reliable enough to meet your organization's growing needs?
- Will the solution need to be integrated with an existing identity solution?
- Will adding new functionality require considerable engineering effort?
- How will the solution help to increase the organization's security posture? Consider controls to manage sensitive data, prevent data breaches, and specify granular access to the right resources.

As you draft an idea of the type of solution your organization needs, you will hopefully get to expand on what an IAM solution is capable of. It involves having the right framework of policies and mechanisms to ensure that only authorized people or user types have the appropriate access level to the resources needed to perform and operate in their daily job, while maintaining a log for audit and compliance.

The aforementioned translates into leveraging features such as password management, role-based access control, governance, auto-provisioning the capabilities of accounts, single sign-on, audit, and compliance.

Understanding Azure AD (now Microsoft Entra ID)

Authentication and **authorization** are the two pillars of an IAM solution, and therefore, organizations need to adopt a compelling identity framework that helps them improve their security posture.

Identity has become the core control plane of security for organizations to securely access resources on-premises and in the cloud. Microsoft was named a leader in the 2022 Gartner Magic Quadrant for Access Management for Microsoft Azure AD (now Microsoft Entra ID).

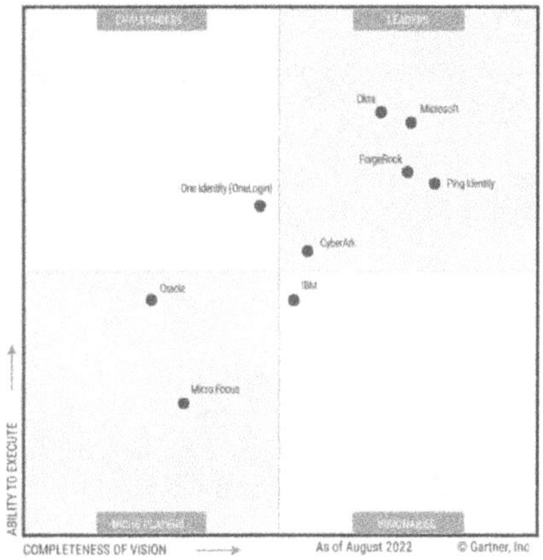

Okta
Microsoft
ForgeRock
Ping Identity
One Identity (OneLogin)
CyberArk
Oracle
IBM
Micro Focus
ABILITY TO EXECUTE
COMPLETENESS OF VISION
As of August 2022
© Gartner, Inc

Figure 2.1 – The Magic Quadrant for Access Management

Microsoft Entra packages Microsoft's rich IAM features, including Azure AD, permissions management, identity governance, workload identities, and Microsoft Entra Verified ID.

Azure AD, a cloud-based service for IAM, is suitable for different types of users, such as the following:

- IT admins looking to control access to applications while protecting user identities and credentials to meet governance requirements
- App developers, who can leverage Microsoft Entra ID capabilities and integration with Microsoft Entra ID APIs to provide personalized app experiences
- Subscribers to multiple services such as Microsoft 365, Dynamics 365, or Microsoft Azure

The following diagram highlights a high-level overview of the integration of AD (now Microsoft Entra ID).

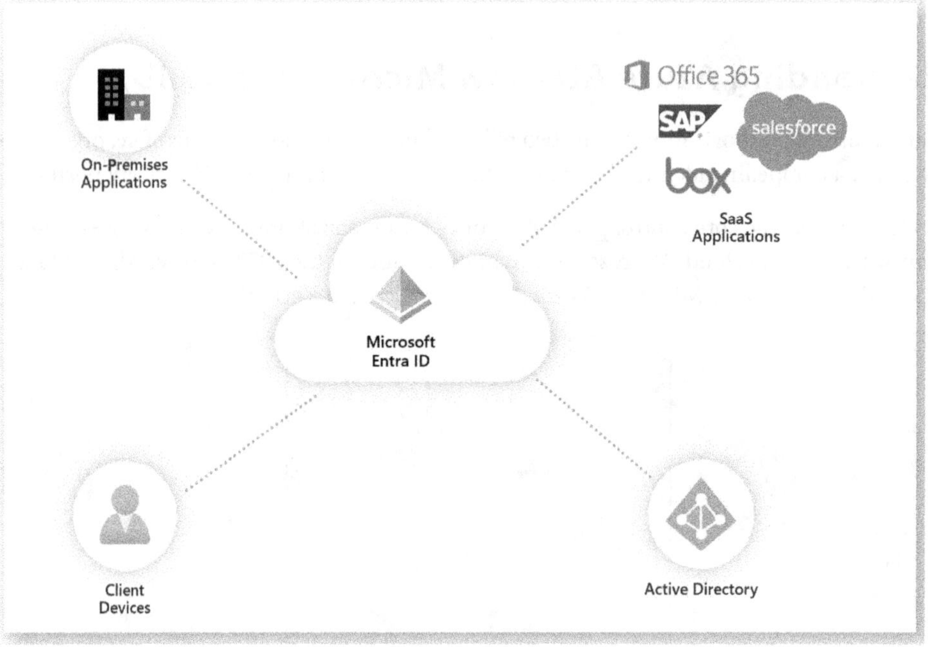
On-Premises Applications
Office 365
SAP
salesforce
box
SaaS Applications
Microsoft Entra ID
Client Devices
Active Directory

Figure 2.2 – Azure AD

It comprises a very rich set of features for a comprehensive IAM solution, including core directory services, application access management, and advanced identity protection.

As with any other similar solution on the market, Microsoft Entra ID is available in four editions – **Free**, **Microsoft Entra ID Governance**, **Premium P1**, and **Premium P2**.

Exploring the Microsoft Entra ID editions

In this section, we'll discuss the Azure AD editions available and which one will best suit your needs.

Azure AD Free

Whenever you purchase an online subscription from services such as Azure, Dynamics 365, Intune, Power Platform, or Microsoft 365, you will be entitled to a dedicated Azure AD service instance; the Azure AD tenant usually represents an organization in Azure AD. You will be able to leverage cloud authentication, including pass-through authentication and *password hash synchronization*.

Password hash synchronization serves as one of the sign-in approaches employed to achieve hybrid identity. With Azure AD Connect, a hash of a user's password from an on-premises AD is synchronized to a cloud-based Azure AD instance.

This capability extends the directory synchronization feature provided by Azure AD Connect sync. By utilizing this functionality, you gain the ability to access Azure AD services, including Microsoft 365, using the same password used for signing in to your on-premises Active Directory instance.

Additional benefits such as federated authentication (Active Directory Federation Services or federation with other identity providers), single sign-on, **Multi-Factor Authentication** (**MFA**), and even passwordless (Windows Hello for Business, Microsoft Authenticator, and FIDO2 security key integrations) are included in the Azure AD Free edition.

An Azure AD tenant can also be created through the Azure portal, in which you can configure and do all the administrative tasks related to Azure AD. In addition, you can make use of the Microsoft Graph API to access Azure AD resources. And now, you can leverage a simpler, integrated experience to manage all your IAM resources using the Microsoft Entra admin center.

The Microsoft Entra admin center can be accessed through the following URL: `entra.microsoft.com`.

Microsoft Entra ID Premium 1

Consider a scenario in which you need to permit users that are part of your organization access across both on-premises and cloud resources; for this requirement, Microsoft Entra ID Premium P1 is the best fit.

This edition also enables administration and hybrid identity benefits such as advanced user and group management. This means you can leverage dynamic groups. Think of dynamic groups as a mechanism that allows you to create rules to determine group membership, based on the user or device properties in Azure AD.

For example, suppose employees of your company are part of the sales department. If so, you can use specific business rules to group all users of the same department as part of the group, instead of manually assigning them to it.

When using dynamic groups, you can create a new user in Azure AD, and based on the user properties, they will be automatically assigned to the specific group. In this case, they will be assigned to the sales group.

> **Note**
>
> Microsoft Entra ID Premium 1 or Premium 2 licenses are required in order to implement dynamic groups.

Consider using Microsoft Entra ID Premium P1 when you want to enable hybrid environments and scale and automatically assign users to groups. Your organization would need to enable identity management and cloud write-back capabilities to allow your on-premises users to perform a self-service password reset, as shown in the following figure:

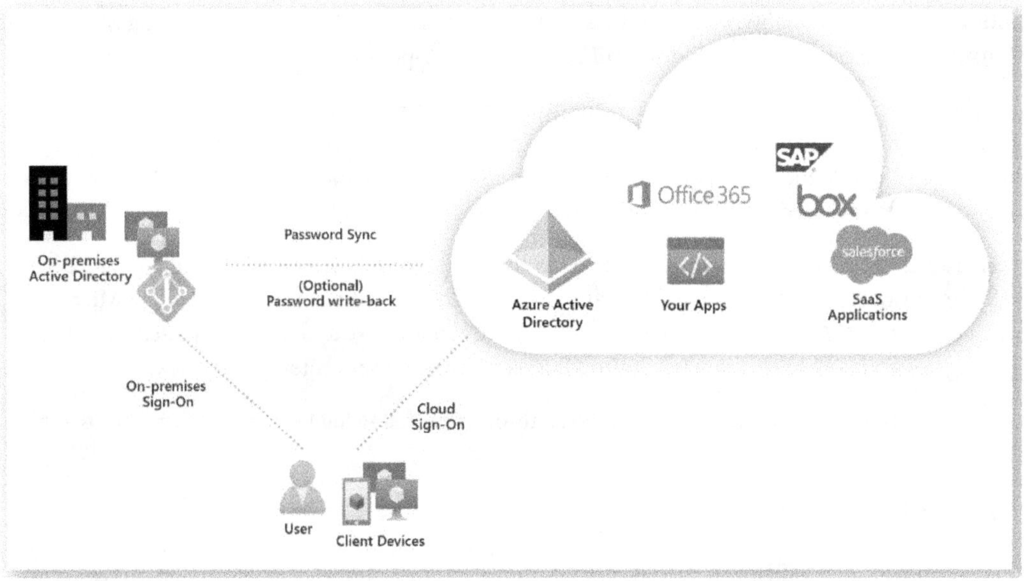
On-premises Active Directory
Password Sync
(Optional) Password write-back
On-premises Sign-On
Cloud Sign-On
User
Client Devices
Azure Active Directory
Office 365
Your Apps
SAP
box
salesforce
SaaS Applications

Figure 2.3 – Azure Active Directory hybrid scenario

Now, let's understand the difference between Azure AD Premium P1 and Premium P2.

Microsoft Entra ID Premium P2

This edition adds on top of Microsoft Entra ID Premium P1 by providing a set of benefits focused on identity protection (risky sign-ins, risky users, and risk-based conditional access), event logging and reporting, and identity governance (access certifications and reviews, entitlements management, **Privileged Identity Management** (**PIM**), and just-in-time access).

Consider the adoption of Microsoft Entra ID Premium P2 when an organization needs to leverage security capabilities such as risk management, based on the identity protection capabilities of Microsoft Entra ID.

The detection of a potential risk is covered by Microsoft Entra ID Premium P2 and allows you to identify suspicious actions related to a user's accounts in the directory. Identity protection can help you to quickly respond to suspicious actions.

These risks can be detected in real time and offline and will help you prove that a user is who they say they are. Also, while some risks are available in the Free and Microsoft Entra ID Premium P1 editions, others are only available in the Microsoft Entra ID Premium P2 edition.

Examples of *Premium* sign-in risk detections include the following:

- Atypical travel
- An anomalous token
- A token issuer anomaly
- A malware-linked IP address
- A suspicious browser
- A malicious IP address

You can find the complete list of *Premium* sign-in risk detections at the following URL:

`https://learn.microsoft.com/en-us/azure/active-directory/identity-protection/concept-identity-protection-risks#sign-in-risk`

Having gained a comprehensive understanding of how Azure AD functions, we will now delve into a more hands-on approach and demonstrate how you can integrate Azure AD into your organization.

Understanding the capabilities of Microsoft Entra ID

To better understand Azure AD's potential (now Microsoft Entra ID), we will perform a series of exercises with practical examples that you can follow step by step. Despite the recent renaming of the product, we will stick to the name Azure AD for the following exercises. We will perform the following tasks:

- **Task 1**: Create a new Azure AD tenant using the Azure portal and activate the Azure AD Premium P2 offer
- **Task 2**: Create and configure Azure AD users
- **Task 3**: Create an Azure AD group with dynamic membership

Let's begin!

Task 1 – creating a new Azure AD tenant using the Azure portal

To create a new tenant, take the following steps:

1. First, we will go to the Azure portal and select **Azure Active Directory**. Then, we will create a new Azure AD tenant, as shown here:

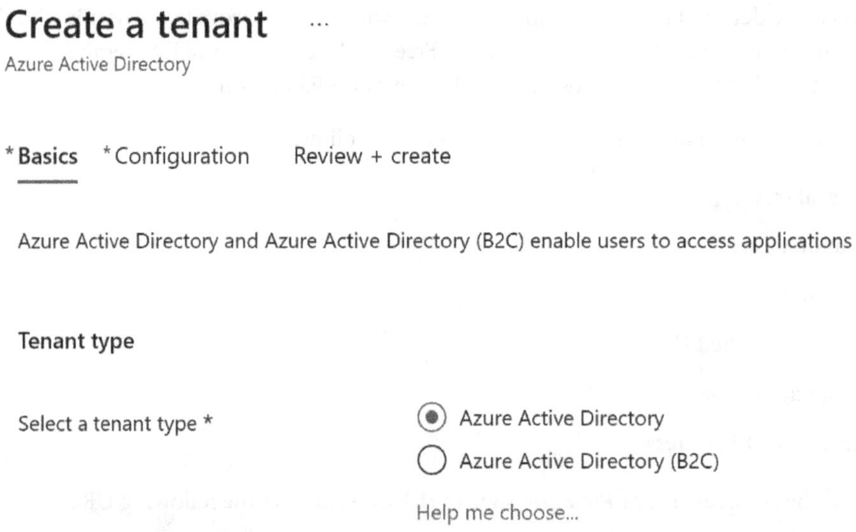
Create a tenant ...
Azure Active Directory
* Basics
* Configuration
Review + create
Azure Active Directory and Azure Active Directory (B2C) enable users to access applications
Tenant type
Select a tenant type *
Azure Active Directory
Azure Active Directory (B2C)
Help me choose...

Figure 2.4 – Create a tenant – the Basics tab

2. Now, we will provide the details for the configuration of the Azure AD tenant:

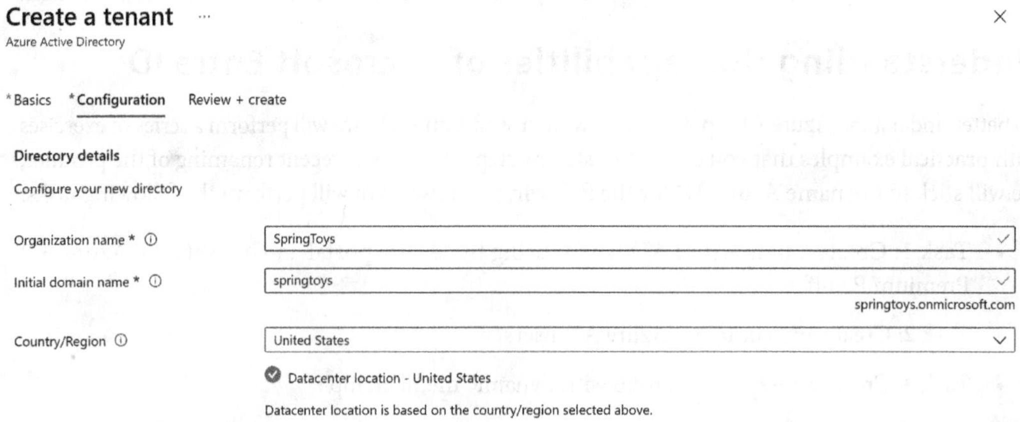
Create a tenant ...
Azure Active Directory
* Basics
* Configuration
Review + create
Directory details
Configure your new directory
Organization name *
SpringToys
Initial domain name *
springtoys
springtoys.onmicrosoft.com
Country/Region
United States
Datacenter location - United States
Datacenter location is based on the country/region selected above.

Figure 2.5 – Create a tenant – the Configuration tab

3. Next, we will proceed to create the Azure AD tenant:

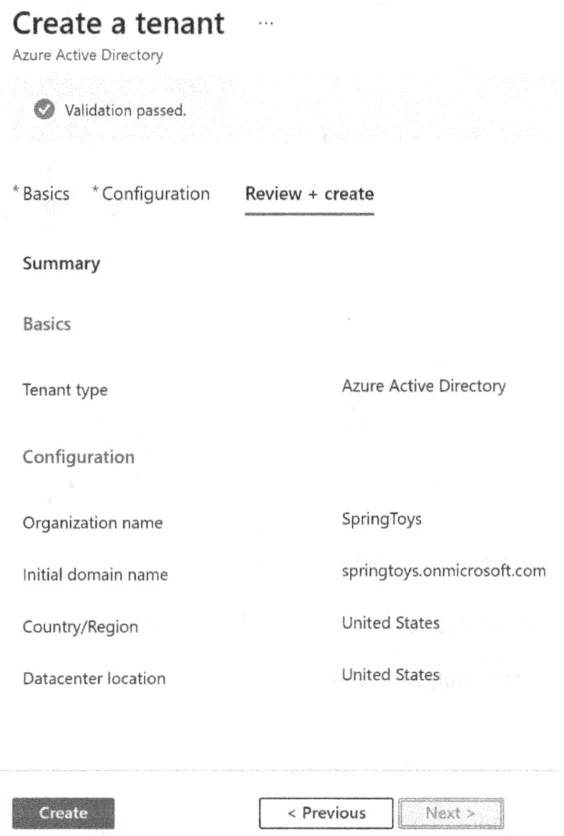
Create a tenant
Azure Active Directory
Validation passed.
* Basics
* Configuration
Review + create
Summary
Basics
Tenant type
Azure Active Directory
Configuration
Organization name
SpringToys
Initial domain name
springtoys.onmicrosoft.com
Country/Region
United States
Datacenter location
United States
Create
< Previous
Next >

Figure 2.6 – The Review + create tab

Note that the creation process for the new Azure AD tenant might take a few minutes.

4. After a few minutes, the new Azure AD tenant should be ready, and you will see a notification, as shown here:

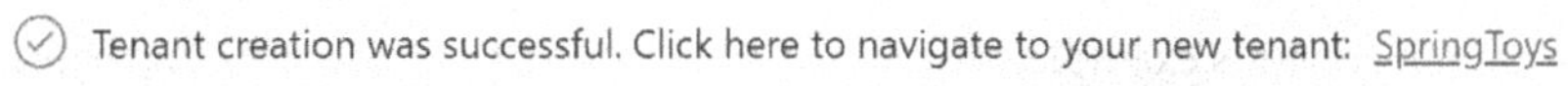
Tenant creation was successful. Click here to navigate to your new tenant: SpringToys

Figure 2.7 – Tenant creation notification

5. Once the new Azure AD tenant is ready, go to the **Configuration** tab, and select the **Licenses** option. From there, choose the **All products** option, then **Try / Buy**, and *activate* the Azure AD Premium P2 offer, as shown in the following figure.

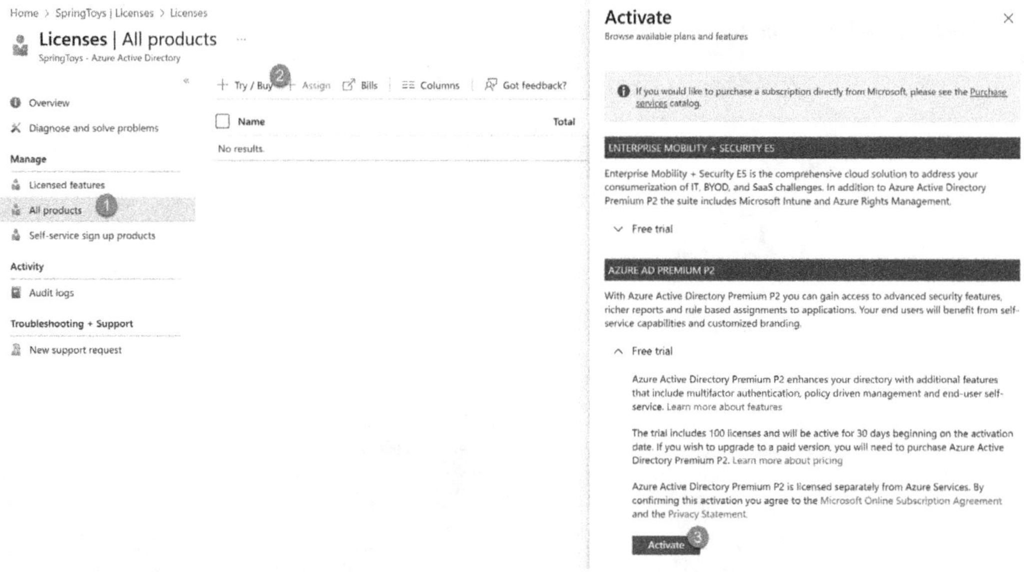
Home > SpringToys | Licenses > Licenses
Licenses | All products
SpringToys - Azure Active Directory
Overview
Diagnose and solve problems
Manage
Licensed features
All products
Self-service sign up products
Activity
Audit logs
Troubleshooting + Support
New support request
+ Try / Buy
+ Assign
Bills
Columns
Got feedback?
Name
Total
No results.
Activate
Browse available plans and features
If you would like to purchase a subscription directly from Microsoft, please see the Purchase services catalog.
ENTERPRISE MOBILITY + SECURITY E5
Enterprise Mobility + Security E5 is the comprehensive cloud solution to address your consumerization of IT, BYOD, and SaaS challenges. In addition to Azure Active Directory Premium P2 the suite includes Microsoft Intune and Azure Rights Management.
Free trial
AZURE AD PREMIUM P2
With Azure Active Directory Premium P2 you can gain access to advanced security features, richer reports and rule based assignments to applications. Your end users will benefit from self-service capabilities and customized branding.
Free trial
Azure Active Directory Premium P2 enhances your directory with additional features that include multifactor authentication, policy driven management and end-user self-service. Learn more about features
The trial includes 100 licenses and will be active for 30 days beginning on the activation date. If you wish to upgrade to a paid version, you will need to purchase Azure Active Directory Premium P2. Learn more about pricing
Azure Active Directory Premium P2 is licensed separately from Azure Services. By confirming this activation you agree to the Microsoft Online Subscription Agreement and the Privacy Statement.
Activate

Figure 2.8 – Activate Azure AD Premium P2

After a few seconds, you should see a notification that the offer has been activated, as shown here:

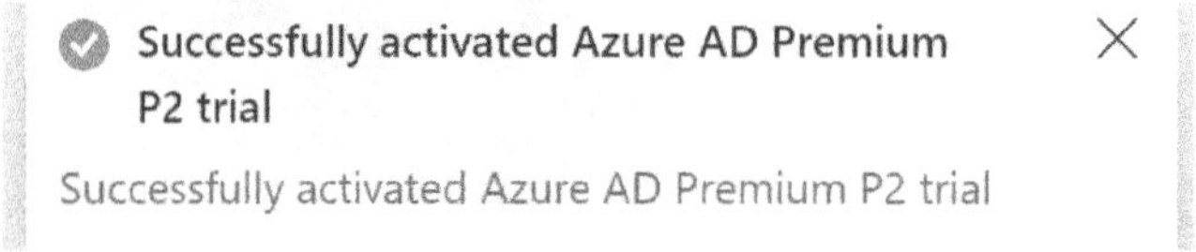
Successfully activated Azure AD Premium P2 trial
Successfully activated Azure AD Premium P2 trial

Figure 2.9 – The Azure AD Premium P2 trial

This process completes task 1. In this task, we created a new Azure AD tenant and activated the Azure AD Premium P2 trial.

Now that the Azure AD tenant is ready, let's create and configure Azure AD users.

Task 2 – creating and configuring Azure AD users

Following on from our previous example, we need to assign users from the sales department to a sales group in the Azure AD tenant. Therefore, we will create and configure two users in the Azure AD tenant with a few properties:

1. In the Azure portal, navigate to **Azure Active Directory**, and under the **Manage** section, select the **All users (preview)** option. Then, select the **+ New user** option.

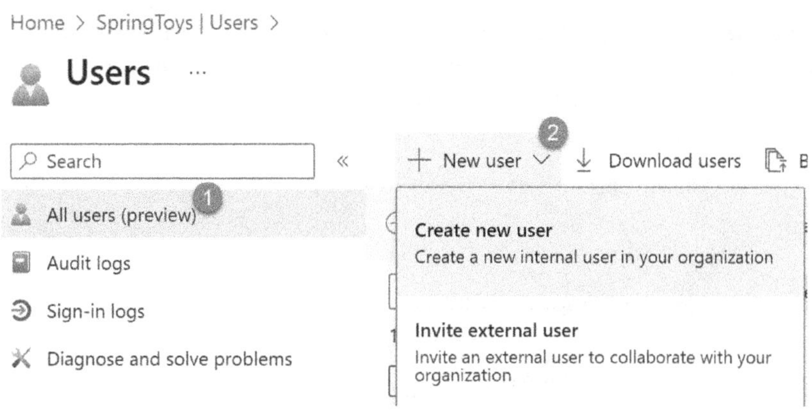
Home > SpringToys | Users >
Users
Search
All users (preview)
Audit logs
Sign-in logs
Diagnose and solve problems
New user
Download users
Create new user
Create a new internal user in your organization
Invite external user
Invite an external user to collaborate with your organization

Figure 2.10 – Create new user

2. Now, on the **Create user** page, we will initially provide the details for the identity and password for this employee:

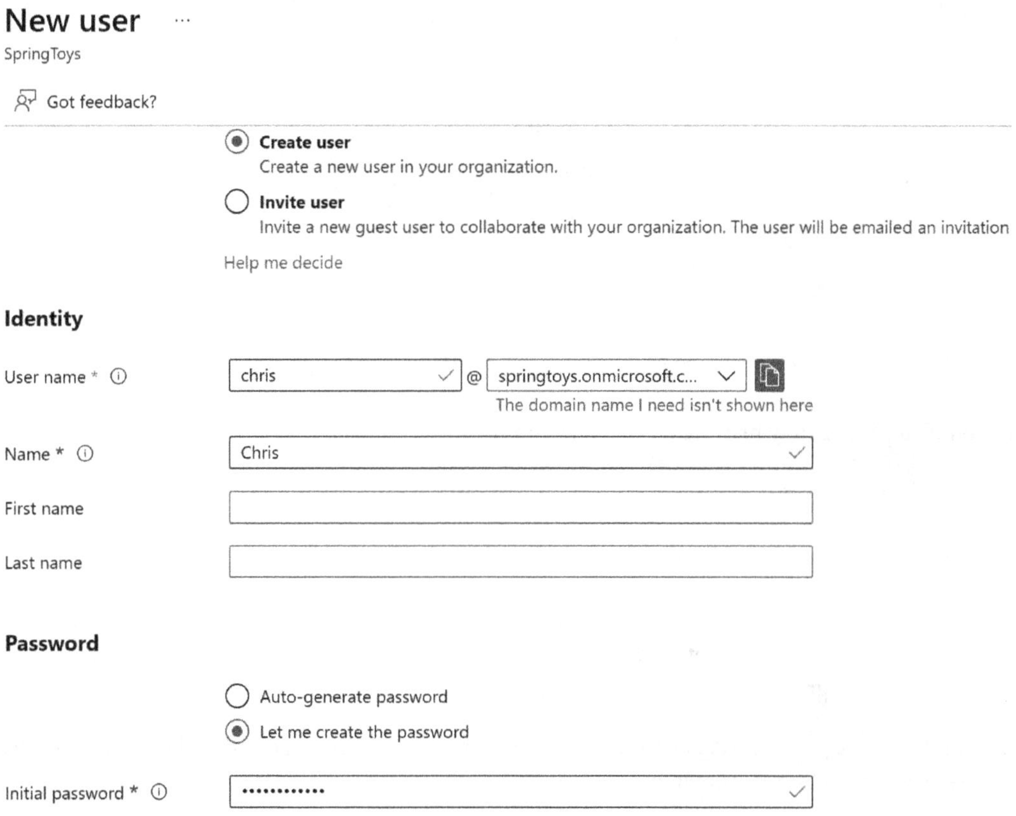
New user
SpringToys
Got feedback?
Create user
Create a new user in your organization.
Invite user
Invite a new guest user to collaborate with your organization. The user will be emailed an invitation
Help me decide
Identity
User name *
chris
springtoys.onmicrosoft.c...
The domain name I need isn't shown here
Name *
Chris
First name
Last name
Password
Auto-generate password
Let me create the password
Initial password *

Figure 2.11 – Creating a new user – identity and password

3. Then, provide additional properties for this user. This includes the ability to specify whether this user should be part of a specific group, role, or location. We can also define properties related to the job information. In this case, we will fill in the job title, department, and company name, as shown here:

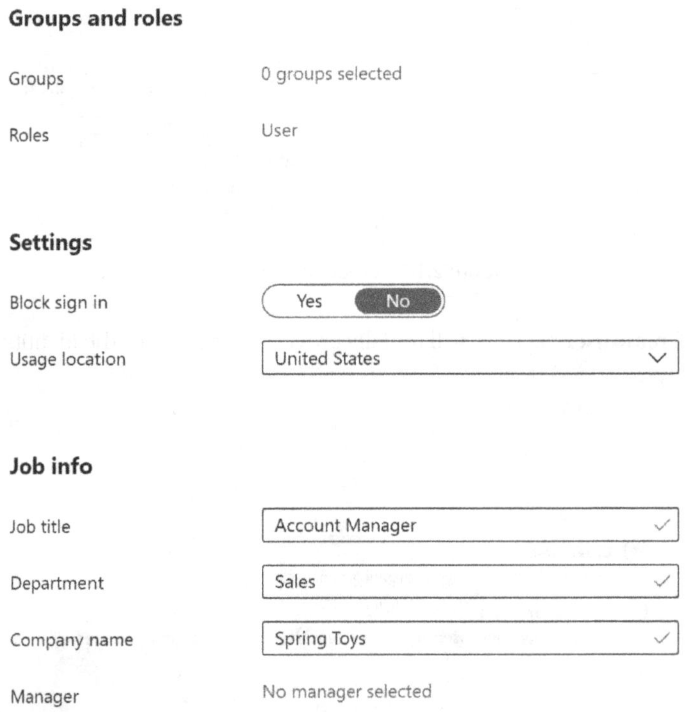
Groups and roles
Groups
0 groups selected
Roles
User
Settings
Block sign in
Yes
No
Usage location
United States
Job info
Job title
Account Manager
Department
Sales
Company name
Spring Toys
Manager
No manager selected

Figure 2.12 – Creating a new user – groups and roles, settings, and job info

This completes task 2, in which we created and configured a new user in the Azure AD tenant.

Now, let's proceed to task 3.

Task 3 – creating an Azure AD group with dynamic membership

As the final step to achieve the automated assignation of users to a specific group, we will leverage the dynamic membership assignation available in Azure AD Premium P1 and P2. Therefore, we will create an Azure AD group with dynamic membership:

1. In the Azure portal, navigate to Azure AD, and select the **Groups** option. Then, select the + **New group** option, as shown here:

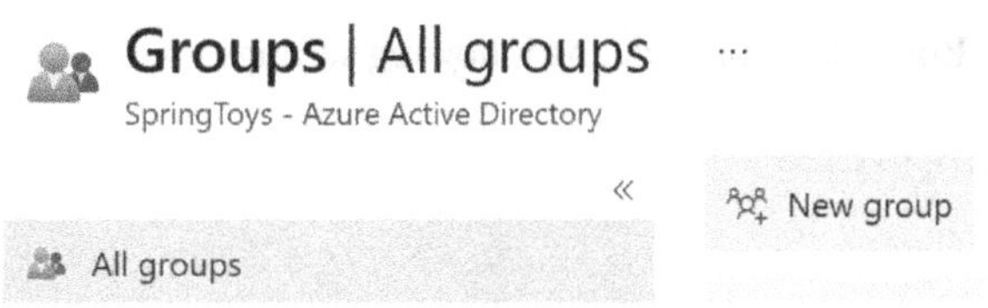
Groups | All groups
SpringToys - Azure Active Directory
New group
All groups

Figure 2.13 – Create a new group

2. Now, we will provide the details for the creation of the new group. We will use the following properties:

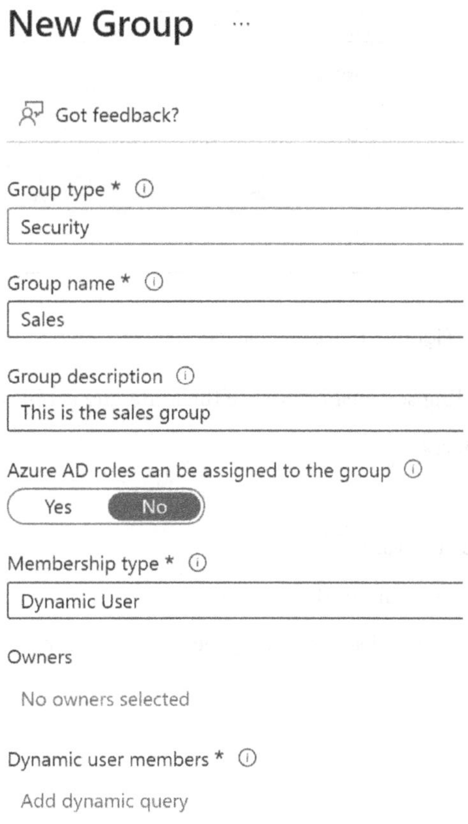
New Group
Got feedback?
Group type *
Security
Group name *
Sales
Group description
This is the sales group
Azure AD roles can be assigned to the group
Yes
No
Membership type *
Dynamic User
Owners
No owners selected
Dynamic user members *
Add dynamic query

Figure 2.14 – New group settings

3. Before moving forward, we will add a dynamic query. Select the **Add dynamic query** option. Here, we will configure the business rules to appropriately assign users to the group.

4. On the **Configure Rules** page, provide the following settings.

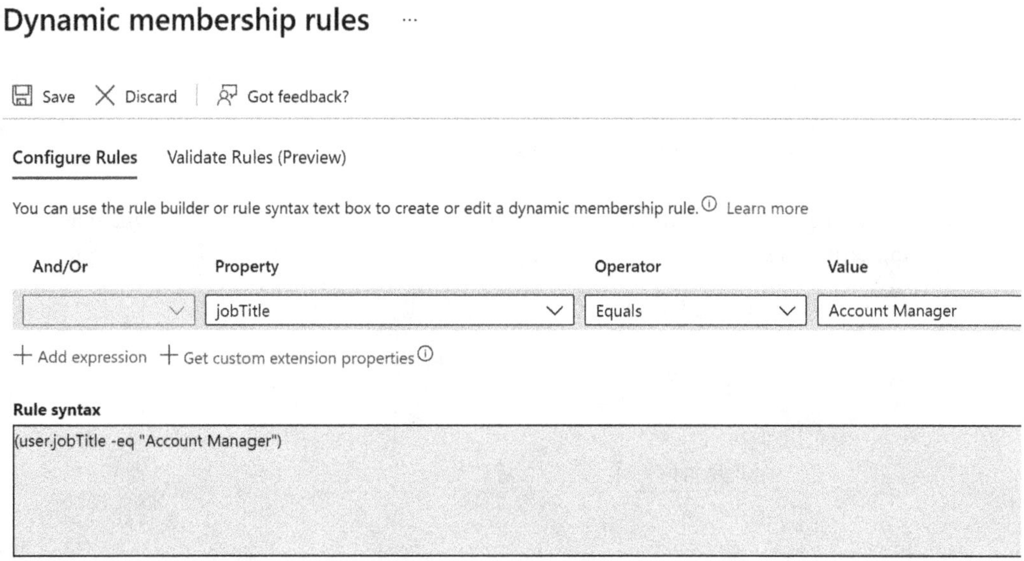
Dynamic membership rules
Save
Discard
Got feedback?
Configure Rules
Validate Rules (Preview)
You can use the rule builder or rule syntax text box to create or edit a dynamic membership rule.
Learn more
And/Or
Property
Operator
Value
jobTitle
Equals
Account Manager
Add expression
Get custom extension properties
Rule syntax
(user.jobTitle -eq "Account Manager")

Figure 2.15 – Dynamic membership rules

As shown in the preceding screenshot, we use the following properties:

- **jobTitle** as the property
- **Equals** as the operator
- **Account Manager** as the value

Then, proceed to save the configuration and create the group.

After a few seconds, you should see the new group in the Azure portal, as shown here:

Name ↑
Object Id
Group type
Membership type
SA
Sales
764bac44-3536-462e-852b-8f5f2eca818c
Security
Dynamic

Figure 2.16 – The group in the Azure AD tenant

5. Now, go to the **Sales** group and then choose **Members**, and note that the user we previously created in *task 2* has been automatically added to the **Sales** group, as shown in the following figure:

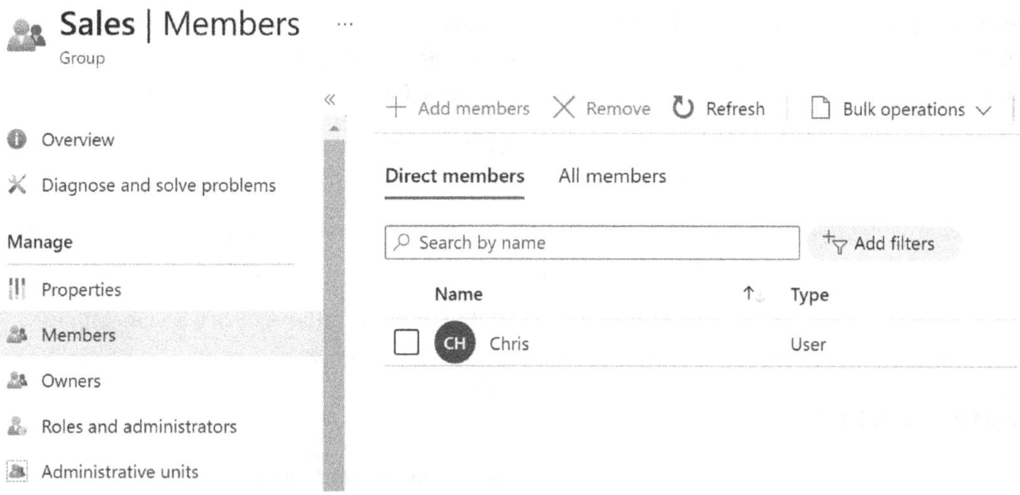
Sales | Members
Group
Overview
Diagnose and solve problems
Manage
Properties
Members
Owners
Roles and administrators
Administrative units
Add members
Remove
Refresh
Bulk operations
Direct members
All members
Search by name
Add filters
Name
Type
CH
Chris
User

Figure 2.17 – The Sales group – membership

> **Note**
>
> When a group membership rule is applied, the user and device properties are evaluated for matches with the membership rule.

This completes *task 3*, in which we reviewed how to work with dynamic groups.

Working with dynamic groups is just one of the variety of benefits included in Azure AD Premium P1. Hopefully, you now better understand Azure AD and some of its capabilities.

So far, we have reviewed the core need to adopt an IAM solution and the benefits of leveraging Azure AD. And very often, organizations might already have an identity management solution on-premises, such as a traditional Active Directory.

Now, let's understand how you can leverage the capabilities from your local on-premises directories with cloud-based directories, through the utilization of tools such as Azure AD Connect sync and cloud sync.

Hybrid identity – integrating your on-premises directories (Azure AD Connect sync and cloud sync)

Organizations can leverage their existing Active Directory on-premises and Azure AD to modernize their identity infrastructure. Integrating on-premises directories with Azure AD gives organizations a common identity to access cloud and on-premises resources.

A prevalent practice is establishing directory synchronization, which helps organizations synchronize their identities or objects, including users, groups, contacts, and devices, between different directories. This is typically configured between an on-premises Active Directory environment and Azure AD.

This approach, commonly referred to as a **hybrid identity**, helps organizations provide users with a common identity across on-premises or cloud-based services.

Microsoft provides two main tools to achieve the configuration of a hybrid identity approach – **Azure AD Connect Sync** and **Azure AD Connect Cloud Sync**.

By leveraging hybrid identity, organizations can extend their footprint to Azure, while maintaining seamless access for their users, by synchronizing their existing identities to the cloud.

Azure AD Connect sync

Azure AD Connect sync can simplify the synchronization process and manage complex topologies. While it was built using traditional architecture, it has helped many organizations bridge the gap between on-premises and the cloud.

The benefits of using Azure AD Connect Sync include the following:

- Leveraging existing on-premises infrastructure, such as Windows Server Active Directory, and enabling a common hybrid identity across on-premises and cloud-based services
- Enabling features from Azure AD, such as Conditional Access, to provide access based on application resource, location, and MFA
- Allowing developers to build applications using a common identity model

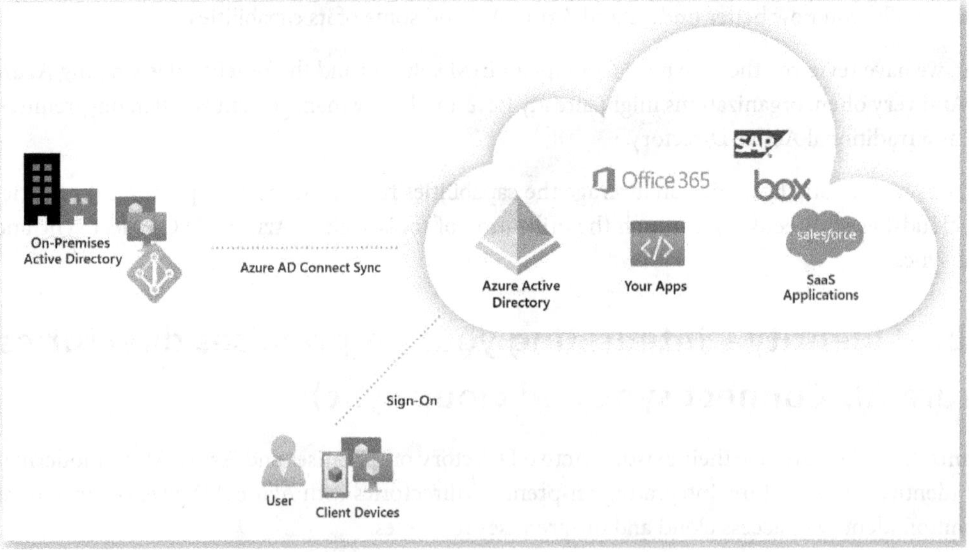
On-Premises
Active Directory
Azure AD Connect Sync
Office 365
Azure Active
Directory
Your Apps
SaaS
Applications
Sign-On
User
Client Devices

Figure 2.18 – An Azure AD Connect diagram

Azure AD Connect provides multiple capabilities:

- Synchronization
- Password hash synchronization
- Pass-through authentication
- Federation integration
- Health monitoring

When enabling a hybrid identity solution, it is crucial to decide on the correct authentication method that will be leveraged. As part of the selection process for the authentication method, there are major factors to consider, such as what the existing infrastructure is, the level of complexity, cost, and time.

There are three main scenarios that you can leverage as a baseline to select the most appropriate authentication method:

- Cloud authentication with Azure AD password hash synchronization
- Cloud authentication with Azure AD pass-through authentication
- Federated authentication

Let's look at them in detail next.

Cloud authentication with Azure AD password hash synchronization

This is the simplest approach to enable hybrid identity, as organizations can enable authentication for on-premises Active Directory objects in Azure AD.

Using this method, users are able to use the existing username and password they use on-premises.

Consider using this method if your organization needs to only enable users to sign in to SaaS applications, Microsoft 365, and other Azure AD resources.

In this case, Azure AD Connect sync will enable password hash synchronization and execute the sync process every two minutes.

Using advanced capabilities from Azure AD, such as Identity Protection, included in the Azure AD Premium P2 edition, organizations can leverage insights provided by Azure AD to get a report of leaked credentials. Microsoft finds leaked credentials on public sites such as `pastebin.com` and `paste.ca` and through law enforcement agencies and dark web research.

These leaked credentials are processed and published in batches, usually multiple times per day.

The following diagram shows a high-level overview of cloud authentication with Azure AD password hash synchronization.

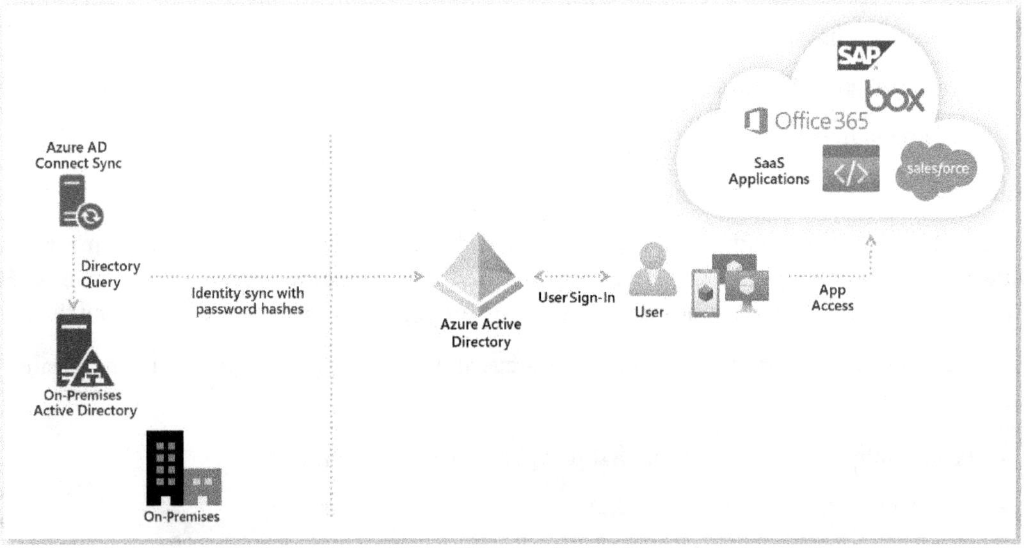
SAP
box
Office 365
SaaS Applications
salesforce
Azure AD Connect Sync
Directory Query
Identity sync with password hashes
On-Premises Active Directory
On-Premises
Azure Active Directory
User Sign-In
User
App Access

Figure 2.19 – A diagram of cloud authentication with Azure AD password hash synchronization

Cloud authentication with Azure AD password hash synchronization means that when you create a password on your on-premises server, it is synchronized with the cloud-based directory. This allows you to use the same password to access both your on-premises and cloud-based resources, simplifying the login process. Additionally, password hash synchronization helps to keep your login credentials secure by encrypting them before they are sent to the cloud.

Cloud authentication with Azure AD pass-through authentication

The second approach to enable hybrid identity is by using pass-through authentication. In this method, organizations can install a software agent that runs on the on-premises servers. Then, the servers validate users directly with the on-premises Active Directory. This way, the password validation process will not happen in the cloud. This method provides a simple password validation.

Consider using this method if your organization needs to enforce password policies or on-premises user account states for security purposes. When using this approach, we strongly recommend using at least an additional pair of pass-through authentication agents for the high availability of the authentication requests.

The next figure shows a high-level overview of cloud authentication with Azure AD pass-through authentication. When a user logs in to a cloud-based service, the username and password are verified by the on-premises server instead of being stored in the cloud. This helps to keep login credentials secure while still allowing you to access cloud-based services.

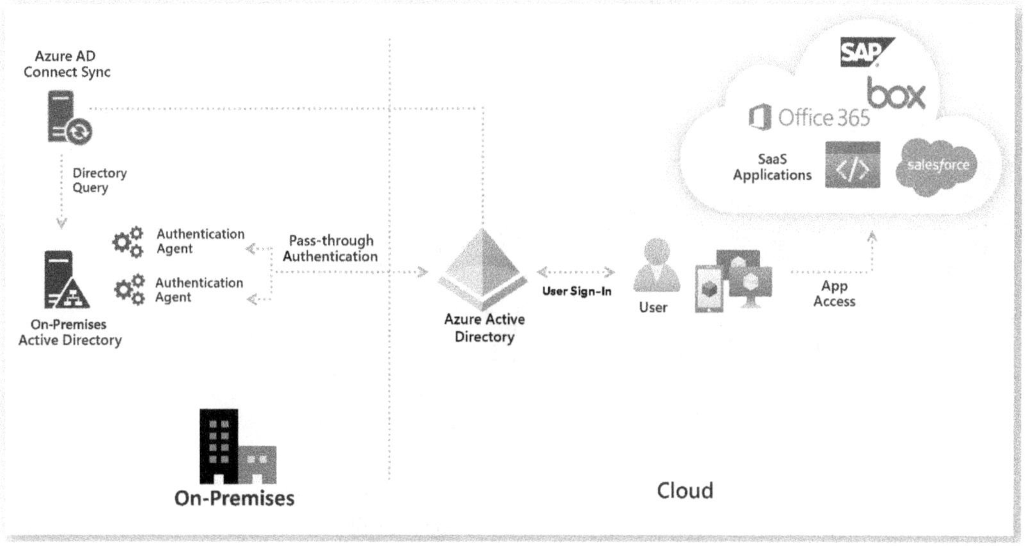
Azure AD
Connect Sync
Directory
Query
Authentication
Agent
Authentication
Agent
Pass-through
Authentication
On-Premises
Active Directory
Azure Active
Directory
User Sign-In
User
App
Access
SAP
box
Office 365
SaaS
Applications
salesforce
On-Premises
Cloud

Figure 2.20 – A diagram of cloud authentication with Azure AD pass-through authentication

Cloud authentication with Azure AD pass-through authentication means that when you sign in to a cloud-based application or service, your login credentials are checked by a server located on-premises, rather than in the cloud. This provides an extra layer of security for your login information while still allowing you to access cloud-based resources.

Federated authentication

With this approach, the authentication process happens through a separate trusted authentication system. This can be the on-premises Active Directory Federation Services or any other external trusted system, and it will validate the user's password.

Azure AD delegates the authentication process, and the organization is responsible for maintaining the federated system and for the security configuration to properly authenticate users.

Consider using this method if your organization has requirements that are not fully covered by Azure AD capabilities – for example, a sign-in that requires additional attributes instead of a **User Principal Name (UPN)**.

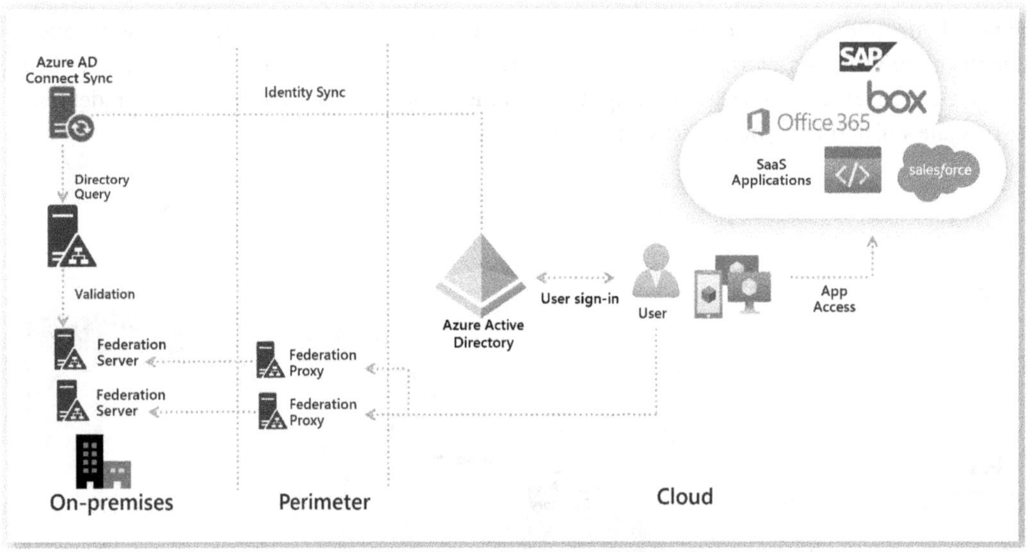
Azure AD
Connect Sync
Identity Sync
Directory
Query
Validation
Federation
Server
Federation
Server
Federation
Proxy
Federation
Proxy
SAP
box
Office 365
SaaS
Applications
salesforce
Azure Active
Directory
User sign-in
User
App
Access
On-premises
Perimeter
Cloud

Figure 2.21 – A diagram of federated authentication

Now, you can see that Azure AD Connect sync offers a rich set of features to enable a hybrid identity solution. However, in some cases, it can be limited, as it doesn't fully leverage the advantages of the cloud. That's where the second tool comes into play – Azure AD Connect cloud sync.

Azure AD Connect cloud sync

Azure AD Connect cloud sync is built on a new architecture and aims to reduce customers' on-premises footprints, as it focuses on a cloud-first approach.

How does it work? An organization can deploy lightweight agents in their on-premises environment to connect Active Directory and Azure AD. Then, all the configurations are completed in the cloud.

The lightweight agents installed on-premises are the bridge from Active Directory to Azure AD, and multiple agents can be leveraged for highly available deployments, or when organizations need to use password hash synchronization.

There are different capabilities that can be used with Azure AD Connect cloud sync and Azure AD Connect sync. Here are the most relevant features from Azure AD Connect cloud sync:

- Connectivity to multiple Active Directory forests
- High availability by using multiple agents
- Password write-back support
- Support for multiple objects including user, group, and contact objects

For a full comparison between Azure AD Connect sync and Azure AD Connect cloud sync, review the following URL: `https://bit.ly/azuread-cloud-sync-comparison`.

Azure AD Connect cloud sync will help you achieve the synchronization process through lightweight agents instead of using the Azure AD Connect application.

By using agents, there are more flexible configurations that can be achieved, such as synchronizing to an Azure AD tenant from a multi-forest on-premises Active Directory.

When using Azure AD Connect cloud sync, the provisioning process to Azure AD is done through Microsoft online services.

On the other hand, we need to consider how cloud or remote users who need access to internal on-premises resources can authenticate and be authorized to perform their daily work. We will address how to achieve this by using Azure AD Application Proxy.

Azure AD Application Proxy

Azure AD offers another capability to protect users, applications, and data on-premises. One of these features is **Azure AD Application Proxy** (**Azure AD App Proxy**). This feature can help organizations publish their on-premises applications outside of the local network.

By leveraging Azure AD App Proxy, external users can access internal applications securely. Azure AD App Proxy supports the publishing of internal applications to the external world without needing a perimeter network, also known as a DMZ, and maintains the benefits of single sign-on, MFA, and centralized control of identity.

Configuration of Azure AD App Proxy is done through the Azure portal. You can publish the external HTTPS URL endpoint in Azure. Then, the endpoint connects to the internal local application server URL. This way, users are able to access the on-premises applications just as they access any other SaaS application.

The components included in this mechanism include the following:

- **The Application Proxy service**: This service runs in the cloud
- **The Application Proxy connector**: This is an agent installed on an on-premises server
- **Azure AD**: This acts as the identity provider

The user signs in, and Azure AD App Proxy provides remote access and single sign-on to the services.

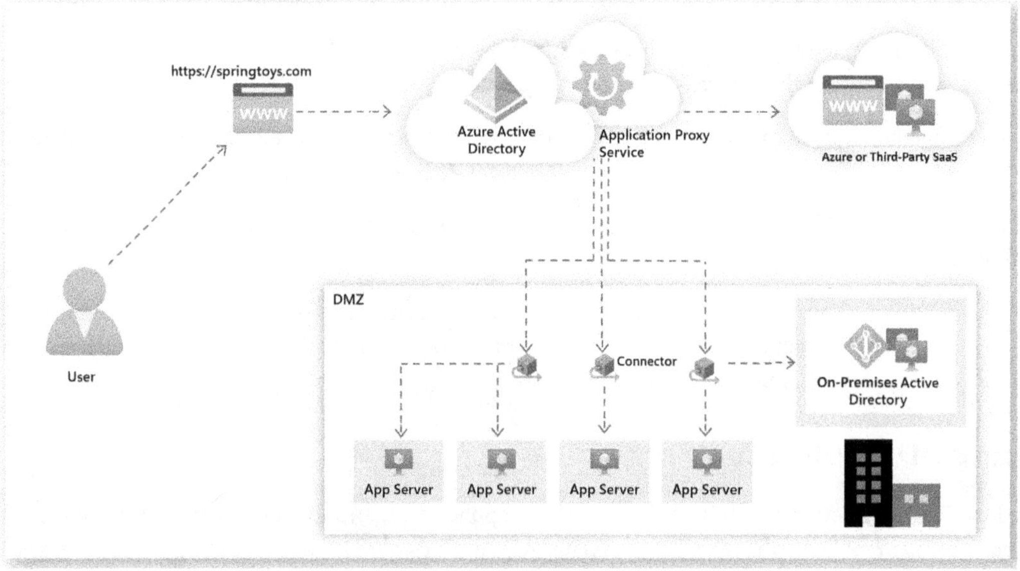
https://springtoys.com
WWW
Azure Active Directory
Application Proxy Service
WWW
Azure or Third-Party SaaS
DMZ
User
Connector
On-Premises Active Directory
App Server
App Server
App Server
App Server

Figure 2.22 – An Azure AD App Proxy diagram

Azure AD App Proxy improves the security posture of organizations by taking advantage of its features, including the following:

- **Pre-authentication**: To ensure that only authenticated users are eligible to access resources.
- **Conditional access**: Organizations can specify restrictions for allowed traffic to the application.
- **Traffic termination**: The Application Proxy service terminates all traffic in the cloud and re-establishes sessions with the server.
- **Outbound access**: The Application Proxy connectors manage the outbound connections to the Application Proxy service, typically using ports `80` and `443`. Therefore, there's no need to manage firewall ports.
- **Security analytics**: By using Azure AD Premium P2, organizations can leverage Identity Protection and identify compromised accounts in real time.

As organizations extend their workloads in the cloud, balancing security for remote access has been one of the main challenges in the past few years. Making resources accessible to the right people at the right moment has been of utmost importance.

Azure AD brings another benefit to the table by leveraging identity-driven signals by using Conditional Access so that organizations can easily take access control decisions.

Azure AD Conditional Access

Conditional Access is the core of the identity-driven control plane included in Azure AD. This is a premium feature that allows organizations to leverage a policy-based mechanism to deny or permit access to resources in the cloud or on-premises.

Consider using Conditional Access if your organization has complex security requirements. Think of Conditional Access as an `if-then` statement; you can use signals available in Azure AD such as users and groups, locations, applications, devices, and risk. Based on these signals, when users try to access a resource, they will be prompted to complete an action.

This way, you will be able to enforce access requirements if a specific condition is met.

The following figure shows a high-level overview of Conditional Access:

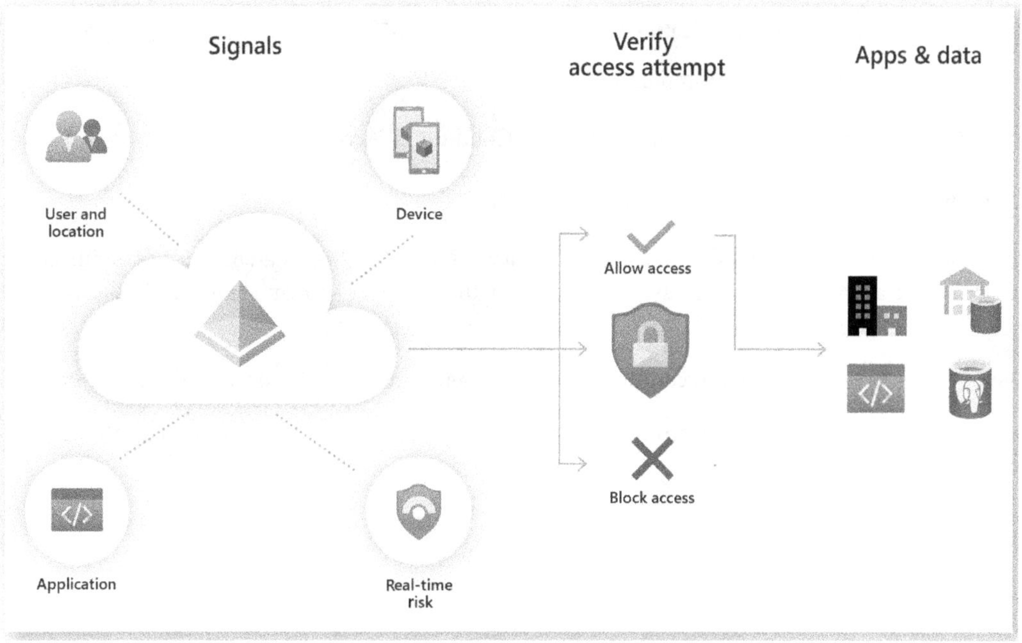
Signals
Verify
access attempt
Apps & data
User and
location
Device
Allow access
Block access
Application
Real-time
risk

Figure 2.23 – Conditional Access

Some of the signals you can use for evaluation in policy decision-making include the following:

- User or group membership
- The IP location
- The device

- The application
- Real-time and calculated risk detection

Decisions can be the following:

- **Block access**: The user is blocked.
- **Grant access**: There are multiple options if access is granted. Organizations can require an additional action before providing access to the resource, such as one of the following:
    - MFA
    - A device to be compliant
    - A hybrid Azure AD-joined device
    - An approved client application
    - An application protection policy

Next, let's review how you can configure Azure AD Conditional Access.

Azure AD Conditional Access configuration

We previously mentioned that Conditional Access acts as an `if-then` statement; *if* a condition is met, *then* we can apply specific access controls. Configuring the `if-then` statement is referred to as the Conditional Access policy.

When you create a Conditional Access policy, you can configure the *assignments* and *access controls*.

Think of *assignments* as the conditions that are to be met in the policy and *access controls* as the action that will be performed if the condition is met.

The following screenshot shows the initial configuration of an Azure AD Conditional Access policy.

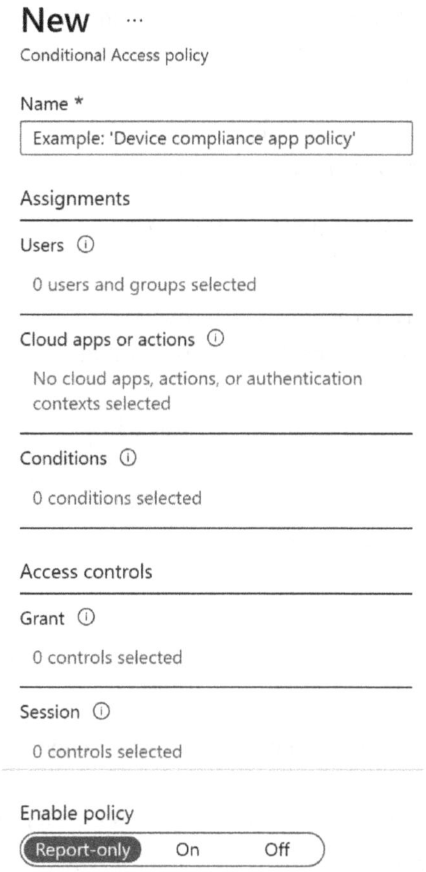
New ...
Conditional Access policy
Name *
Example: 'Device compliance app policy'
Assignments
Users
0 users and groups selected
Cloud apps or actions
No cloud apps, actions, or authentication contexts selected
Conditions
0 conditions selected
Access controls
Grant
0 controls selected
Session
0 controls selected
Enable policy
Report-only
On
Off

Figure 2.24 – Conditional Access – a new policy

Note that under the **Assignments** section, we configure the users and groups that will be impacted by the Conditional Access policy, and we can configure the cloud apps, actions, or authentication contexts to which the policy applies. Conditions are also defined under this section.

The **Access controls** section will allow you to specify whether access should be granted or blocked. In the session configuration, you define how to control access to cloud apps.

In short, the **Assignments** section will help you control the who, what, and where of the policy, the access controls, and how the policy is enforced.

> **Note**
>
> You'll need to use Azure AD Premium P1 and Microsoft 365 E3 for Conditional Access with standard rules, or Azure AD Premium P2 to use the risky sign-in, risky users, and risk-based sign-in features included in Identity Protection.

Exercise – creating a Conditional Access policy

Now, we will create a Conditional Access policy and assign it to a group of users. We will enforce the use of MFA when a user signs in to the Azure portal:

1. First, go to the Azure portal using an account with the Global Administrator role. Then, from the top search bar, look for the Azure AD Conditional Access service, as shown in the following figure:

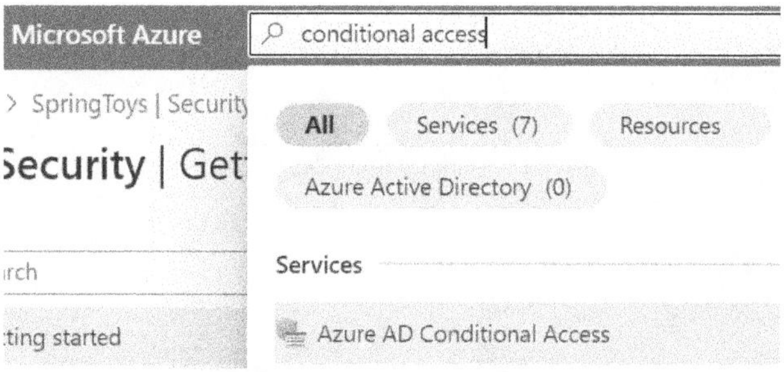
Microsoft Azure
conditional access
> SpringToys | Security
Security | Get
All
Services (7)
Resources
Azure Active Directory (0)
Services
Azure AD Conditional Access

Figure 2.25 – The Azure search bar – Azure AD Conditional Access

2. From there, you can select the + **New Policy** option. Now, we will proceed to configure the policy.
3. Provide a name for the policy, `MFA enforcement`. Then, select the **Users** option under **Assignments** and choose the **Select users and groups** option. Then, we will choose the IT group, as shown here:

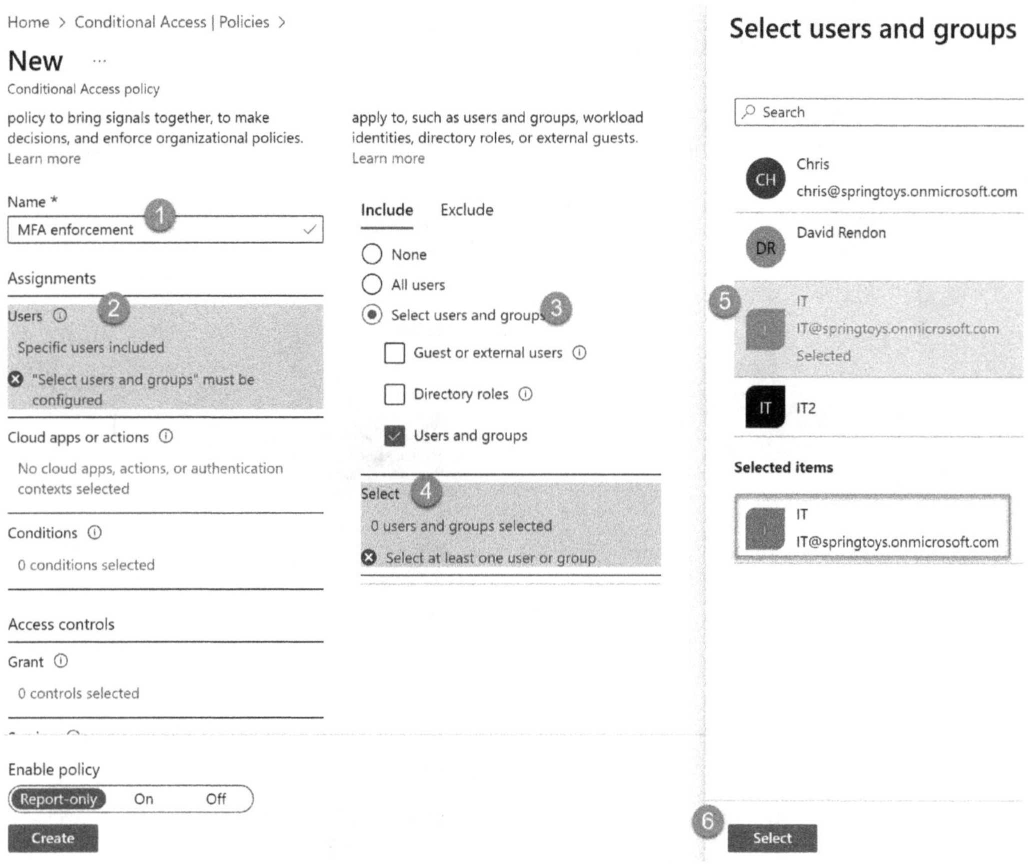
Home > Conditional Access | Policies >
New
Conditional Access policy
policy to bring signals together, to make decisions, and enforce organizational policies. Learn more
Name *
MFA enforcement
Assignments
Users
Specific users included
"Select users and groups" must be configured
Cloud apps or actions
No cloud apps, actions, or authentication contexts selected
Conditions
0 conditions selected
Access controls
Grant
0 controls selected
Enable policy
Report-only
On
Off
Create
apply to, such as users and groups, workload identities, directory roles, or external guests. Learn more
Include
Exclude
None
All users
Select users and groups
Guest or external users
Directory roles
Users and groups
Select
0 users and groups selected
Select at least one user or group
Select users and groups
Search
Chris
chris@springtoys.onmicrosoft.com
David Rendon
IT
IT@springtoys.onmicrosoft.com
Selected
IT2
Selected items
IT
IT@springtoys.onmicrosoft.com
Select
1
2
3
4
5
6

Figure 2.26 – Configuring the new Conditional Access policy

Next, we will configure the conditions to enforce MFA.

> **Note**
>
> The `IT Spring Toys Group` is a security group created for this exercise.

4. Under the section called **Cloud apps or actions**, select the **Select what this policy applies to** option. Then, we will select **All cloud apps**, as shown here:

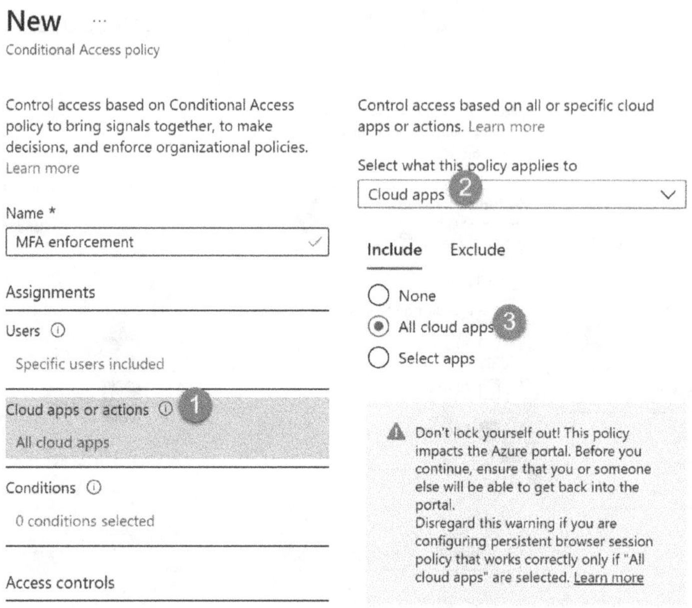
New
Conditional Access policy
Control access based on Conditional Access policy to bring signals together, to make decisions, and enforce organizational policies. Learn more
Name *
MFA enforcement
Assignments
Users
Specific users included
Cloud apps or actions
1
All cloud apps
Conditions
0 conditions selected
Access controls
Control access based on all or specific cloud apps or actions. Learn more
Select what this policy applies to
Cloud apps
2
Include
Exclude
None
All cloud apps
3
Select apps
Don't lock yourself out! This policy impacts the Azure portal. Before you continue, ensure that you or someone else will be able to get back into the portal.
Disregard this warning if you are configuring persistent browser session policy that works correctly only if "All cloud apps" are selected. Learn more

Figure 2.27 – Cloud apps or actions configuration

The next step is to configure the access controls. This will allow organizations to define the specific requirements for a user to be blocked or granted access.

5. Go to the **Access controls** section and select the **Grant** option. From there, we will grant access to users and select the **Require multifactor authentication** checkbox, as shown here:

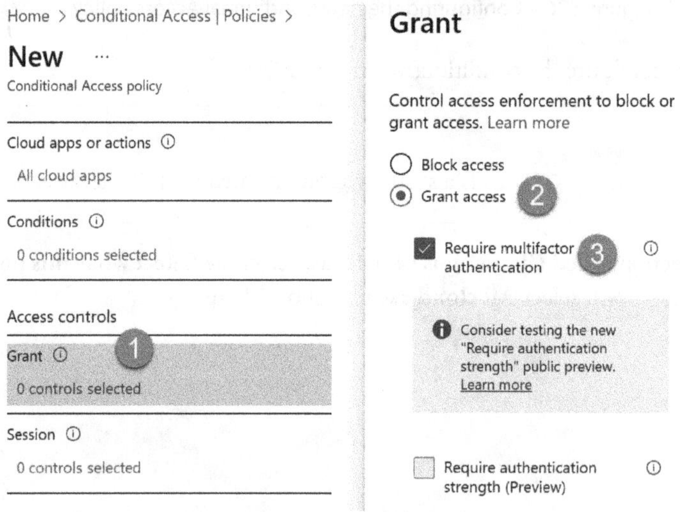
Home > Conditional Access | Policies >
New
Conditional Access policy
Cloud apps or actions
All cloud apps
Conditions
0 conditions selected
Access controls
Grant
1
0 controls selected
Session
0 controls selected
Grant
Control access enforcement to block or grant access. Learn more
Block access
Grant access
2
Require multifactor authentication
3
Consider testing the new "Require authentication strength" public preview.
Learn more
Require authentication strength (Preview)

Figure 2.28 – Configuring the Access controls section in Conditional Access

Once the **Access controls** section is configured, we can proceed to activate the policy.

There are multiple options to activate your policy. Before impacting users, you can leverage the **Report-only** option to see how the policy would affect users. You can set it to **On** or **Off** as needed.

Enable policy
Report-only
On
Off
Create

Figure 2.29 – Enable policy in Conditional Access

Administrators in an organization might need specific elevated access to resources to perform certain privileged operations, and while Conditional Access might help, we want to reduce as much as possible the number of people who access secure information and resources.

For those scenarios, organizations can leverage PIM to protect data and resources. PIM is one of the services included in Azure AD to manage, control, and monitor access to important resources in the organization.

Azure AD PIM

PIM enables time-based and approval-based role activation so that users can access and perform operations during a specific timeframe with a specific role. This helps reduce the risk of misused or unnecessary access to resources.

In short, PIM is a service to help you manage access to privileged resources. This feature is included in Azure AD Premium P2 or Enterprise Mobility + Security E5 licenses.

What do we mean by access to privileged resources? This refers to resources included in Azure, Azure AD, and Microsoft online services such as Microsoft Intune and Microsoft 365.

The most relevant capabilities of PIM include the following:

- **Enabling just-in-time access to Azure and Azure AD resources**: This can help you protect your Azure resources – for example, protecting Azure VMs from unauthorized access. Just-in-time allows access to the VMs only when needed, and you can specify through which ports and for how long.
- **Requiring approval to activate privileged roles**: In PIM, a user who is a Global Administrator or Privileged Role Administrator can manage assignments for other administrators. You can enable this feature so that users can gain approval to activate a role in Azure PIM.
- **Mail notifications**: This can be configured when privileged roles are activated.
- **Audit history**: This resource audit provides you with a view of all the activity related to Azure AD roles.

PIM can help you improve the protection of your resources in Azure. The first step to enable protection of your resources is to define the resources that should be protected with PIM.

> **Note**
>
> Ensure that you elevate your access to all subscriptions and management groups. This can be done using the Azure portal or the Azure AD admin center as a Global Administrator. Then, select **Properties** under the **Manage** section, toggle the **Access management for Azure resources** option to **Yes**, and select **Save**, as shown in the following figure.

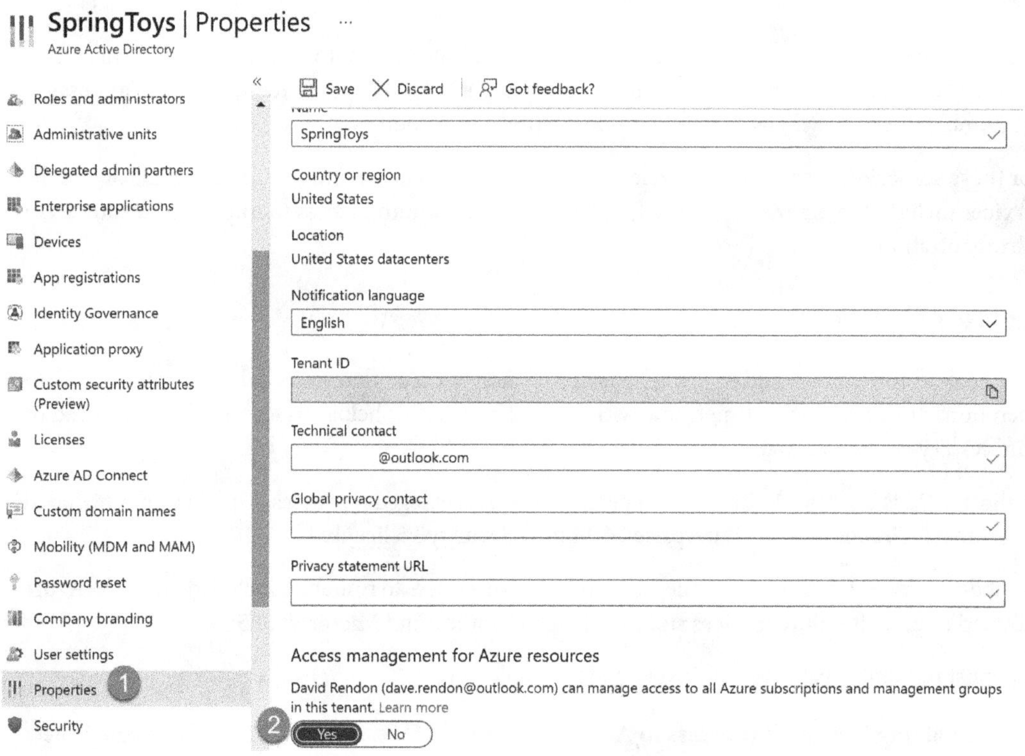
SpringToys | Properties
Azure Active Directory
Save
Discard
Got feedback?
Roles and administrators
Administrative units
Delegated admin partners
Enterprise applications
Devices
App registrations
Identity Governance
Application proxy
Custom security attributes (Preview)
Licenses
Azure AD Connect
Custom domain names
Mobility (MDM and MAM)
Password reset
Company branding
User settings
Properties
Security
SpringToys
Country or region
United States
Location
United States datacenters
Notification language
English
Tenant ID
Technical contact
@outlook.com
Global privacy contact
Privacy statement URL
Access management for Azure resources
David Rendon (dave.rendon@outlook.com) can manage access to all Azure subscriptions and management groups in this tenant. Learn more
1
2
Yes
No

Figure 2.30 – Elevated access for a global admin user

Now, let's see how you can assign roles in PIM.

Assigning roles in PIM

1. In the Azure portal, search for `Privileged Identity Management`, and under the **Manage** menu, select the **Roles** option, as shown in the following figure:

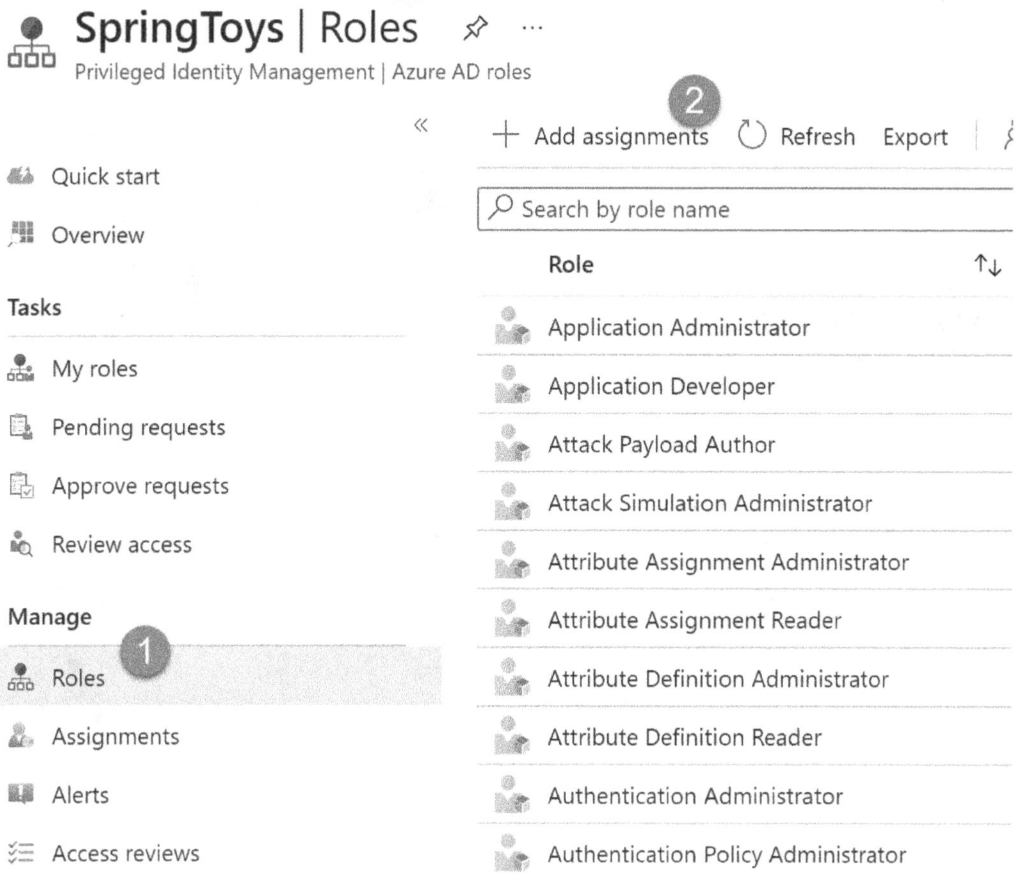
SpringToys | Roles
Privileged Identity Management | Azure AD roles
2
Add assignments
Refresh
Export
Quick start
Overview
Tasks
My roles
Pending requests
Approve requests
Review access
Manage
1
Roles
Assignments
Alerts
Access reviews
Search by role name
Role
Application Administrator
Application Developer
Attack Payload Author
Attack Simulation Administrator
Attribute Assignment Administrator
Attribute Assignment Reader
Attribute Definition Administrator
Attribute Definition Reader
Authentication Administrator
Authentication Policy Administrator

Figure 2.31 – Role assignments in PIM

2. Now, proceed to add a new assignment. You can select the role needed and then add the member or group, as shown here:

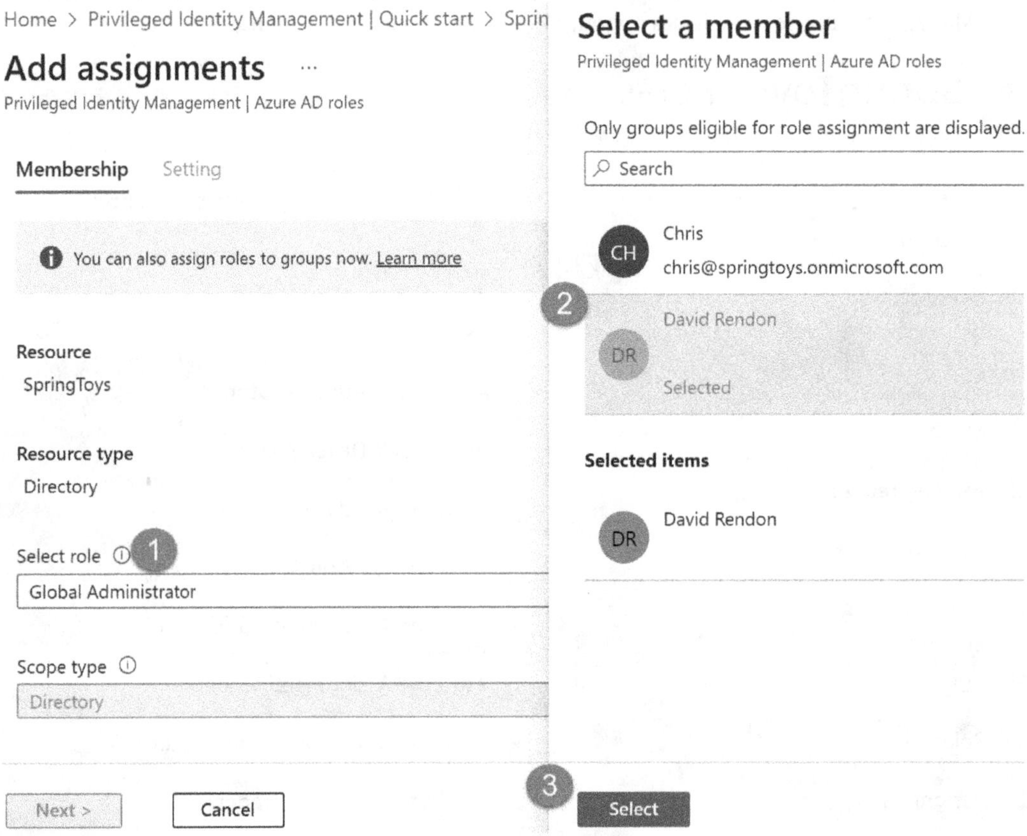
Home > Privileged Identity Management | Quick start > Sprin
Add assignments ...
Privileged Identity Management | Azure AD roles
Membership
Setting
You can also assign roles to groups now. Learn more
Resource
SpringToys
Resource type
Directory
Select role
Global Administrator
Scope type
Directory
Next >
Cancel
Select a member
Privileged Identity Management | Azure AD roles
Only groups eligible for role assignment are displayed.
Search
CH
Chris
chris@springtoys.onmicrosoft.com
DR
David Rendon
Selected
Selected items
DR
David Rendon
Select
1
2
3

Figure 2.32 – Add assignments in PIM

3. Once you select the user or group, the next step is to define the *assignment type*. Assignment types can be **Eligible** or **Active**. *Eligible* means that the member needs to activate the role before using it.

   An *active* assignment means that the member does not need to activate the role before using it. Note that you can also specify the assignment duration, as shown in the following figure:

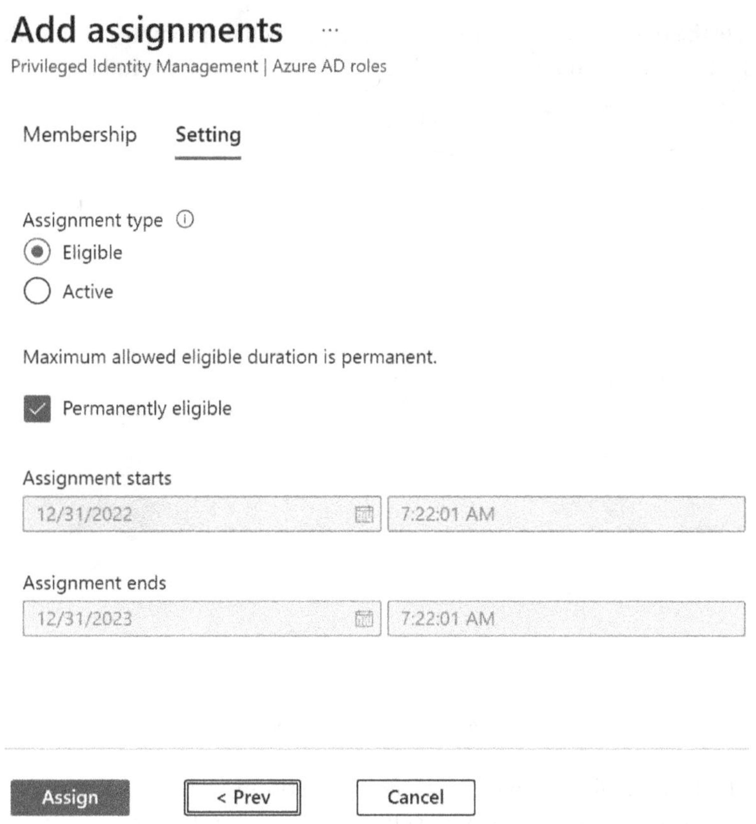
Add assignments ...
Privileged Identity Management | Azure AD roles
Membership
Setting
Assignment type
Eligible
Active
Maximum allowed eligible duration is permanent.
Permanently eligible
Assignment starts
12/31/2022
7:22:01 AM
Assignment ends
12/31/2023
7:22:01 AM
Assign
< Prev
Cancel

Figure 2.33 – Assignments in PIM

4. Back on the main PIM page, go to the **My roles** option, and you should see now the role we just assigned, as shown in the following figure:

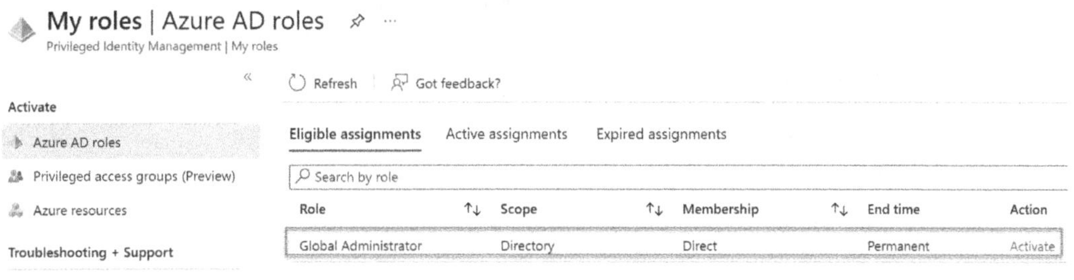
My roles | Azure AD roles
Privileged Identity Management | My roles
Refresh
Got feedback?
Activate
Azure AD roles
Privileged access groups (Preview)
Azure resources
Troubleshooting + Support
Eligible assignments
Active assignments
Expired assignments
Search by role
Role
Scope
Membership
End time
Action
Global Administrator
Directory
Direct
Permanent
Activate

Figure 2.34 – Azure AD roles in PIM

5. Let's activate the role. When you activate the role, you will be prompted to a blade where you can specify the duration in hours in which this role can perform operations.

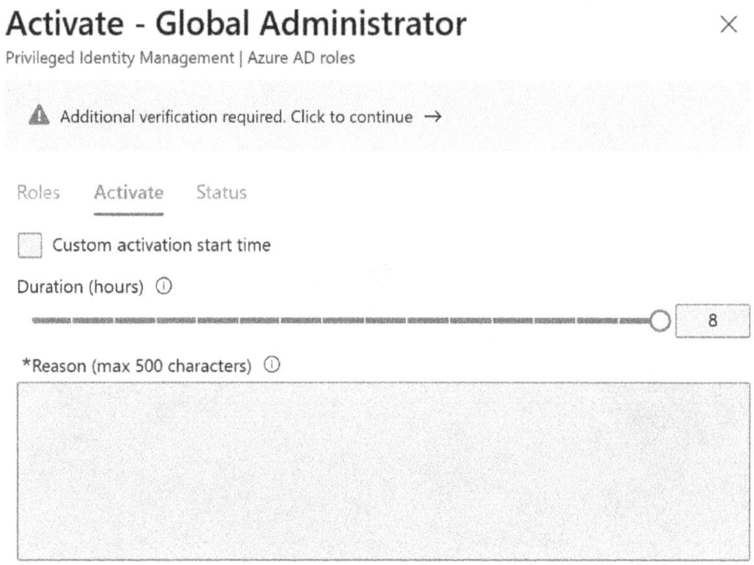
Activate - Global Administrator
Privileged Identity Management | Azure AD roles
Additional verification required. Click to continue →
Roles
Activate
Status
Custom activation start time
Duration (hours)
8
*Reason (max 500 characters)

Figure 2.35 – Activating roles in PIM

Note that, for this specific example, we will need to perform an additional verification process. This includes the use of a verification process through the Microsoft Authenticator app or a phone.

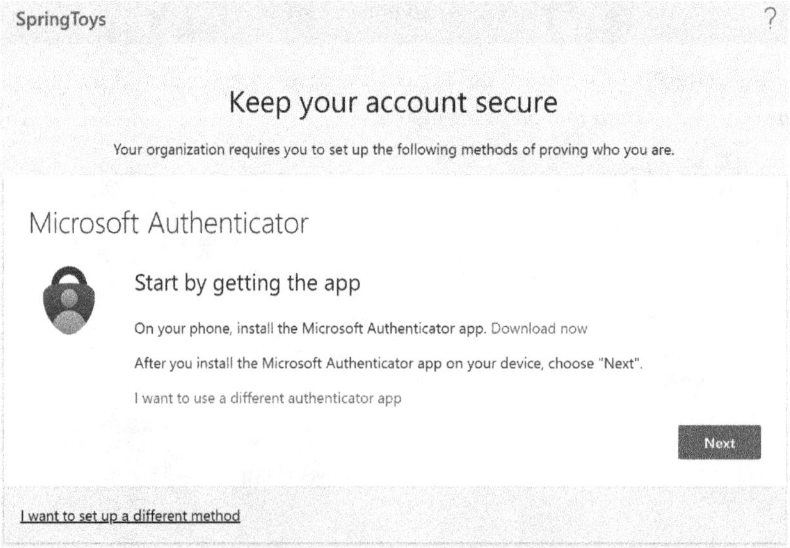
SpringToys
?
Keep your account secure
Your organization requires you to set up the following methods of proving who you are.
Microsoft Authenticator
Start by getting the app
On your phone, install the Microsoft Authenticator app. Download now
After you install the Microsoft Authenticator app on your device, choose "Next".
I want to use a different authenticator app
Next
I want to set up a different method

Figure 2.36 – Verification options for PIM

The following screenshot shows the verification using SMS.

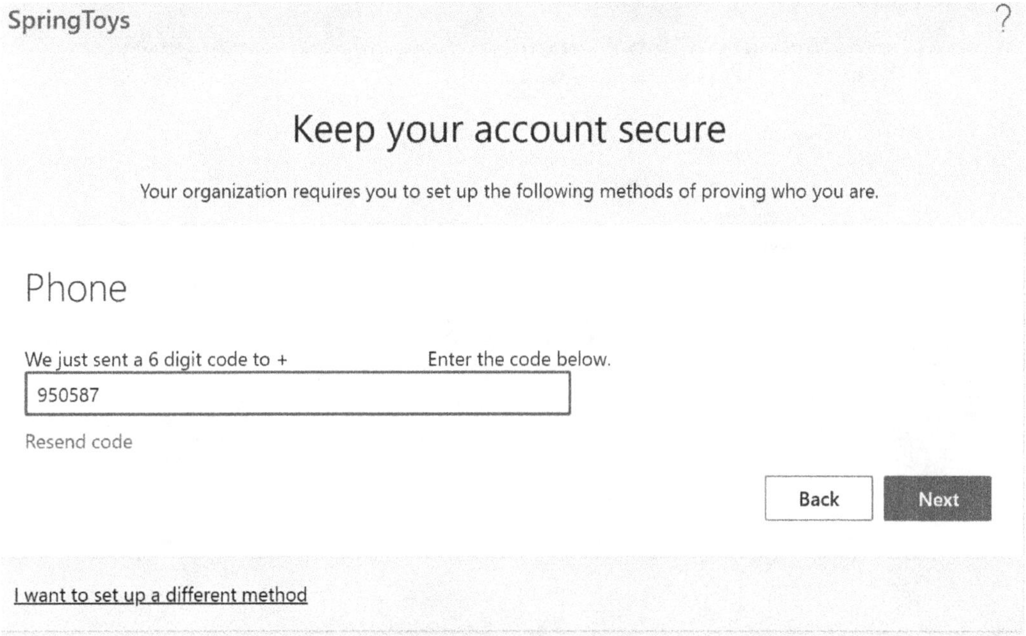
SpringToys
?
Keep your account secure
Your organization requires you to set up the following methods of proving who you are.
Phone
We just sent a 6 digit code to +
Enter the code below.
950587
Resend code
Back
Next
I want to set up a different method

Figure 2.37 – The verification process in PIM

6. Once verified, a notification will appear, saying that the phone was registered successfully.

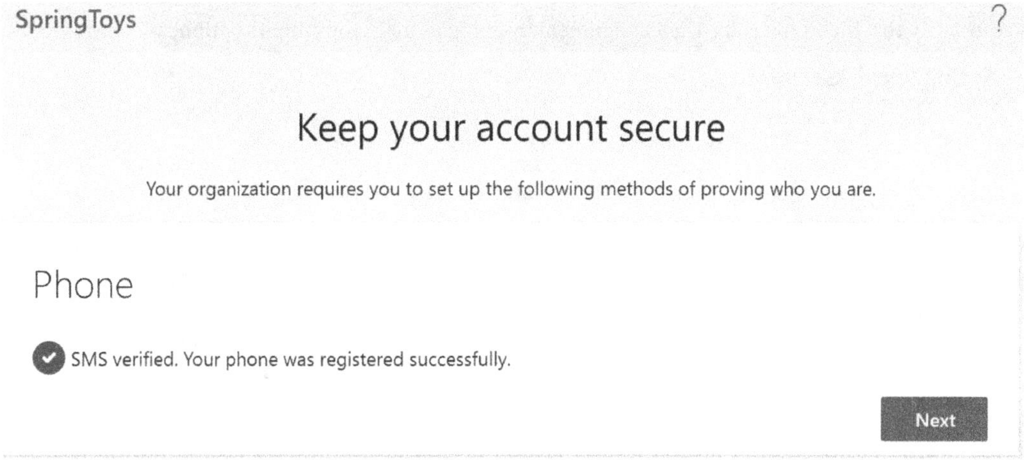
SpringToys
?
Keep your account secure
Your organization requires you to set up the following methods of proving who you are.
Phone
SMS verified. Your phone was registered successfully.
Next

Figure 2.38 – Phone verification in PIM

7. Now, you will be redirected to the previous page, and you will be able to provide justification to activate the role. Then, activate the role:

Activate - Global Administrator
Privileged Identity Management | Azure AD roles
Roles
Activate
Status
Custom activation start time
Duration (hours)
8
*Reason (max 500 characters)
Perform privileged operations
Activate
Cancel

Figure 2.39 – Activating a role in PIM

You can verify that the user has the right assignment by going to the **Assignments** option, as shown here:

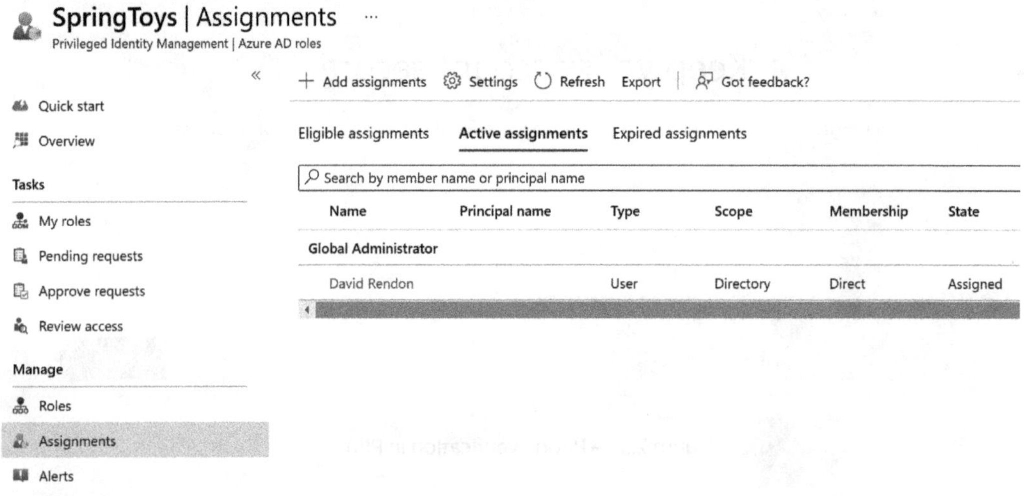
SpringToys | Assignments
Privileged Identity Management | Azure AD roles
Add assignments
Settings
Refresh
Export
Got feedback?
Quick start
Overview
Tasks
My roles
Pending requests
Approve requests
Review access
Manage
Roles
Assignments
Alerts
Eligible assignments
Active assignments
Expired assignments
Search by member name or principal name
Name
Principal name
Type
Scope
Membership
State
Global Administrator
David Rendon
User
Directory
Direct
Assigned

Figure 2.40 – Assignments in PIM

Now, you can utilize this user to manage, control, and monitor access to resources in your organization across Azure AD, Azure, Microsoft 365, or Intune.

Summary

In this chapter, we reviewed the most relevant features of the IAM capabilities you can leverage using Azure AD. You should now have a better understanding of how you can implement mechanisms to protect access to resources and manage users inside and outside your organization.

In the next chapter, we will address security controls that can be implemented to protect your organizations from potential attackers, who use accounts to gain access to sensitive resources in your network.

3

Using Microsoft Sentinel to Mitigate Lateral Movement Paths

This chapter explains Microsoft Sentinel's capabilities to detect and investigate advanced security threats, compromised identities, and potentially malicious actions in our organization.

Lateral movement is a technique used by cyber attackers to move across a network once they have gained access to one device. Microsoft Sentinel is a tool that helps to detect and respond to cyber threats.

In this chapter, we will review how organizations can identify suspicious activity and prevent lateral movement by setting up alerts and automated responses to potential threats using Microsoft Sentinel, helping to protect a network from cyberattacks and keep sensitive information safe.

By using Microsoft Sentinel to mitigate lateral movement paths, you can detect and prevent attackers from moving from one device to another within a network. This is important because once an attacker gains access to one device, they can use that device to move to other devices and potentially access sensitive information or cause damage.

In this chapter, we'll cover the following main topics:

- Understanding the Zero Trust strategy
- Understanding lateral movements
- Leveraging Microsoft Sentinel to improve your security posture
- Enabling Microsoft Sentinel
- Mitigating lateral movements

Let's get started!

Understanding the Zero Trust strategy

Identity protection has become a central part of adopting a Zero Trust strategy in organizations looking to improve their security posture. Zero Trust, a security model, describes an approach to designing and implementing systems to protect organizations better.

Zero Trust responds to modern enterprise trends that enable remote users, bring-your-own-device policies, and access to cloud-based resources from multiple locations.

Zero Trust principles are *verified explicitly*, *use least-privilege access*, *assume breach*, and focus on protecting resources, including assets, services, workflows, and network accounts. Therefore, a **Zero Trust Architecture** (**ZTA**) leverages these zero-trust principles to plan enterprise infrastructure and workflows.

A Zero Trust model provides a holistic security control plane, segmented into multiple layers of defense:

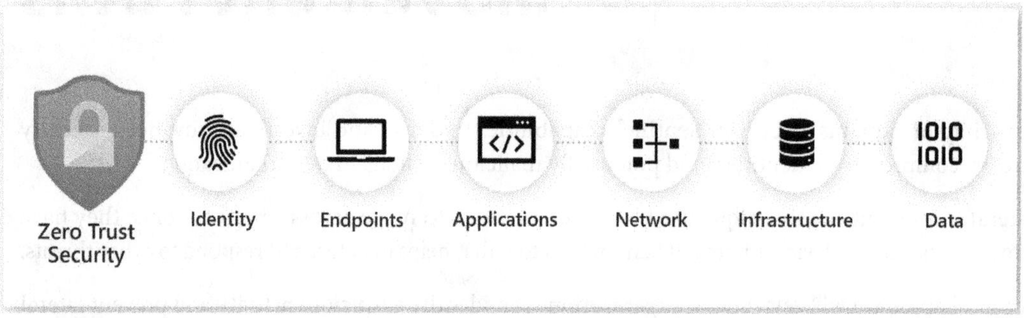
Zero Trust Security
Identity
Endpoints
</>
Applications
Network
Infrastructure
1010
1010
Data

Figure 3.1 – A Zero Trust model and its layers of defense

Let's examine each one in brief:

- **Identity**: As previously mentioned, identity has become the control plane to manage access to resources in a modern workplace. This includes establishing stronger authentication and authorization methods with controls and access policies with granular permissions to resources in the organization.
- **Endpoints**: This refers to devices that access your organization's resources; the goal is to verify the security of the devices that will access data and applications, enforcing least-privilege policies and implementing mechanisms to handle security breaches.
- **Applications**: As the core productivity tools for employees, it is critical to improve the visibility and control of applications, how internal and external users access them, and how we can enforce policies to enhance the security of the various applications in an organization.

- **Network**: As organizations expand their footprint in the cloud, access to resources and data is ultimately done through networks. This means that organizations should segment their networks and implement controls to allow access only to the right users at the right moment and improve traffic visibility. Implementing real-time threat protection, end-to-end encryption, monitoring, and analytics are examples of controls that can be used.
- **Infrastructure**: This layer focuses on verifying access to infrastructure resources explicitly and enforcing compliance requirements for cloud and on-premises resources.
- **Data**: All organizational data should be protected; ensuring data safety is of utmost importance, and controls such as data classification, labeling, and encryption should be implemented.

Now that we have understood the benefits of adopting layered defense and leveraging identity controls, let's understand the fundamental concept of lateral movement.

Understanding lateral movement

Threat actors or cyber attackers leverage several techniques to search for sensitive data and assets. **Lateral movement** refers to the technique of gaining initial access to organizational assets and extending access to other hosts or applications in an organization.

After gaining access to a compromised endpoint, the attacker can maintain access, move through the compromised environment, and search for sensitive data and other assets. The attacker can impersonate a legitimate user and access other network resources.

Imagine there's an employee in your organization, Chris, who opens an email with a malicious attachment. Chris's computer is compromised, so the threat actor can already start performing enumeration operations and gathering information about the internal systems.

Now, the threat actor can perform reconnaissance or credential or privilege gathering, and gain access to other assets in the network. Detecting and preventing lateral movement in your organization is of utmost importance.

Microsoft's response to take preemptive action on your organization to prevent threat actors from succeeding at lateral movements for credential dumping is by leveraging Microsoft Defender for Identity, Advanced Threat Analytics, and Microsoft Sentinel.

These solutions are designed to detect, investigate, and respond to various cyber threats, providing comprehensive protection for organizations' digital environments:

- **Microsoft Defender for Identity**: A security solution that helps detect and investigate identity-based threats across on-premises and cloud environments
- **Advanced Threat Analytics**: A solution that uses behavioral analytics to detect and alert you to suspicious activities and advanced threats within an organization's network

- **Microsoft Sentinel**: A cloud-native **Security Information and Event Management** (**SIEM**) system that provides intelligent security analytics and threat response capabilities for quick detection and response to cyber threats

Now, let's understand how Microsoft Sentinel helps organizations successfully mitigate lateral movements.

Leveraging Microsoft Sentinel to improve your security posture

Microsoft Sentinel, a unified **Security Operations** (**SecOps**) platform, focuses primarily on two fronts – SIEM and **Security Orchestration, Automation, and Response** (**SOAR**).

Microsoft Sentinel allows data collection across your organization; it detects threats while minimizing false positives by leveraging Microsoft's analytics and threat intelligence. Organizations can investigate threats, hunt for suspicious activities, and accelerate their response to incidents by using the built-in orchestration and automation components available in Sentinel.

Through Microsoft Sentinel, organizations can protect their critical assets by gaining visibility into security data and performing searches across all their data, including archive logs, investigating historical data, and then transforming data by enriching and filtering it as needed. Microsoft Sentinel provides the right tools for all members of your security team.

> **Note**
>
> Microsoft Sentinel is built on top of the Azure Monitor Log Analytics platform, a tool you can use through the Azure portal to edit and run log queries from data collected by Azure Monitor logs and interactively analyze results.

Log Analytics is designed to store and analyze massive amounts of data. By using the **Kusto Query Language** (**KQL**), a specific language to join data from multiple tables, aggregate large sets of data, and perform complex operations, security operation professionals and researchers can quickly analyze, investigate, and remediate incidents using automated tools.

Let's take a look at the four main areas of Microsoft Sentinel:

- Collecting data
- Detecting threats
- Investigation
- Response

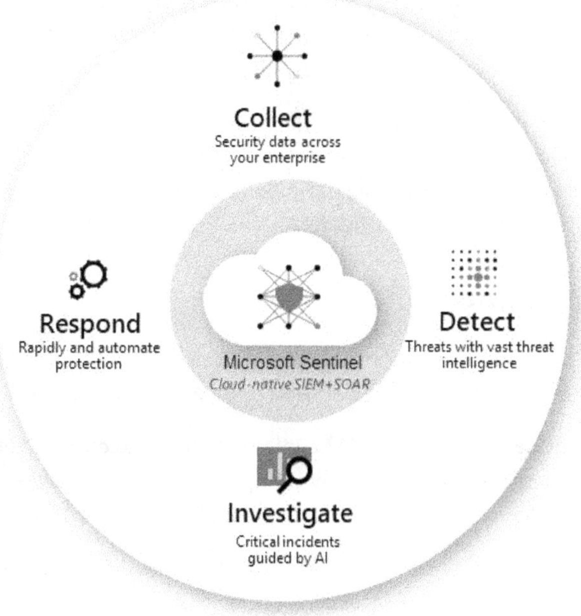
Collect
Security data across your enterprise
Respond
Rapidly and automate protection
Microsoft Sentinel
Cloud-native SIEM+SOAR
Detect
Threats with vast threat intelligence
Investigate
Critical incidents guided by AI

Figure 3.2 – The main focus areas in Microsoft Sentinel

Let's discuss these stages in detail.

Collecting data

As with other SIEM solutions, Microsoft Sentinel collects data from different sources such as devices, users, applications, on-premises, and multiple cloud providers. Once you enable Microsoft Sentinel in your environment, you can leverage built-in connectors or use third-party connectors to collect data.

In the case of physical servers and virtual machines, the Log Analytics agent can be installed, collect the logs, and forward them to Microsoft Sentinel. You can also leverage Linux syslog servers, install the Log Analytics agent, and forward the logs to Microsoft Sentinel.

You can find the latest available connectors at `https://bit.ly/sentinel-data-connectors`.

Detecting threats

Microsoft Sentinel includes analytics capabilities that you can use to detect potential threats and vulnerabilities across your organization. Security analysts can benefit from this feature and perform complex investigations faster.

You will be able to detect threats and reduce false positives by leveraging the threat intelligence and analytics capabilities included in Microsoft Sentinel.

Microsoft Sentinel Analytics will help you detect, investigate, and remediate potential threats. You can configure *analytics rules* and queries to detect issues. So, what can you do with Microsoft Sentinel Analytics? You can analyze data from various sources, such as servers, firewalls, sensors, and networking devices, and identify correlations and anomalies.

Analytic rules are a key concept in Microsoft Sentinel. Think of an analytical rule as a query that contains specific logic to discover threats and anomalous behaviors in your environment, and while Microsoft provides built-in analytics rules and detections, you can customize your analytics rules.

Investigating anomalies

Proactive security analysts handle security threats and translate data into meaningful events. This requires standardizing the processes and lists of tasks to investigate and remediate an incident.

Given that Microsoft Sentinel leverages Log Analytics for query execution, your security team can search for long periods in large datasets by simply using a search job. You can execute a search job on any type of log. You can also restore data to perform your queries.

Microsoft Sentinel also provides a compelling dashboard to handle incidents and includes powerful search and query tools to hunt for threats across an organization's data sources.

Once you connect data sources to your Microsoft Sentinel solution, you can investigate incidents, and each one is created based on the analytics rules you configure. Think of an incident as an aggregation of relevant evidence for a specific investigation, including single or multiple alerts.

Now, picture a scenario in which security analysts perform a hunt, or an investigation; as these operations become more complex, security analysts need a way to simplify data analysis and visualizations.

Microsoft Sentinel includes a feature called **notebooks** to help security analysts simplify how they work with data, the code executed in the data, results, and visualizations of a specific hunt or investigation. So, if an investigation becomes too complex, notebooks can help you remember and view details and save the queries and results of the investigation.

Once the investigation is complete, it is time to leverage mechanisms to respond to security threats.

Responding to incidents

Microsoft Sentinel includes SOAR capabilities, as one of the primary purposes of Sentinel is to automate any response and remediation tasks that are part of the responsibilities of a security operations center.

Therefore, you will find controls to help you manage incident handling and response automation, such as *automation rules* and *playbooks*.

Automation rules are different from analytics rules. Automation rules will help centrally manage the automation of incident handling, so when there's an incident, you can automate responses from multiple analytics rules and tag, assign, or close incidents. Simply put, *automation rules* simplify the workflows of a threat response orchestration process.

Conversely, think of *playbooks* as a collection of actions that can be executed in Sentinel as a routine. Playbooks contain a set of response and remediation actions and logic steps. You can run a playbook automatically or manually on-demand in response to alerts or incidents. Playbooks are based on workflows built in Azure Logic Apps. You can find out more about Logic Apps at the following URL: `https://bit.ly/az-logic-apps`.

Enabling Microsoft Sentinel

In this section, we will demonstrate how to configure Microsoft Sentinel, and we will take the following steps:

1. List the prerequisites
2. Enable Microsoft Sentinel using the Bicep language
3. Enable Microsoft Sentinel using the Azure portal
4. Set up data connectors

Let's review the components we need to configure before enabling Microsoft Sentinel.

Global prerequisites

To successfully enable Microsoft Sentinel, we need the following:

- An active Azure subscription
- A Log Analytics workspace
- A user with contributor role permissions where the Sentinel workspace resides

Let's begin.

Enabling Microsoft Sentinel using the Bicep language

In this section, we will enable Microsoft Sentinel in your environment using infrastructure as code with Azure Bicep. Azure Bicep is a **Domain-Specific Language** (**DSL**) that uses a declarative syntax to deploy Azure resources.

Think of Bicep as an abstraction on top of **Azure Resource Manager** (**ARM**) templates to define Azure resources using declarative infrastructure as code.

The following are the prerequisites for this exercise:

- An active Azure subscription
- Azure Bicep installed on your local machine – you can install it from here: `https://github.com/Azure/bicep/releases`
- Azure PowerShell – you can install it from here: `https://learn.microsoft.com/en-us/powershell/scripting/install/installing-powershell?view=powershell-7.3`
- A resource group in your Azure subscription
- Visual Studio Code with the Bicep extension

One of the main benefits of leveraging infrastructure as code is the ability to simplify the deployment of resources by describing the resource type and its properties through code. Bicep truly simplifies how you can declare your environment's resource types and properties.

We will author a Bicep template that creates a Log Analytics workspace and enables Microsoft Sentinel. We will create the following two files:

- `main.bicep`: This is the Bicep template
- `azuredeploy.parameters.json`: This parameter file contains the values to use to deploy your Bicep template

Now, let's work on the `main.bicep` file:

1. First, we will define the following parameters in our file:

```
param location string = resourceGroup().location
param sentinelName string

@minValue(30)
@maxValue(730)
param retentionInDays int = 90w
```

2. Then, we will add two variables, one for the name of the workspace and another for the Sentinel workspace:

```
var workspaceName = '${location}-${sentinelName}
${uniqueString(resourceGroup().id)}'
var solutionName = 'SecurityInsights(${sentinelWorkspace.name})'
```

3. Finally, we will define the actual resources – `sentinelWorkspace` and `sentinelSolution`:

```
resource sentinelWorkspace 'Microsoft.OperationalInsights/
workspaces@2022-10-01' = {
  name: workspaceName
  location: location
  properties: {
    sku: {
      name: 'PerGB2018'
    }
    retentionInDays: retentionInDays
  }
}

resource sentinelSolution 'Microsoft.OperationsManagement/
solutions@2015-11-01-preview' = {
  name: solutionName
  location: location
  properties: {
    workspaceResourceId: sentinelWorkspace.id
  }
  plan: {
    name: solutionName
    publisher: 'Microsoft'
    product: 'OMSGallery/SecurityInsights'
    promotionCode: ''
  }
}
```

This completes the definition of the `main.bicep` file. Now, we will create our second file, `azuredeploy.parameters.json`. This file will include the parameters' values, and we will define only the name for the Sentinel solution:

```
{
    "$schema": "https://schema.management.azure.com/
schemas/2019-04-01/deploymentParameters.json#",
    "contentVersion": "1.0.0.0",
    "parameters": {
        "sentinelName": {
            "value": "azinsider"
        }
    }
}
```

The next step is to use the following script to deploy the files that will create the resources:

```
$date = Get-Date -Format "MM-dd-yyyy"
$rand = Get-Random -Maximum 1000
$deploymentName = "AzInsiderDeployment-"+"$date"+"-"+"$rand"

New-AzResourceGroupDeployment -Name $deploymentName -ResourceGroupName
azinsider_demo -TemplateFile .\main.bicep -TemplateParameterFile .\
azuredeploy.parameters.json -c
```

In the preceding code, we define the new resource group deployment and pass the `main.bicep` and `parameters` files. Also, we use the `-c` flag to get a preview of the deployment operation.

The following figure shows a preview of the deployment operation:

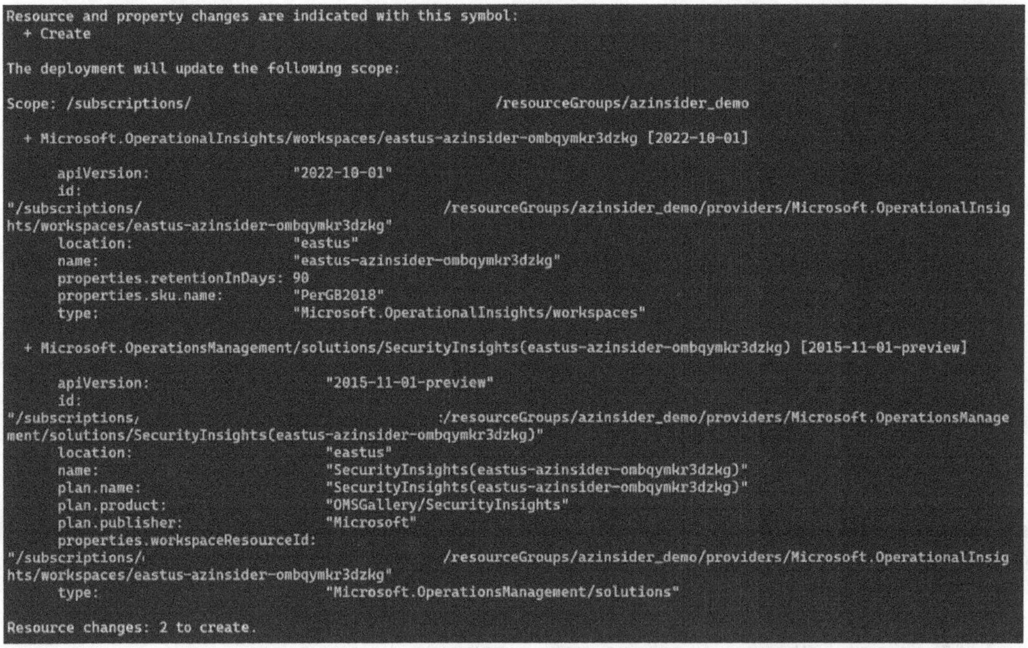
Resource and property changes are indicated with this symbol:
+ Create
The deployment will update the following scope:
Scope: /subscriptions/ /resourceGroups/azinsider_demo
+ Microsoft.OperationalInsights/workspaces/eastus-azinsider-ombqymkr3dzkg [2022-10-01]
apiVersion: "2022-10-01"
id:
"/subscriptions/ /resourceGroups/azinsider_demo/providers/Microsoft.OperationalInsights/workspaces/eastus-azinsider-ombqymkr3dzkg"
location: "eastus"
name: "eastus-azinsider-ombqymkr3dzkg"
properties.retentionInDays: 90
properties.sku.name: "PerGB2018"
type: "Microsoft.OperationalInsights/workspaces"
+ Microsoft.OperationsManagement/solutions/SecurityInsights(eastus-azinsider-ombqymkr3dzkg) [2015-11-01-preview]
apiVersion: "2015-11-01-preview"
id:
"/subscriptions/ :/resourceGroups/azinsider_demo/providers/Microsoft.OperationsManagement/solutions/SecurityInsights(eastus-azinsider-ombqymkr3dzkg)"
location: "eastus"
name: "SecurityInsights(eastus-azinsider-ombqymkr3dzkg)"
plan.name: "SecurityInsights(eastus-azinsider-ombqymkr3dzkg)"
plan.product: "OMSGallery/SecurityInsights"
plan.publisher: "Microsoft"
properties.workspaceResourceId:
"/subscriptions/ /resourceGroups/azinsider_demo/providers/Microsoft.OperationalInsights/workspaces/eastus-azinsider-ombqymkr3dzkg"
type: "Microsoft.OperationsManagement/solutions"
Resource changes: 2 to create.

Figure 3.3 – Microsoft Sentinel using the Bicep language – the deployment preview

Note that we can review the details of the resources that will be provisioned before they are actually deployed in the environment.

Next, we will execute the deployment. The following screenshot shows the deployment output:

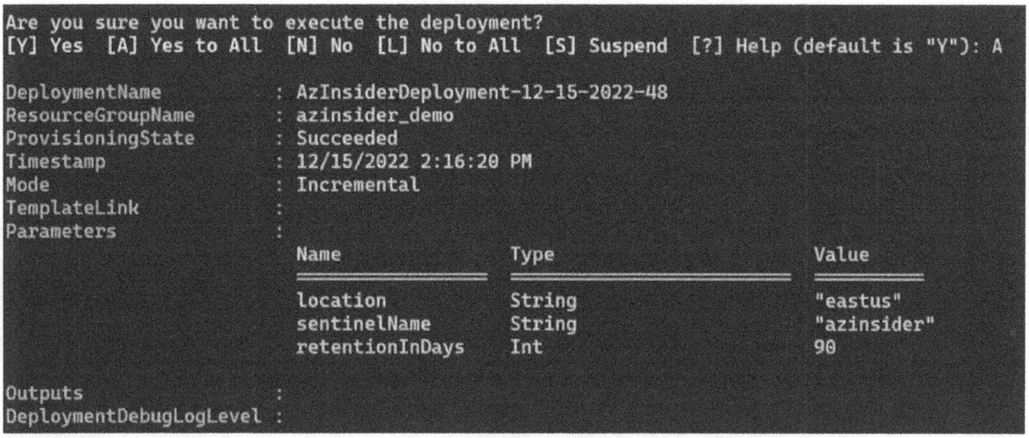
Are you sure you want to execute the deployment?
[Y] Yes [A] Yes to All [N] No [L] No to All [S] Suspend [?] Help (default is "Y"): A
DeploymentName : AzInsiderDeployment-12-15-2022-48
ResourceGroupName : azinsider_demo
ProvisioningState : Succeeded
Timestamp : 12/15/2022 2:16:20 PM
Mode : Incremental
TemplateLink :
Parameters :
Name Type Value
location String "eastus"
sentinelName String "azinsider"
retentionInDays Int 90
Outputs :
DeploymentDebugLogLevel :

Figure 3.4 – Microsoft Sentinel deployment output using the Bicep language

You can download the complete solution at `https://bit.ly/sentinel-bicep`.

Enabling Microsoft Sentinel using the Azure portal

Now, let's review how we can enable Microsoft Sentinel using the Azure portal:

1. Go to the Azure portal and search for `log analytics`, as shown in the following figure:

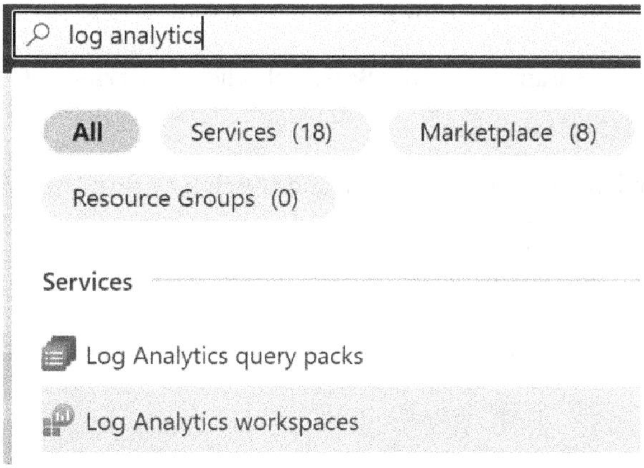
log analytics
All
Services (18)
Marketplace (8)
Resource Groups (0)
Services
Log Analytics query packs
Log Analytics workspaces

Figure 3.5 – A Log Analytics search

2. Then, provide the details to provision the Log Analytics workspace, as shown here:

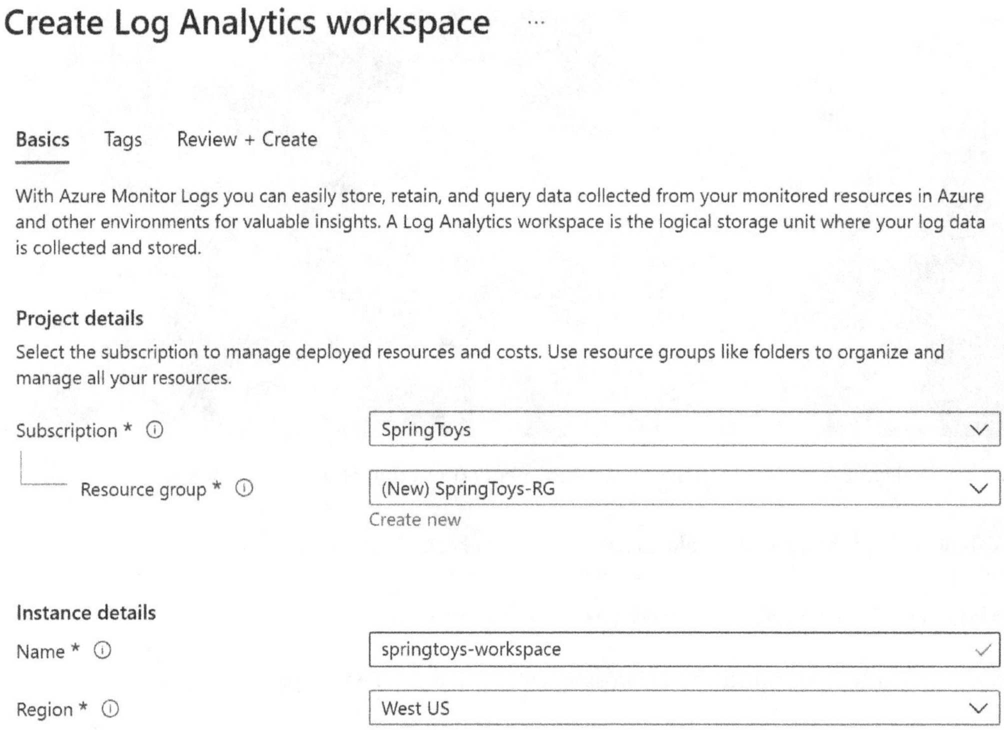
Create Log Analytics workspace
Basics
Tags
Review + Create
With Azure Monitor Logs you can easily store, retain, and query data collected from your monitored resources in Azure and other environments for valuable insights. A Log Analytics workspace is the logical storage unit where your log data is collected and stored.
Project details
Select the subscription to manage deployed resources and costs. Use resource groups like folders to organize and manage all your resources.
Subscription *
SpringToys
Resource group *
(New) SpringToys-RG
Create new
Instance details
Name *
springtoys-workspace
Region *
West US

Figure 3.6 – Log Analytics workspace configuration

Once you provide the parameters in the **Basics** tab, click on **Review + Create**.

Next, we will enable Microsoft Sentinel.

3. Go to the top search bar of the Azure portal and search for `sentinel`, as shown here:

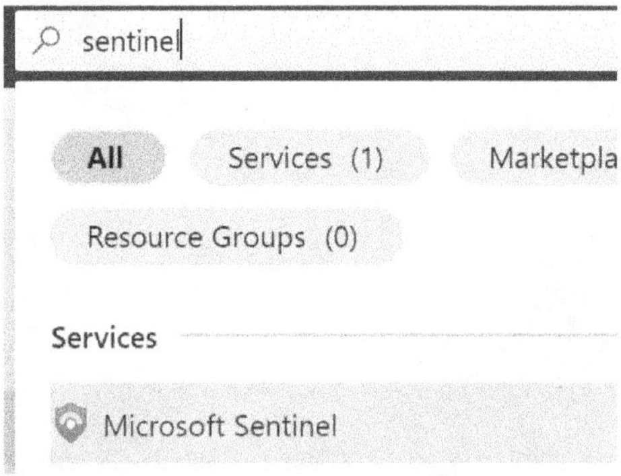
sentinel
All
Services (1)
Marketpla
Resource Groups (0)
Services
Microsoft Sentinel

Figure 3.7 – A Microsoft Sentinel search

4. Click on **Add** after selecting the workspace to enable Microsoft Sentinel:

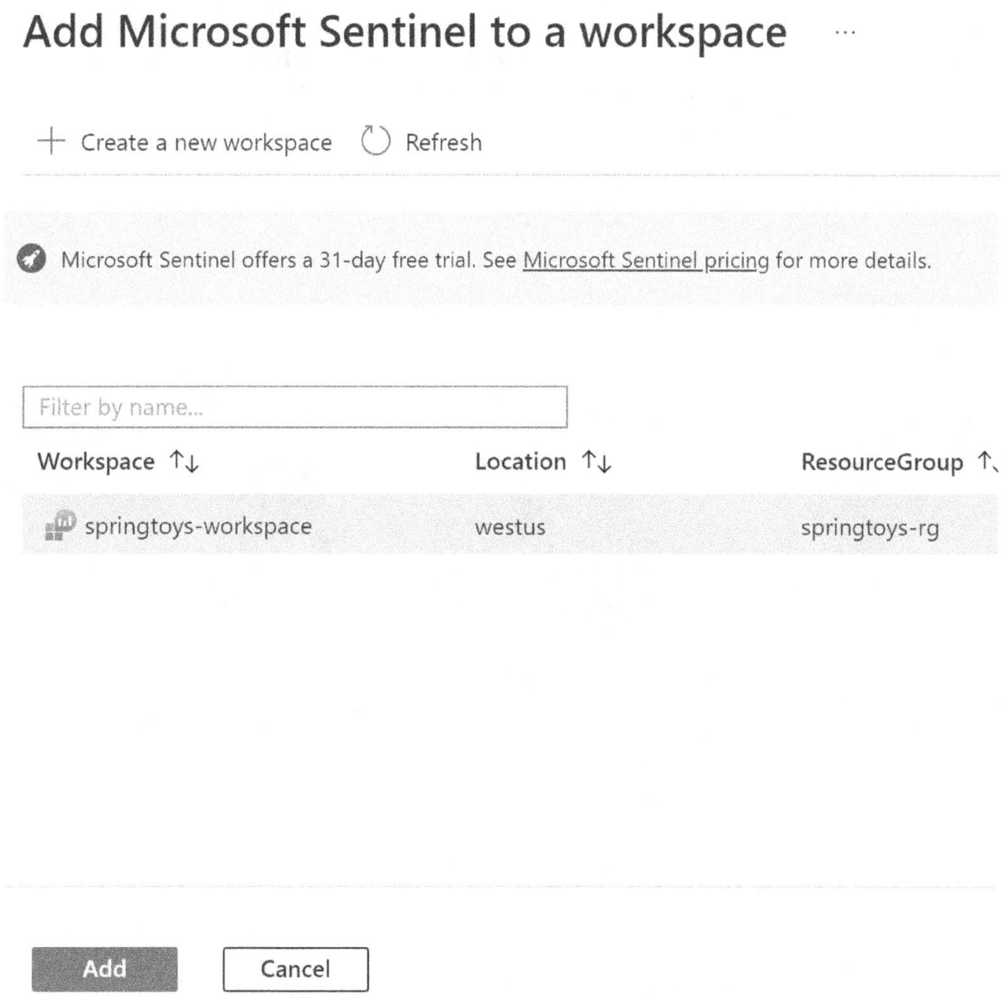
Add Microsoft Sentinel to a workspace
Create a new workspace
Refresh
Microsoft Sentinel offers a 31-day free trial. See Microsoft Sentinel pricing for more details.
Filter by name...
Workspace
Location
ResourceGroup
springtoys-workspace
westus
springtoys-rg
Add
Cancel

Figure 3.8 – Microsoft Sentinel configuration

> **Note**
>
> Microsoft Sentinel comes with a 31-day free trial, after which it is a paid service. Up to 10 GB per day is free for both Microsoft Sentinel and Log Analytics during the trial period.

The next step is to connect data sources to collect data in Microsoft Sentinel.

Setting up data connectors

Organizations can use data connectors to ingest data into Microsoft Sentinel. Some of these connectors are built-in connectors for Microsoft services, such as the Microsoft 365 Defender connector, which integrates data from different services such as Azure Active Directory, Office 365, Microsoft Defender for Identity, and Microsoft Defender for Cloud Apps.

Let's start by enabling the Microsoft 365 Defender connector.

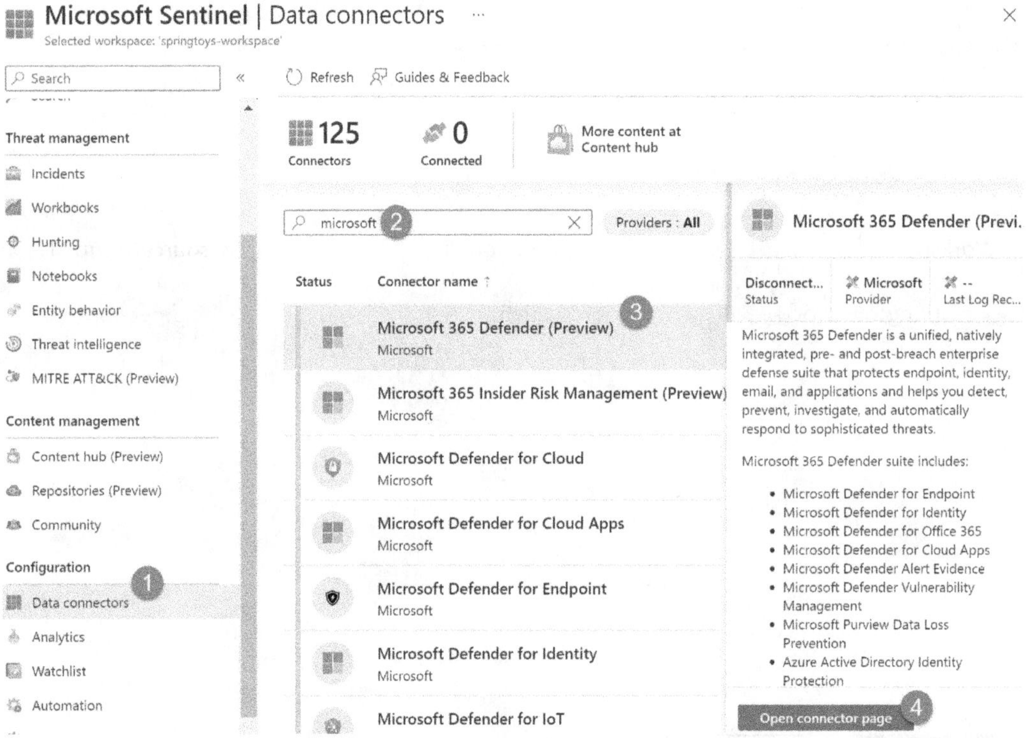
Microsoft Sentinel | Data connectors
Selected workspace: 'springtoys-workspace'
Search
Refresh
Guides & Feedback
125
Connectors
0
Connected
More content at Content hub
Threat management
Incidents
Workbooks
Hunting
Notebooks
Entity behavior
Threat intelligence
MITRE ATT&CK (Preview)
Content management
Content hub (Preview)
Repositories (Preview)
Community
Configuration
Data connectors
Analytics
Watchlist
Automation
microsoft
Providers : All
Status
Connector name
Microsoft 365 Defender (Preview)
Microsoft
Microsoft 365 Insider Risk Management (Preview)
Microsoft
Microsoft Defender for Cloud
Microsoft
Microsoft Defender for Cloud Apps
Microsoft
Microsoft Defender for Endpoint
Microsoft
Microsoft Defender for Identity
Microsoft
Microsoft Defender for IoT
Microsoft 365 Defender (Previ.
Disconnect...
Status
Microsoft
Provider
--
Last Log Rec...
Microsoft 365 Defender is a unified, natively integrated, pre- and post-breach enterprise defense suite that protects endpoint, identity, email, and applications and helps you detect, prevent, investigate, and automatically respond to sophisticated threats.
Microsoft 365 Defender suite includes:
Microsoft Defender for Endpoint
Microsoft Defender for Identity
Microsoft Defender for Office 365
Microsoft Defender for Cloud Apps
Microsoft Defender Alert Evidence
Microsoft Defender Vulnerability Management
Microsoft Purview Data Loss Prevention
Azure Active Directory Identity Protection
Open connector page
1
2
3
4

Figure 3.9 – The Microsoft 365 Defender connector

Once you open the connector page, you will see the log types that can be ingested in Microsoft Sentinel. The following figure shows the logs that can be ingested using the Microsoft 365 Defender connector:

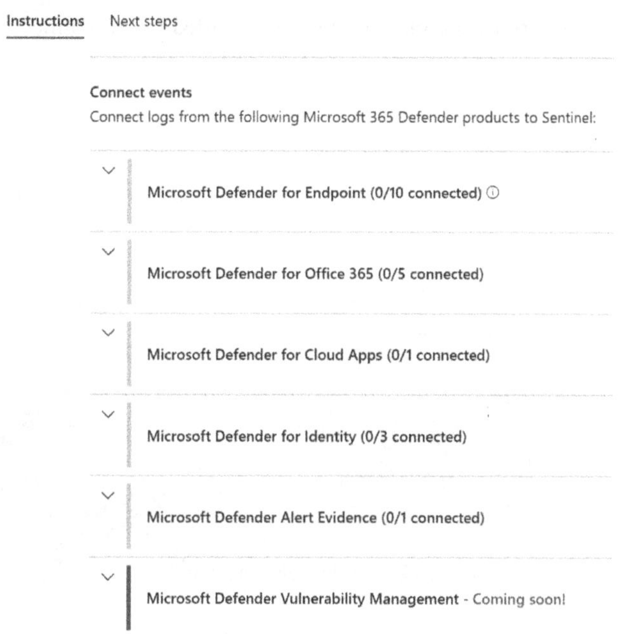
Instructions
Next steps
Connect events
Connect logs from the following Microsoft 365 Defender products to Sentinel:
Microsoft Defender for Endpoint (0/10 connected)
Microsoft Defender for Office 365 (0/5 connected)
Microsoft Defender for Cloud Apps (0/1 connected)
Microsoft Defender for Identity (0/3 connected)
Microsoft Defender Alert Evidence (0/1 connected)
Microsoft Defender Vulnerability Management - Coming soon!

Figure 3.10 – Microsoft 365 Defender connector

You can select the specific log event you want to collect for each product. For example, if you choose **Microsoft Defender for Endpoint** and expand the options, you will see the particular event types, as shown here:

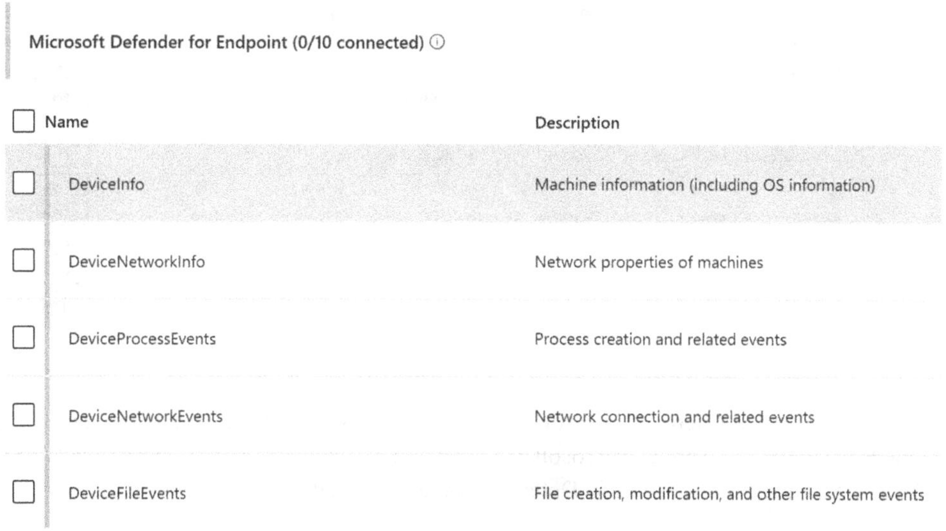
Microsoft Defender for Endpoint (0/10 connected)
Name
Description
DeviceInfo
Machine information (including OS information)
DeviceNetworkInfo
Network properties of machines
DeviceProcessEvents
Process creation and related events
DeviceNetworkEvents
Network connection and related events
DeviceFileEvents
File creation, modification, and other file system events

Figure 3.11 – The event types to collect

You can also review the workbooks and built-in queries included in this connector, as shown here:

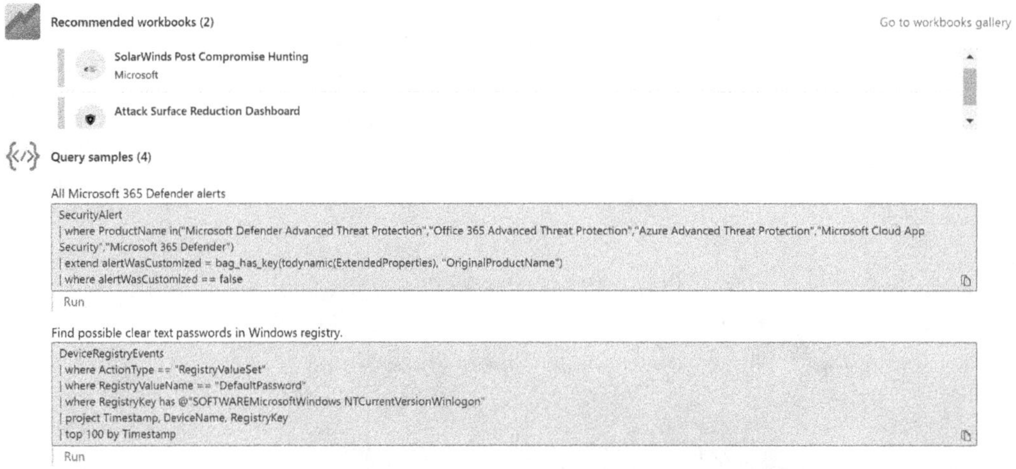
Recommended workbooks (2)
Go to workbooks gallery
SolarWinds Post Compromise Hunting
Microsoft
Attack Surface Reduction Dashboard
Query samples (4)
All Microsoft 365 Defender alerts
SecurityAlert
| where ProductName in("Microsoft Defender Advanced Threat Protection","Office 365 Advanced Threat Protection","Azure Advanced Threat Protection","Microsoft Cloud App Security","Microsoft 365 Defender")
| extend alertWasCustomized = bag_has_key(todynamic(ExtendedProperties), "OriginalProductName")
| where alertWasCustomized == false
Run
Find possible clear text passwords in Windows registry.
DeviceRegistryEvents
| where ActionType == "RegistryValueSet"
| where RegistryValueName == "DefaultPassword"
| where RegistryKey has @"SOFTWAREMicrosoftWindows NTCurrentVersionWinlogon"
| project Timestamp, DeviceName, RegistryKey
| top 100 by Timestamp
Run

Figure 3.12 – Workbooks and queries

In addition, the Microsoft 365 Defender connector also provides you with relevant analytics templates that you can use as required:

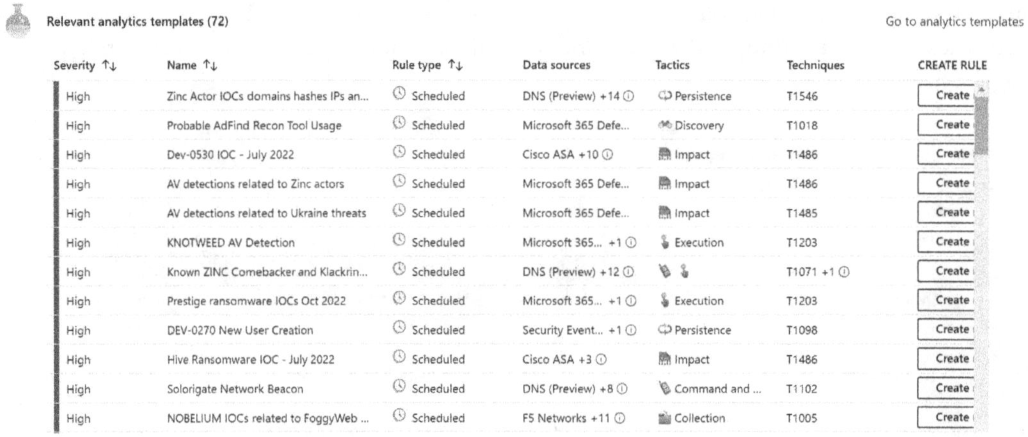
Relevant analytics templates (72)
Go to analytics templates
Severity
Name
Rule type
Data sources
Tactics
Techniques
CREATE RULE
High
Zinc Actor IOCs domains hashes IPs an...
Scheduled
DNS (Preview) +14
Persistence
T1546
Create
High
Probable AdFind Recon Tool Usage
Scheduled
Microsoft 365 Defe...
Discovery
T1018
Create
High
Dev-0530 IOC - July 2022
Scheduled
Cisco ASA +10
Impact
T1486
Create
High
AV detections related to Zinc actors
Scheduled
Microsoft 365 Defe...
Impact
T1486
Create
High
AV detections related to Ukraine threats
Scheduled
Microsoft 365 Defe...
Impact
T1485
Create
High
KNOTWEED AV Detection
Scheduled
Microsoft 365... +1
Execution
T1203
Create
High
Known ZINC Comebacker and Klackrin...
Scheduled
DNS (Preview) +12
T1071 +1
Create
High
Prestige ransomware IOCs Oct 2022
Scheduled
Microsoft 365... +1
Execution
T1203
Create
High
DEV-0270 New User Creation
Scheduled
Security Event... +1
Persistence
T1098
Create
High
Hive Ransomware IOC - July 2022
Scheduled
Cisco ASA +3
Impact
T1486
Create
High
Solorigate Network Beacon
Scheduled
DNS (Preview) +8
Command and ...
T1102
Create
High
NOBELIUM IOCs related to FoggyWeb ...
Scheduled
F5 Networks +11
Collection
T1005
Create

Figure 3.13 – Analytics templates

So far, we have configured Microsoft Sentinel and the Microsoft 365 Defender data connector. You can find more documentation about Microsoft 365 Defender at the following URL: `https://bit.ly/ms365-defender`. Microsoft 365 Defender includes the following:

- Microsoft Defender for Endpoint
- Microsoft Defender for Identity

- Microsoft Defender for Office 365
- Microsoft Defender for Cloud Apps
- Microsoft Defender Alert Evidence
- Microsoft Defender Vulnerability Management
- Microsoft Purview Data Loss Prevention
- Azure Active Directory Identity Protection

Hopefully, this gives you a better understanding of how you can enable Microsoft Sentinel, connect data sources to ingest data, and use them for analysis.

Now, let's focus on our goal to remediate lateral movements. The Microsoft 365 Defender connector will help us mitigate lateral movements, as we will see in the next section.

Mitigating lateral movements

We can use Microsoft Defender for Identity, which includes a mechanism to detect lateral movements, and a powerful alternate method to remediate lateral movements is by configuring Fusion in Microsoft Sentinel, a correlation engine to detect multistage attacks automatically.

Fusion is a powerful engine integrated with Microsoft Sentinel that identifies combinations of anomalous behaviors and malicious activities observed at different kill chain stages.

Microsoft Sentinel can generate incidents with multiple alerts or activities, and Fusion will correlate all signals from various products and detect advanced attacks. Fusion detections will be shown as Fusion incidents on the **Microsoft Sentinel Incidents** page. Fusion incidents are stored in the **SecurityIncident** table in **Logs**.

How can we enable Fusion? It is enabled as an analytics rule and can help us cover a variety of scenarios, such as the following:

- Compute resource abuse
- Credential access
- Credential harvesting
- Crypto-mining
- Data destruction
- Data exfiltration
- Denial of service
- Lateral movement

- Malicious administrative activity
- Malicious execution with the legitimate process
- Malware C2 or download
- Persistence
- Ransomware
- Remote exploitation
- Resource hijacking

For lateral movement mitigation, Fusion can be utilized on two fronts:

- An Office 365 impersonation following a suspicious Azure AD sign-in
- Suspicious inbox manipulation rules set following a suspicious Azure AD sign-in

An Office 365 impersonation following a suspicious Azure AD sign-in

The Fusion approach involves the identification of suspicious sign-ins from an Azure AD account that follow an unusual number of impersonation actions.

These alerts are triggered by various scenarios, including impossible travel to an unusual location, sign-ins from unfamiliar locations, infected devices, anonymous IP addresses, and users with leaked credentials, all leading to Office 365 impersonation.

To enable this control, Microsoft Defender for Cloud Apps and Azure AD Identity Protection data connectors can be utilized in Microsoft Sentinel.

Suspicious inbox manipulation rules set following suspicious Azure AD sign-in

The second approach to mitigate lateral movement is to have Fusion detect anomalous inbox rules set in a user's inbox, followed by a suspicious sign-in to an Azure AD account. The alerts relate to the following:

- Impossible travel to an atypical location, leading to a suspicious inbox manipulation rule
- A sign-in event from an unfamiliar location, leading to a suspicious inbox manipulation rule
- A sign-in event from an infected device, leading to a suspicious inbox manipulation rule
- A sign-in event from an anonymous IP address, leading to a suspicious inbox manipulation rule
- A sign-in event from a user with leaked credentials, leading to a suspicious inbox manipulation rule

How we do enable this second approach? In Microsoft Sentinel, we can use the Microsoft Defender for Cloud Apps and Azure AD Identity Protection data connectors. We use the same data connectors for both options.

Summary

In this chapter, we reviewed how organizations can benefit from adopting the Zero Trust model and protect their assets by providing the right tools for their security analysts, such as Microsoft Sentinel. We also reviewed different methods to enable Microsoft Sentinel in your environment, and we saw how you can work with the Microsoft 365 Defender connector and leverage Fusion to mitigate lateral movement in your organization.

In the next chapter, we will discuss the various Azure data solutions offered by Microsoft that are designed to help organizations manage and analyze their data more effectively. The solutions are flexible and scalable and can be used to process and store data of various types and sizes, from structured to unstructured, from batch to real time, and from small to very large.

Part 2 – Architecting Compute and Network Solutions

Architecting compute and network solutions is a fundamental aspect of designing and implementing robust IT infrastructure. As organizations increasingly rely on cloud computing and network technologies to power their operations, the importance of strategically planning and deploying compute and network solutions becomes paramount.

This part will demonstrate how an effective architecture involves designing and optimizing the allocation of compute resources, such as servers and virtual machines, to efficiently handle workloads and meet performance requirements.

This part has the following chapters:

- *Chapter 4, Understanding Azure Data Solutions*
- *Chapter 5, Migrating to the Cloud*
- *Chapter 6, End-to-End Observability in Your Cloud and Hybrid Environments*
- *Chapter 7, Working with Containers in Azure*
- *Chapter 8, Understanding Networking in Azure*
- *Chapter 9, Securing Access to Your Applications*

4

Understanding Azure Data Solutions

When building solutions, at some point, you need to be able to read and write data.

There are many kinds of data we may need to consider interacting with when designing and building solutions. Pieces of data aren't just records in a database – images, documents, and even binary files are all data.

In technical terms, all these different types of data can be grouped into *structured*, *semi-structured*, and *unstructured* data.

Once we have decided what type of data we need to store, we must make choices based on cost, performance, availability, and resilience.

How we plan to use our data can also impact our choice of technology – for example, a web application that simply needs to display images or allow a user to download a document would have very different requirements to a data processing solution that needs to quickly read thousands of small files and output a report.

We must also consider the sensitivity of data. Personal information, especially **personally identifiable information** (**PII**), must be handled very carefully. Not only must sensitive data be properly secured and encrypted, but you also need to consider the lifetime of that information as many laws restrict how long you can keep it after it is no longer required.

To help us understand and answer these questions and more, this chapter is going to delve into the following main topics:

- Understanding Azure storage types
- Understanding Azure database options

Technical requirements

This chapter will use the Azure portal (`https://portal.azure.com`) throughout for the examples.

Understanding Azure storage types

Before we can choose an Azure data solution, we need to understand what our data is. As mentioned previously, data can be grouped into *structured*, *semi-structured*, and *unstructured* data:

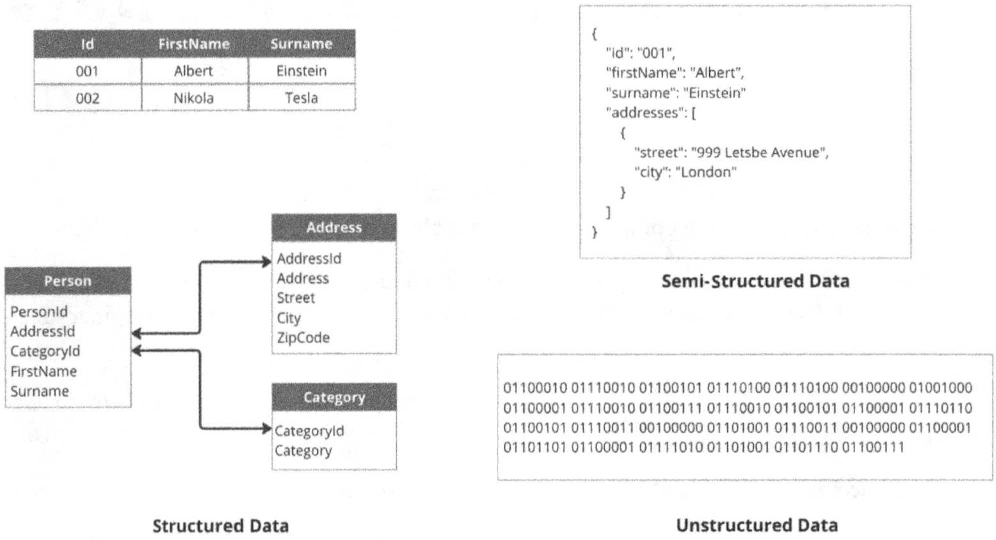
Id
FirstName
Surname
001
Albert
Einstein
002
Nikola
Tesla
Address
AddressId
Address
Street
City
ZipCode
Person
PersonId
AddressId
CategoryId
FirstName
Surname
Category
CategoryId
Category
{
"id": "001",
"firstName": "Albert",
"surname": "Einstein"
"addresses": [
{
"street": "999 Letsbe Avenue",
"city": "London"
}
]
}
Semi-Structured Data
01100010 01110010 01100101 01110100 01110100 00100000 01001000
01100001 01110010 01100111 01110010 01100101 01100001 01110110
01100101 01110011 00100000 01101001 01110011 00100000 01100001
01101101 01100001 01111010 01101001 01101110 01100111
Structured Data
Unstructured Data

Figure 4.1 – Examples of different types of data

Let's look at them in detail.

Structured data

Structured data is data that has a predefined definition or schema and is managed in a column and row format – that is, rows of data separated into columns.

This type of data is sometimes stored as a comma-separated or other text-based file that conforms to a set format. By this, we mean it should be easy to read the data a row at a time and parse each bit of data or column. So, in the case of a comma-separated file, each row would be one record, and each column would be separated by a comma.

The key feature here is that we should know ahead of time what shape the data is in – that is, each field is in the same column.

For example, in the following comma-separated data, each row has the same data in the same field so that we always know that in field 1, we expect an ID, in field 2, we expect a name, and so on:

```
001, Albert, Einstein
002, Nikola, Tesla
```

On the other hand, the following code would not be valid because there is different data on each row, and there is more data on row 2 than on row 1:

```
001, Albert, Einstein
Fred, true, 25, Blue
```

SQL databases also fall into this area. For example, if you want to store a `customer` record, you must define the `customer` schema – that is, you must state the pieces of individual data you want to capture as fields (or columns), such as name, age, and so on. For each field, we might also define the type of data, such as text, a number, or a Boolean value, along with a maximum length.

Fields are then grouped into a table, and each table usually covers a particular aspect of the object you are trying to record. This can include personal details such as name and age or address.

One key aspect of this type of data is that it can be relational. This means that you don't store all the data in a single table; instead, you break the data up into different tables and then link those tables. Again, as an example, a `customer` record may store personal details in a `person` table, but their address will be in a separate `address` table, with the two linked on a special field called a **key** (often denoted as an ID).

As you can see, we call it structured for a reason – the data should always follow a predetermined pattern.

Unstructured data

On the opposite side of the spectrum, we have unstructured data, and as its name suggests, it does not follow any kind of format. Also called raw data, it is often used as a proprietary file type that needs a particular program to read it, such as a Word or Excel document.

Images, application files, and even disk images used by **virtual machines** (**VMs**) are all raw data files – essentially, these are files that can't be read by a human to make any sense of them.

Text files can also be technically unstructured if they don't follow any format that would make them structured or semi-structured. But, generally speaking, when we talk about unstructured data, we mean images and application files.

Semi-structured data

Semi-structured data means it has some structure; however, it is usually a bit more flexible.

HTML, JSON, and XML files are great examples of semi-structured data. *Figure 4.1* shows an example JSON file.

Semi-structured data, as in the JSON example, still has a structure, and if we know the rules, we can work out what it means. In the case of JSON, data is stored as key-value pairs separated by a colon. However, the flexible part of JSON is that a value can contain more data as key-value pairs.

The following snippet highlights a simple data block. Since we know the rules of JSON, we can see that the data consists of an ID, a first name, and a surname:

```
{
      "id" : "001",
      "firstName" : "Brett",
      "surname" : "Hargreaves",
}
```

A key difference with semi-structured data is that we can easily mix and match data within records – in other words, we don't need to predefine the structure or schema. In the preceding example, we stated the following would be invalid:

```
001, Albert, Einstein
Fred, true, 400, Blue
```

In a semi-structured JSON document, the following is perfectly fine:

```
{
"id" : 001,
"firstName" : "Albert",
"surname" : "Einstein"
},
{
"name" : "Fred",
"likesCheese" : true,
"age" : 25,
"favouriteColor" : "Blue"
}
```

As you can see, semi-structured data is incredibly flexible yet still allows us to be able to parse the information relatively easily.

As mentioned previously, JSON configuration files are a common use case for this type of data, as are NoSQL databases.

Now that you understand what data you have, we can start to consider which might be the best fit for our solution. There are three main options – they are **Azure SQL**, **Azure Cosmos DB**, and **Azure storage accounts**. Your solution may even use all three. Next, we will examine each one and learn about their configuration options and what use cases fit them best.

Azure storage accounts

We will start with Azure storage accounts because they are the most versatile. Whereas Azure SQL and Azure Cosmos DB are just databases, storage accounts are several different services all rolled into one. At a high level, an Azure storage account offers the following features:

- **Blob Storage**: This is a massively scalable storage for any type of file. However, file access is controlled using RBAC and must be performed over HTTPS.
- **File shares**: This is another scalable store for any type of file. However, file shares are like a more traditional Windows file server and support NTFS permissions, and they can be accessed over NTFS, SMB, and HTTPS.
- **Storage queues**: Technically, this is not really storage. Queues are for applications or other components to write small messages on them. They are held until another process reads the message and then deletes it.
- **Tables**: A table is a simple, lightweight but scalable key-value database. Tables are a great option for storing very simple structured data.
- **Data Lake Storage (Gen2)**: This is similar to Blob Storage – in fact, Data Lake Storage is an option you can enable on Blob Storage. Data Lake Storage uses hierarchical namespaces, which are like file directories, but they are specifically built for quickly processing files in data analytics solutions.

Creating a storage account is simple enough, and except for Data Lake Storage (Gen 2), all services are available by default.

In the Azure portal, do the following:

1. From the **Azure** menu, click + **Create a resource**.
2. In the search box, type `storage account`.
3. Select **Storage account** from the list of options.
4. Click **Storage account** in the list of marketplace items.
5. Click **Create**.

6. Fill in some basic details, including the following:

   - Select your subscription
   - Select a resource group or click **Create new** to create a new one
   - Give your storage account a name – this needs to be unique within the region and must be in lowercase and only use alphanumeric characters, hyphens, and underscores
   - Choose a region:

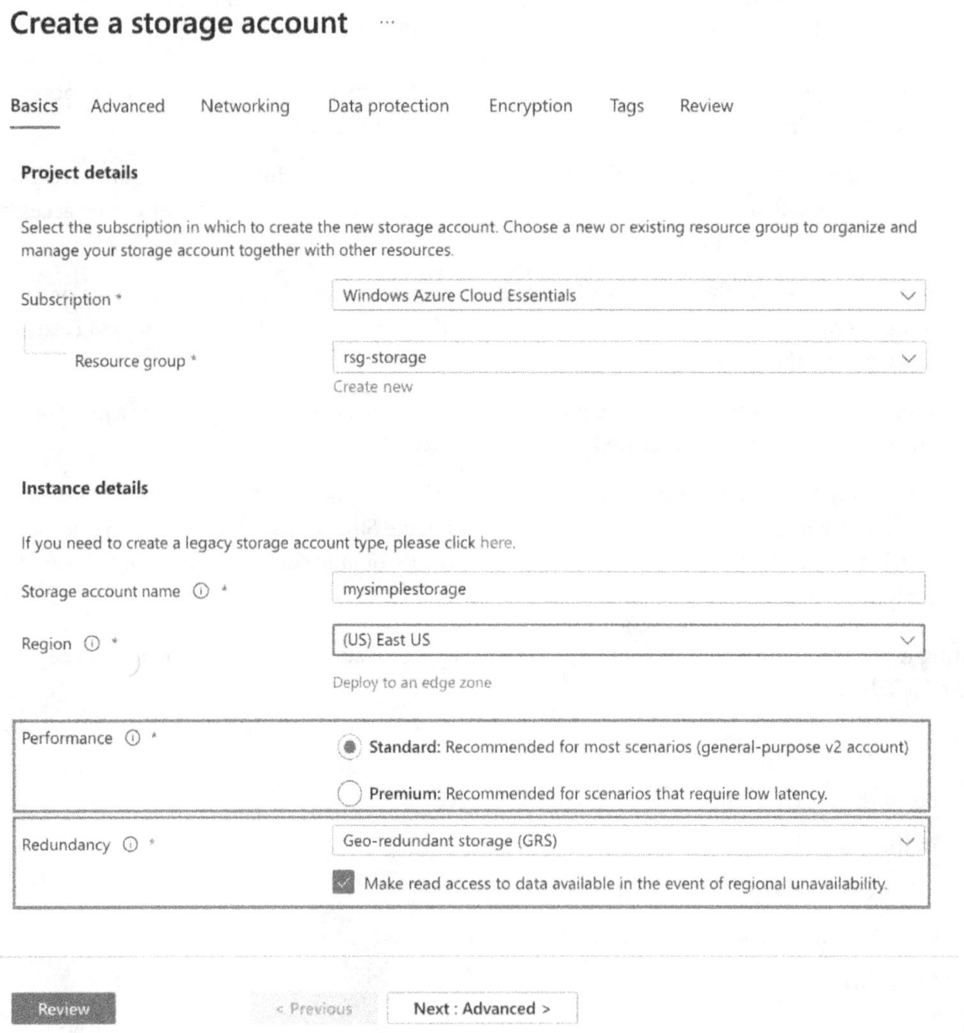
Create a storage account
Basics
Advanced
Networking
Data protection
Encryption
Tags
Review
Project details
Select the subscription in which to create the new storage account. Choose a new or existing resource group to organize and manage your storage account together with other resources.
Subscription *
Windows Azure Cloud Essentials
Resource group *
rsg-storage
Create new
Instance details
If you need to create a legacy storage account type, please click here.
Storage account name *
mysimplestorage
Region *
(US) East US
Deploy to an edge zone
Performance *
Standard: Recommended for most scenarios (general-purpose v2 account)
Premium: Recommended for scenarios that require low latency.
Redundancy *
Geo-redundant storage (GRS)
Make read access to data available in the event of regional unavailability.
Review
< Previous
Next : Advanced >

Figure 4.2 – Creating a storage account

Figure 4.2 shows an example screen. Two options have been highlighted here that we need to examine in more detail – **Performance** and **Redundancy**.

Performance tiers

As the text in the portal says, for most of your scenarios, you would set the performance to *Standard*. This option offers blobs (including Data Lake), queues, and table and file storage and offers all levels of redundancy, which we will cover shortly.

Premium is for those occasions when your solution demands lower latency and faster read/write times. If you select this option, you must select a Premium account type, for which there are three further options:

- **Block Blob**: This is particularly suited to applications that require high numbers of transactions or use lots of small objects.
- **File Shares**: For enterprise-class file server solutions that require high performance. Premium is also required if you want to use the NFS protocol instead of SMB.
- **Page Blobs**: Page blobs are a particular type of blob that suits larger files. Block blobs (the default type) are optimized for text and binary files. Page blobs use a bigger page size, which makes them better suited for larger files that have lots of random read/write operations. The use case is a **virtual hard disk** (**VHD**) file that would be used by a VM.

As already stated, unless there is a specific need for your solution, in most cases, the recommended choice is Standard. Another aspect to consider is that the Premium performance tier does not support all the redundancy options.

Block blobs and file shares only provide **locally redundant storage** (**LRS**) and **zone-redundant storage** (**ZRS**), and page blobs only support LRS.

We will discuss what the different options are next.

Redundancy

The Standard performance tier allows several different redundancy options. When you write data to a storage account, that data is replicated to at least two other copies. In this way, if there is ever a physical failure that would affect your data, you always have another copy. If there is a physical outage, you would not even be aware of it, and the Azure platform will simply read data from one of the replicas.

I specifically said *read* because, when dealing with local replicas, the data is always written **synchronously** – that is, your data is written to each replica *at the same time*. This is important because it also means that at any moment in time, all three copies contain the same data – so there can be no data loss in the event of a failure.

There are four different redundancy options to choose from; we will look at each in turn.

Locally redundant storage (LRS)

Here, your data is written synchronously to three copies within the same region within the same availability zone. In effect, this means you are protected from a single physical disk or rack failure within a data center. LRS is the cheapest option, but you are not protected if something happens to the availability zone you are in.

> **Note**
>
> An availability zone can be thought of as a single data center. Within any one region, you have at least three availability zones that are 10-30 miles apart. You rarely get direct visibility of which availability zone you are in, and for production applications, you should always opt for a solution that spans availability zones (at least).

Zone-redundant storage (ZRS)

Your data is written synchronously to three copies across three availability zones within a single region. This means you are protected from disk, rack, and data center failures.

Figure 4.3 shows the difference between LRS and ZRS:

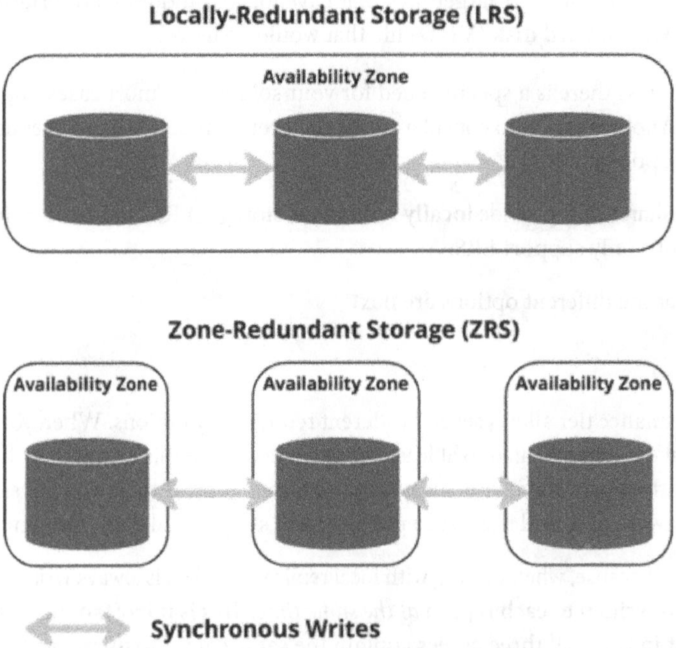
Locally-Redundant Storage (LRS)
Availability Zone
Zone-Redundant Storage (ZRS)
Availability Zone
Availability Zone
Availability Zone
Synchronous Writes

Figure 4.3 – Local and zone redundancy

Protecting your data within a region is fine for many applications, but sometimes, you want the added safety net of having your data copied to another region as well – in other words, to protect you against a full region outage (it does happen!).

In these cases, your data is copied to a paired region. You have no choice in terms of region; each is a pair that Microsoft controls. The copy to the paired region happens asynchronously – this means that it happens after your data has been written to the local replicas, and there is a delay. Depending on the size and type of data, that delay could be as long as 15 minutes. This can have important consequences for your application as you should never assume your copies across regions are in sync.

For this reason, optionally, you can turn off read access to the region replica to prevent out-of-date data from being read accidentally.

If a region fails, Azure DNS is updated to redirect all traffic to the paired region replica, which then becomes read/write.

There are two regional redundant options, and the difference is how they replicate the data in the primary region.

Geo-redundant storage (GRS)

Here, data is written synchronously within a single availability zone within your primary region and then replicated asynchronously to the paired region. Effectively, this means it uses LRS in the primary region.

Geo-zone-redundant storage (GZRS)

In this scenario, data is written synchronously across three availability zones within your primary region and then replicated asynchronously to the paired region. Effectively, this means it uses ZRS in the primary region.

Figure 4.4 shows the differences between these two options:

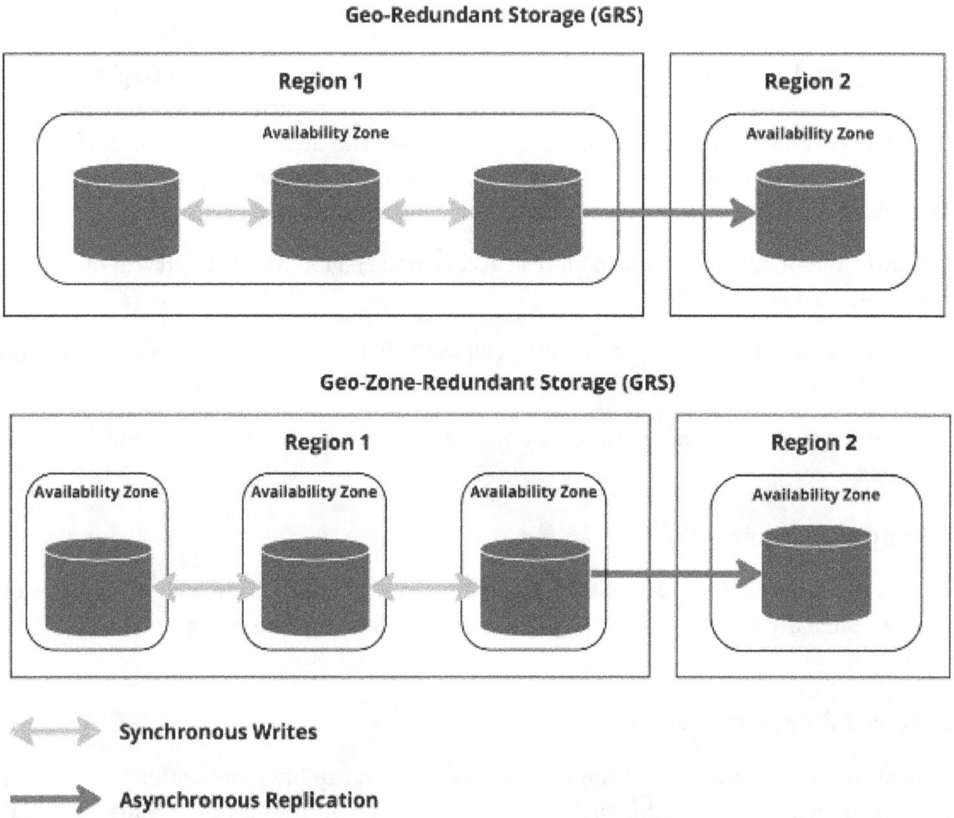
Geo-Redundant Storage (GRS)
Region 1
Availability Zone
Region 2
Availability Zone
Geo-Zone-Redundant Storage (GRS)
Region 1
Availability Zone
Availability Zone
Availability Zone
Region 2
Availability Zone
Synchronous Writes
Asynchronous Replication

Figure 4.4 – Geo and geo-zone redundancy

Once you have decided which performance tier you need, and which replication strategy suits your availability requirements, we can continue configuring the storage account.

Back in the Azure portal, click **Next: Advanced** >. You should see the screen shown in *Figure 4.5*:

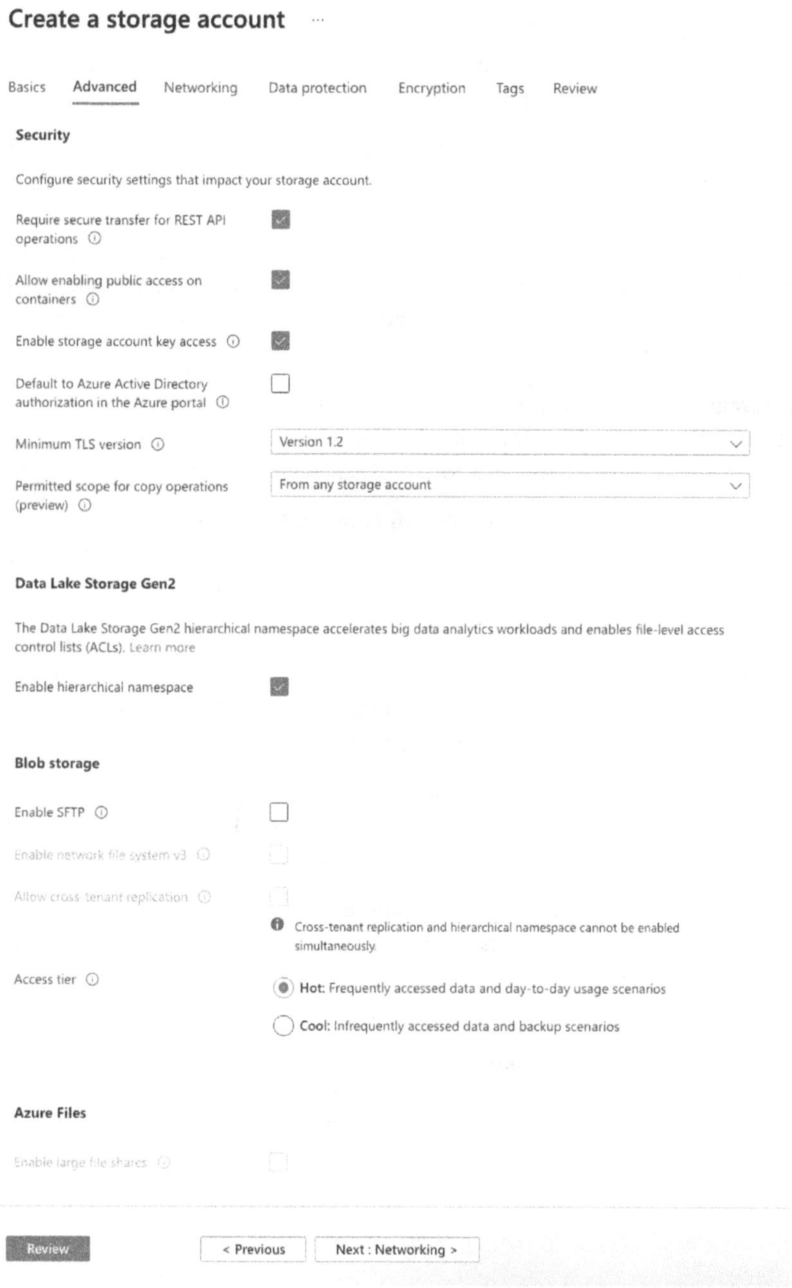
Create a storage account
Basics
Advanced
Networking
Data protection
Encryption
Tags
Review
Security
Configure security settings that impact your storage account.
Require secure transfer for REST API operations
Allow enabling public access on containers
Enable storage account key access
Default to Azure Active Directory authorization in the Azure portal
Minimum TLS version
Version 1.2
Permitted scope for copy operations (preview)
From any storage account
Data Lake Storage Gen2
The Data Lake Storage Gen2 hierarchical namespace accelerates big data analytics workloads and enables file-level access control lists (ACLs). Learn more
Enable hierarchical namespace
Blob storage
Enable SFTP
Enable network file system v3
Allow cross-tenant replication
Cross-tenant replication and hierarchical namespace cannot be enabled simultaneously.
Access tier
Hot: Frequently accessed data and day-to-day usage scenarios
Cool: Infrequently accessed data and backup scenarios
Azure Files
Enable large file shares
Review
< Previous
Next : Networking >

Figure 4.5 – Advanced storage configuration

This screen has options around how to limit access to your account. Let's take a look:

- **Require secure transfer for REST API operations**: Do you want to use HTTP (unsecure) or HTTPS (secure)? The default setting of secure is always best here unless there is a specific need, such as your application not supporting HTTPS (unlikely!).
- **Allow enabling public access on containers**: By default, you allow your storage account to create containers that are publicly accessible with anonymous access. This is useful if you want to store images and documents that a user can access from a web browser on a public website. Your web application can simply point to the file in the container for the user to download directly. This presumes that the storage account has its network open to allow public access (which we will see later). It doesn't mean everything in the account is public, just that you can create public objects.
- **Enable storage account key access**: Objects in Blob Storage, when set to private, require either Active Directory authorization or a **shared access signature** (**SAS**). This allows us to turn off the latter and ensure only Active Directory users can access it.
- **Default to Azure Active Directory authorization in the Azure portal**: By default, storage access keys are used to allow access to accounts when you are accessing them via the portal. You can switch this to require Active Directory authorization instead. This option simply actives that.
- **Minimum TLS version**: Older versions of TLS are not secure and therefore changing this is not recommended. However, some older applications simply do not support TLS 1.2. This option allows you to downgrade it. This essentially makes your account open to exploits and should only be done if you need it and there are no other alternatives.
- **Enable hierarchical namespace**: If you are using a storage account to perform data analytics with tools such as Azure Databricks, Azure Synapse, or Azure Data Factory, enabling this option turns your account into a Hadoop-compatible storage system, such as a Data Lake Gen 2.
- **Enable SFTP**: This is only available if you enable a hierarchical namespace. The option allows you to set up your storage account as an SFTP server. You will set up user accounts within the SFTP configuration blades to allow access to users.
- **Enable large file shares**: This is only available if you set your redundancy to LRS or ZRS. This option allows you to create file shares of up to 100 TiB.

You may have noticed that we missed **Access Tier** – this is because it warrants discussion in more detail. We will cover it shortly in the *Life cycle management* section.

Let's get back to configuring our storage account. Unless you have a specific need, the default options are usually the best. Therefore, don't change any options, or any of the options in the other sections, and just click **Review**.

You will be presented with a last chance to look at your options. Once you're happy, click **Create**.

With that, your storage account will be created. Once complete, in the Azure portal, select your new account. We will examine some more options next.

Life cycle management

When we created our storage account, we set **Access Tier** to **Hot** as this was the default. But what is this?

Storage in Azure is one of the few services that is always running and therefore incurring costs. By comparison, most other services only charge you whilst the service is running. For example, if you have a VM and you shut it down, you stop incurring costs – or rather, you stop incurring compute costs. The disk that is kept on a storage account is still there and therefore you are still being charged for it.

Azure Cloud is also seen as the ideal place for storing data as it is relatively cheap (compared to buying and maintaining your own hardware), and can easily be expanded as required.

The temptation would be to simply push all your data to Azure and forget about it. But not all data is equal – some data is accessed regularly, some infrequently, and some never at all.

For this reason, Azure storage accounts support access tiers. An access tier is a way to move data to different types of disks based on how often you need to access that data. The three tiers are **Hot**, **Cool**, and **Archive**. The biggest differences between the tiers are in terms of performance and how they are charged for:

- **Hot**: This is for normal data that is frequently accessed and is the one you will use most of the time. The cost of storing data here is the most expensive of the three tiers, but read and write operations are the cheapest.
- **Cool**: This is for infrequently accessed data. By this, we mean data that is occasionally accessed, but not regularly. It costs more than the **Hot** tier to read and write data to it, but its storage costs are lower.
- **Archive**: This is for data that is rarely, perhaps never, accessed. A great use case is data that needs to be kept for regulatory reasons but isn't expected to be needed until a point in the future. Archive data is the cheapest to store, but the costliest to read from. You also cannot read data from it straight away – to get at data in the **Archive** tier, you must first move it to the **Hot** tier – and this can take hours!

Even though you can choose the **Hot** or **Cool** tier when creating an account, this just sets the default tier –you can manually move a file between access tiers at any point. This can be done either when you upload a file, or by changing its tier once uploaded. *Figure 4.6* shows an example of changing an existing file's tier:

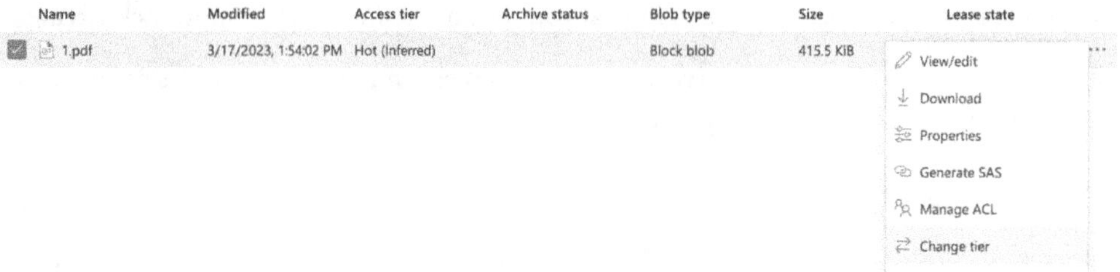
Name
Modified
Access tier
Archive status
Blob type
Size
Lease state
1.pdf
3/17/2023, 1:54:02 PM
Hot (Inferred)
Block blob
415.5 KiB
View/edit
Download
Properties
Generate SAS
Manage ACL
Change tier

Figure 4.6 – Changing the access tier of a document

The real power of document tiers comes into play when we combine them with another feature of storage accounts – life cycle management.

Documents often have a life cycle. During creation and review, they will be accessed frequently, but after a certain period, they will either become reference documents that are infrequently accessed or kept but rarely accessed. A great example is a customer contract.

A customer contract will be used a lot during the onboarding process, which may take a week, but once signed, the document may only be valid for 12 months. During those 12 months, the customer may need to view it occasionally, but after 12 months, it is no longer valid but must be kept to comply with regulations.

The life cycle management feature of storage accounts allows you to define a policy that automatically moves documents between tiers and even deletes them at different periods.

Lifecycle management can be found in the storage blade under **Data management**, as shown in *Figure 4.7*:

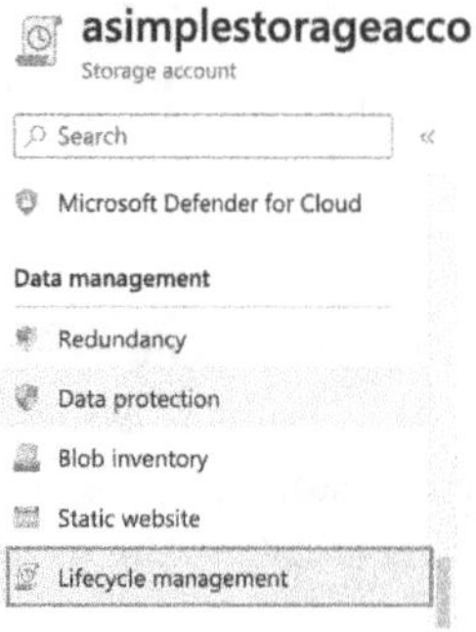
asimplestorageacco
Storage account
Search
Microsoft Defender for Cloud
Data management
Redundancy
Data protection
Blob inventory
Static website
Lifecycle management

Figure 4.7 – Lifecycle management

To create a life cycle policy, from the **Lifecycle management** option, perform the following steps:

1. Click **Add a rule**.
2. Under **Rule name**, enter **Manage Contracts**.
3. Under **Rule scope**, select **Limit blobs with filters**.

4. Leave the other options as-is and click **Next**.
5. Create the first rule:

   - Leave **Last modified** selected
   - Under **More than (days ago)**, enter **7**
   - Under **Then**, choose **Move to cool storage**
   - Click **Add conditions**
   - Leave **Last modified** selected
   - Under **More than (days ago)**, enter **365**
   - Under **Then**, choose **Move to archive storage**

   Your screen should look similar to the one shown in *Figure 4.8*:

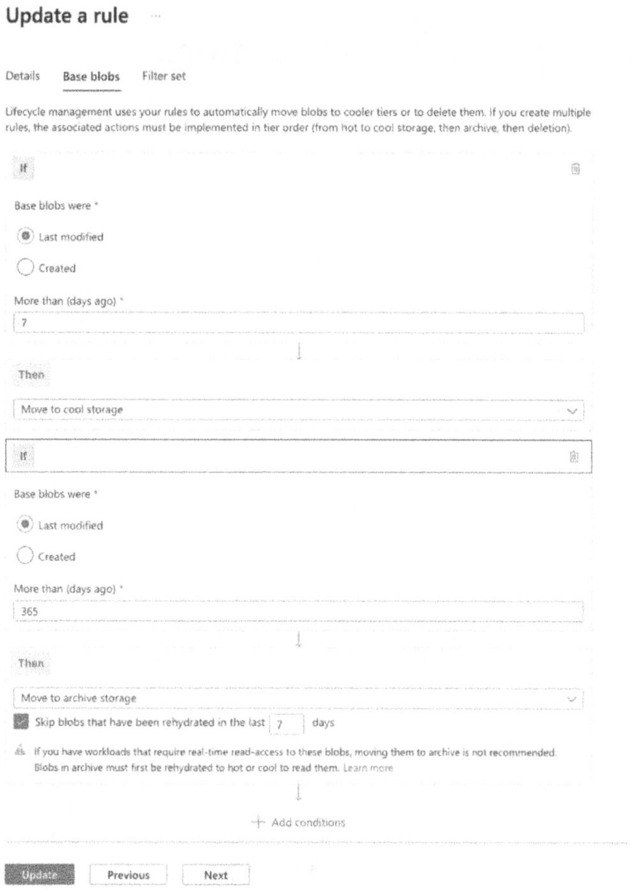
Update a rule
Details
Base blobs
Filter set
Lifecycle management uses your rules to automatically move blobs to cooler tiers or to delete them. If you create multiple rules, the associated actions must be implemented in tier order (from hot to cool storage, then archive, then deletion).
If
Base blobs were *
Last modified
Created
More than (days ago) *
7
Then
Move to cool storage
If
Base blobs were *
Last modified
Created
More than (days ago) *
365
Then
Move to archive storage
Skip blobs that have been rehydrated in the last
7
days
If you have workloads that require real-time read-access to these blobs, moving them to archive is not recommended. Blobs in archive must first be rehydrated to hot or cool to read them. Learn more
Add conditions
Update
Previous
Next

Figure 4.8 – Setting the life cycle rules

If everything looks good, we can finish off by setting a filter.

6. Click **Next**.
7. Under **Blob prefix**, enter `contracts/*.docx`.
8. Click **Add**.

Your rule will now be created, and any documents with a `.docx` extension in a `contracts` container in this storage account will follow the life cycle you've defined.

Storage accounts are incredibly versatile and should be considered for most of your document and image data storage needs. The **Tables** feature can even be considered for basic structured data that doesn't require more complex features.

However, when you do need to store structured and semi-structured data, especially when that data needs to be queried and managed, then a database solution is usually the best option.

Understanding Azure database options

Microsoft has two core database options – Azure SQL, and Azure Cosmos DB – both of which lend themselves to structured or semi-structured data, respectively. Azure SQL has a range of different options. We will consider these first.

Azure SQL

Azure provides three different options for running SQL – a traditional SQL server running on a VM, Azure SQL, and Azure SQL Managed Instance.

The core differences between these three platforms are around how they are managed, and of course how much they cost. Azure SQL on a VM is the most traditional option, and there are specific Azure marketplace images for Windows and Linux that include the cost of the SQL license. If you already have a license, you can still use the marketplace image and apply that license to get a discount, or you can install SQL on a standard VM image.

Azure SQL sits at the opposite end of the **Infrastructure-as-a-Service** (**IaaS**) – **Platform-as-a-Service** (**PaaS**) spectrum. Whereas SQL on a VM is classed as IaaS, Azure SQL is classed as PaaS. The importance of this is in terms of management:

- With **SQL on a VM**, you are responsible for fully patching, maintaining, and upgrading everything. You are also fully responsible for backups, but the Azure Backup solution does include tools that help with this.
- **Azure SQL**, on the other hand, is a fully managed platform. All you need to worry about is setting up the database's schema, indexes, and other database admin stuff. The Azure platform takes care of maintenance, patching, and even automatic tuning if you enable it. The service is

also backed up, and you have choices around some of that configuration. Azure SQL also has a degree of resilience built in that you could only achieve using the IaaS option by running at least two instances of SQL on a VM – which of course doubles your costs! We will examine this in more detail shortly.

- **Azure Managed Instance** sits between SQL on a VM and Azure SQL. It is still classified as a PaaS offering, in that it is managed by the Azure platform, but it isn't quite as flexible as Azure SQL. However, it does have greater feature parity with traditional SQL running on either a VM or physical hardware. This is often a key consideration when migrating databases from an existing on-premises SQL database.

So, how do you choose one or the other? We can get to the answer by asking five questions:

- Are you building a new or migrating an existing database? If you're migrating, is the source database compatible with Azure PaaS?
- Do you need full control over the SQL Server, including defining the precise RAM and CPU requirements?
- Do you want to be responsible for managing all aspects of the service, including resilience and backups?
- Do you need connectivity from on-premises systems, such as over a VPN or ExpressRoute connection?
- How large could a single database grow?

Once you have answered these questions, you can make an informed decision about which option is best for your needs. However, each area has its considerations.

Building a new database or migrating an existing one?

When building a new solution, you would normally opt for Azure SQL. Azure SQL, as a fully managed service, is the most cost-effective and flexible option. You only pay for what you need, and you can scale your databases up to 100 TB.

One of the main reasons not to choose Azure SQL is that of compatibility with older versions. However, when you're building a new solution, this doesn't matter.

Azure SQL offers several different performance tiers, depending on how much resilience you need and how much you want to spend. However, even the lowest service tier offers resilience across availability zones because of how it is built.

However, Azure SQL is not 100% compatible with older versions because of some transact-SQL differences in the data manipulation.

When you're building solutions using SQL running on a VM or physical server, code can be written that uses the `master` database – this is the special database SQL uses for its configuration, and it is possible to query and update it directly. However, in Azure SQL, the `master` database only provides limited access, which means any processes that would try to write to it would fail.

Microsoft provides tools to check your existing databases and recommendations. We will look at these in the next chapter.

Full control or fully managed?

When you create an Azure SQL instance or SQL Managed Instance, you have limited control over the precise configuration of the hardware that runs it compared to when you build SQL on a VM, where you can define the number of vCPUs and RAM.

This might not seem like an issue, but there are advanced situations when you might need this control.

Another consideration, especially with older databases, is that a solution may have additional custom features installed on the server that the database relies on. As managed services, Azure SQL and SQL Managed Instance don't give you that level of access, so in these situations, SQL on a VM might be your only route.

With Azure SQL and SQL Managed Instance, the Azure platform takes care of all the maintenance for you. This includes patching and backups. As mentioned previously, the PaaS versions of SQL also provide a base level of resilience automatically, and you can increase this by choosing different service tiers or creating your own replicas and failover groups.

If you opt for SQL on a VM, you are responsible for creating multiple instances and setting up replicas and failovers. As well as increased complexity and management overhead, your costs will increase as well.

In the cloud or hybrid?

The final question is about how you plan to access your databases once they are in Azure. This becomes important because of how connectivity is managed on the different services.

On-premises applications will be assigned IP addresses on a private range, and when you connect your corporate network to Azure over a VPN or ExpressRoute, you extend that private range into the cloud. Therefore, when you're trying to access anything in Azure over a VPN, the service you connect to also needs an address of that private range.

All VMs in Azure are connected to VNets, which use a private range address, so SQL on a VM is accessible over a VPN. The same is true for SQL Managed Instance – again, the service is built on top of a VNet and therefore receives a private IP address.

However, Azure SQL is not. Because of the underlying architecture of SQL, the IP address is on a public range, plus that IP is not the IP address of your own private SQL in Azure. Instead, it uses a gateway device that connects multiple customers to their own virtual SQL instances. Your virtual instance is identified by the FQDN, which is assigned to your Azure SQL service. This is in the `<servername>.database.windows.net` format.

The FQDN maps to an address that Azure manages and may change. The result is that, by default, SQL Azure cannot be reached on a private address, and there must be a public IP address.

There is, however, a separate service called Private Link IP that does allow you to assign an IP address on that private range. However, the service requires additional setup and other resources in the form of a DNS forwarder and Azure DNS zones. For a complete walkthrough of the service, visit the official Microsoft documentation at `https://learn.microsoft.com/en-us/azure/private-link/tutorial-private-endpoint-sql-portal`.

In conclusion, all three options can be used in a hybrid scenario, but SQL Managed Instance and SQL on a VM are the most straightforward options that don't require additional services to be set up and configured.

Database size

Another point to consider is how big your database is now or may grow. Azure SQL supports a single database of up to 100 TB, SQL Managed Instance can only support up to 16 TB, and SQL on a VM can support up to 256 TB.

So, again, if you expect your databases to grow over 256 TB, Azure SQL may be the best choice. Note, however, that when dealing with databases of this size, it is often common to implement database sharding. This is a technique whereby data is split across many databases and the SQL platform manages that split and exposes the data as though it is a single database.

Choosing a service

There is a final aspect to consider – and that is service levels. Most services in Azure carry a SLA – this is a financially backed agreement that a service will remain running for an amount of time. Azure SQL and Managed Instance have SLAs of up to 99.99%, whereas SQL on Azure is 99.95%. 00.04% may not seem a lot, but for mission-critical services, it could make a big difference.

In summary, we consider Azure SQL when we want a fully managed instance of SQL that is fast to set up, easy to scale, and is a brand-new, modern service.

Managed Instance is ideal when you're migrating databases that use some advanced features not available in Azure SQL PaaS, but when you still want the Azure service and management wrap.

SQL on a VM should be a last resort for those occasions whereby you are migrating old legacy systems that are not compatible with a PaaS offering.

For many, Azure SQL will be the best option, and if you choose that service, there are further decisions to make as there are different service tiers and pricing models. We will delve into those areas next.

Azure SQL pricing models

Azure SQL has two pricing models – **vCore** and **DTU**.

When Azure SQL was first made available, a DTU pricing model was used. A **DTU** is a **database transaction unit** and is essentially a bundled amount of CPU and RAM. It enables you to pay for a particular level of performance – the more you pay, the faster it is!

DTUs allow you to create a functional database at a very low cost – the smallest option is five DTUs and is around just $5 a month. Mapping DTUs to traditional CPU/RAM sizing is very difficult. Some tools will scan the usage of your database and suggest how many DTUs you require, but in practice, they tend to overestimate.

In recent years, Microsoft introduced the vCore pricing model, and this is now the recommended option. As you might expect from the name, the vCore model is based on the number of vCPUs you need for your database to run at an acceptable performance level. The amount of RAM is predefined at 5.1 GB or 10.2 GB per core, depending on the service tier you choose.

The vCore pricing model has a higher price for the lowest available scale – 1 vCore – which is why the DTU model might be better for dev/test environments or smaller use cases when keeping costs low is important.

However, the vCore model introduces a serverless option, whereby the database compute is paused when it's not used for a defined period (the default is 1 hour). Once paused, no compute costs are incurred; you only pay for the storage. Therefore, the serverless model offers potentially better cost efficiencies for dev/test or scenarios where data is only infrequently accessed, such as in archive databases.

SQL service tiers

Azure offers two different service tiers for SQL Managed Instance and three for Azure SQL. Each tier uses a slightly different underlying architecture that affects performance and availability. The two tiers that are available on both are **General** and **Business Critical**; with Azure SQL, the additional tier is **Hyperscale**.

General tier

The General tier is used for generic workloads, or when cost is more important than availability.

The General tier runs a `sqlservr.exe` process (essentially SQL Server) that uses Azure Blob Storage for the database files themselves. Blob Storage can be configured as either LRS or ZRS to protect from local failure or data center failure. If maintenance or a failure occurs, the active node will fail over to another node automatically to keep your service running.

Figure 4.9 shows how this looks:

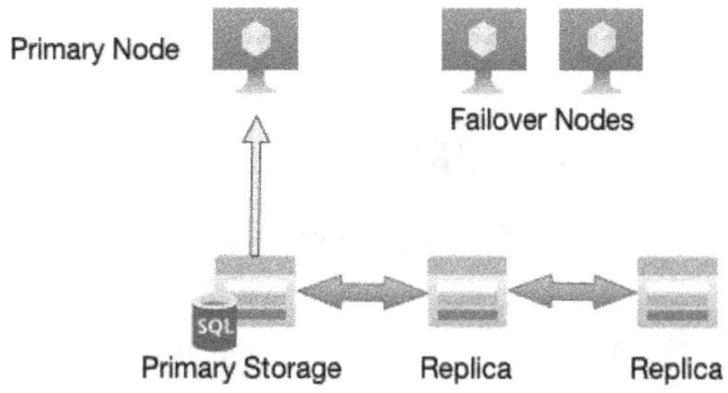
Primary Node
Failover Nodes
SQL
Primary Storage
Replica
Replica

Figure 4.9 – General tier architecture

Therefore, the General tier can provide redundancy with a region but does not protect you from a regional outage.

If you do need a regional outage, you can configure a failover replica of your database in another region. However, this involves creating another Azure SQL instance, which increases costs.

Hyperscale

The Hyperscale tier uses an architecture that is designed to support much larger databases – up to 100 TB.

Traditional database engines run all the management functions on a single process – as we saw in the General tier, where a compute node runs the `sqlservr.exe` process. Hyperscale databases separate the query processing engine and other functions. This enables the service to use a primary compute node, as in the General tier, but also adds the ability to add additional compute nodes that can serve as read-only processors to offload and enable **read scale-out**.

Along with local SSD caches on each node, this allows the Hyperscale tier to manage much bigger databases.

As with the General tier, you can create your own replicas – called **named replicas** – of your database to further enhance readability.

It is the Hyperscale tier that can increase the RAM per vCore to 10.2 GB when required, though this is managed automatically.

Business Critical

The Business Critical tier, as its name suggests, is for databases that cannot afford any downtime, and require the fastest performance.

They always use a replica to ensure high availability, but the underlying architecture differs slightly in that the database files are stored on locally attached SSDs, rather than Blob Storage.

Along with providing greater resilience, the use of locally attached SSDs ensures that latency and **input/output operations per second** (**IOPS**) – which is how fast you can read and write data – is the highest possible:

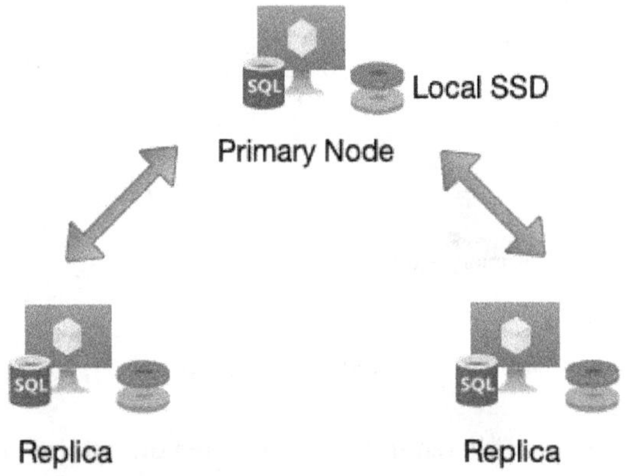
SQL
Local SSD
Primary Node
SQL
Replica
SQL
Replica

Figure 4.10 – Business-critical architecture

Choosing the correct tier for your business needs is very important, but it is possible to upgrade tiers – for example, General to Hyperscale – when required. However, downgrading tiers isn't directly supported in some instances. Despite this, there is always the option to back up a database and restore it to a lower tier.

Creating an Azure SQL database

That's enough theory – let's walk through creating a SQL database. We will choose the serverless tier to keep costs down.

In the Azure portal, perform the following steps:

1. Click **Create a resource** from the menu or main screen.
2. Select **SQL Database**, which is listed as one of the popular services, or type `SQL Database` into the search box.
3. Select your subscription and either select an existing resource group or create a new one by clicking **Create new** under the **Resource group** dropdown.
4. Give your database a name. In my example, I'm using `packt-sqldb`.
5. Under **Server**, click **Create new**.
6. In the new window that appears, give your server a name, such as `packt-sqlsvr`. Names must be unique across Azure.
7. Select a **Location** type.
8. Set **Authentication method** to **Use only Azure Active Directory**.

9. Under **Set Azure AD admin**, click **Set admin**, then select your account from the list of users and click **Select**.
10. Click **OK** to return to the previous screen.
11. Leave **Want to use SQL elastic pool?** set to **No**.
12. Change **Workload environment** to **Development**.
13. Change **Backup storage redundancy** to **Locally-redundant backup storage** to keep costs down. Your screen should look similar to what's shown in *Figure 4.11*:

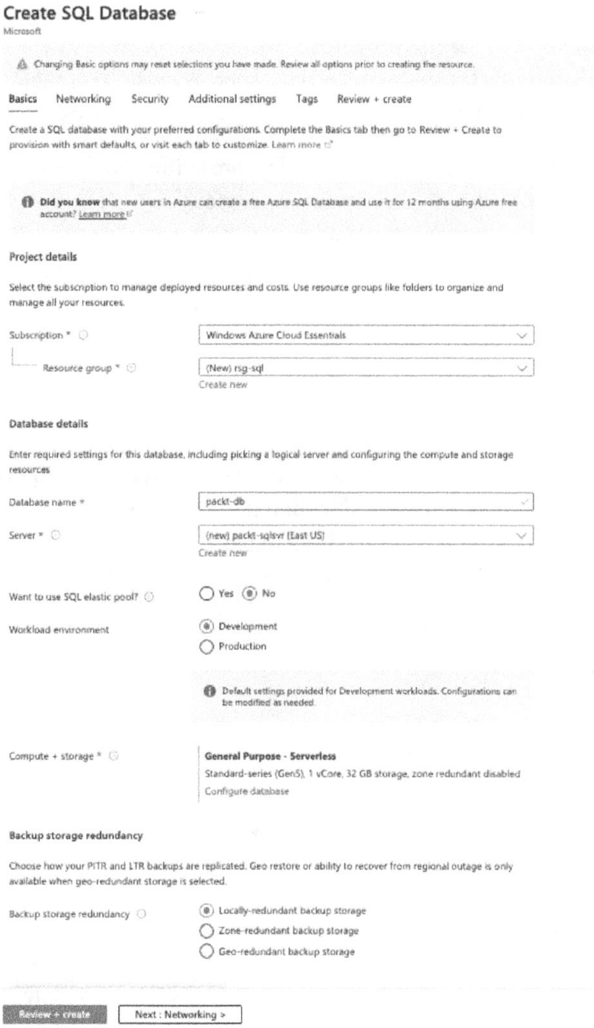
Create SQL Database
Microsoft
Changing Basic options may reset selections you have made. Review all options prior to creating the resource.
Basics
Networking
Security
Additional settings
Tags
Review + create
Create a SQL database with your preferred configurations. Complete the Basics tab then go to Review + Create to provision with smart defaults, or visit each tab to customize. Learn more
Did you know that new users in Azure can create a free Azure SQL Database and use it for 12 months using Azure free account? Learn more
Project details
Select the subscription to manage deployed resources and costs. Use resource groups like folders to organize and manage all your resources.
Subscription *
Windows Azure Cloud Essentials
Resource group *
(New) rsg-sql
Create new
Database details
Enter required settings for this database, including picking a logical server and configuring the compute and storage resources
Database name *
packt-db
Server *
(new) packt-sqlsvr (East US)
Create new
Want to use SQL elastic pool?
Yes
No
Workload environment
Development
Production
Default settings provided for Development workloads. Configurations can be modified as needed.
Compute + storage *
General Purpose - Serverless
Standard-series (Gen5), 1 vCore, 32 GB storage, zone redundant disabled
Configure database
Backup storage redundancy
Choose how your PITR and LTR backups are replicated. Geo restore or ability to recover from regional outage is only available when geo-redundant storage is selected.
Backup storage redundancy
Locally-redundant backup storage
Zone-redundant backup storage
Geo-redundant backup storage
Review + create
Next : Networking >

Figure 4.11 – The Create SQL Database screen

We will leave the other pages of options as-is, so click **Review + create** and then **Create**.

There are two options we must choose from that are worth further discussion: elastic pools and authentication.

Elastic pools

Each database is priced individually, so even though you may use the same SQL Server component in Azure, each database can be configured as different vCores or DTUs, and you are charged *per database* accordingly.

However, at times, you may have databases whose usage patterns differ. For example, a customer database may be used continually throughout the day, whereas a database that is used for overnight processing will only be used out of hours.

A SQL elastic pool allows you to define your DTU and vCore at the pool level, at which point you can add databases to that pool. Regardless of how many databases you have in the pool, you will only be charged for the pool itself. Of course, the databases must all compete for the defined resources in the pool, which is why elastic pools work well with databases that have complimentary usage patterns.

In many ways, an elastic pool is like a traditional SQL server running on a VM – you only pay for the VM, regardless of how many databases you have, but all those databases are sharing the VM's resources.

Authentication

When we created our database, we chose the **Use only Azure Active Directory** option. SQL can use its own authentication, which is completely independent of your user account in AD. However, this requires storing and managing separate usernames and passwords.

In application development, you would normally create users for different applications and assign rights depending on what functions you need the application to perform on the data – for example, reading or writing data.

Along with having to maintain these users and passwords, these details must also be used when you're connecting to the database from an application. This is usually performed by creating a connection string that defines the server's name and the username and password. This can be considered insecure as the connection string is usually plain text.

In Azure, we have key vaults to help protect information like this, but it can still be a management overhead.

An alternative is to use Active Directory accounts. In this way, the account is protected by Azure. In addition, applications that require access can use managed identities or service principals to connect to the database. Using these options removes the need for the security details to be present in a connection string.

Using SQL databases

Now that our database is up and running, we shall look at how we can access and add data to it.

There are several client tools we can use to connect to a database. SQL Studio Manager is probably the most well-known, but another tool is now installed alongside it called Azure Data Studio.

Azure Data Studio can also be installed separately if you download it from `https://github.com/microsoft/azuredatastudio`. If you don't already have it, download and install it to perform the next walkthrough.

Before we can connect with any tool, we must allow access from our computer to the SQL server. This is because when we built the SQL server, the default network rules blocked all connections, forcing us to explicitly define what IP addresses can connect. Let's take a look:

1. To allow access to your computer's IP address, in the Azure portal, go to your new SQL Server database and, on the main page, at the top of the screen, you will see a button called **Set server firewall**, as shown in *Figure 4.12*:

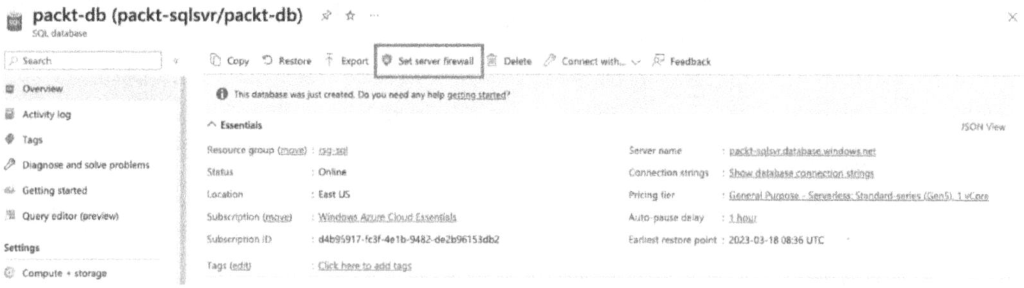
packt-db (packt-sqlsvr/packt-db)
SQL database
Search
Overview
Activity log
Tags
Diagnose and solve problems
Getting started
Query editor (preview)
Settings
Compute + storage
Copy
Restore
Export
Set server firewall
Delete
Connect with...
Feedback
This database was just created. Do you need any help getting started?
Essentials
JSON View
Resource group (move) : rsg-sql
Status : Online
Location : East US
Subscription (move) : Windows Azure Cloud Essentials
Subscription ID : d4b95917-fc3f-4e1b-9482-de2b96153db2
Tags (edit) : Click here to add tags
Server name : packt-sqlsvr.database.windows.net
Connection strings : Show database connection strings
Pricing tier : General Purpose - Serverless: Standard-series (Gen5), 1 vCore
Auto-pause delay : 1 hour
Earliest restore point : 2023-03-18 08:36 UTC

Figure 4.12 – SQL database overview page

2. Click that button to go to the firewall page. Note that the firewall page is for the SQL server – this means any IP addresses or VNets that you allow access from will be set for all databases on this SQL server.
3. First, we must enable public access. Do this by selecting the **Selected networks** option. A range of further options will appear. However, we must save the new selection before making further changes, so click **Save**.
4. We can now allow access in three ways. First, we can allow access from VNets, which means we can provide access from VMs or other VNet-connected services. This is ideal for limiting access to your SQL server from other internal services, such as web apps or containers.
5. Alternatively, at the bottom of the page is a checkbox called **Allow Azure services and resources to access this server**. This can be useful; however, it enables access from all Azure services – not just your subscriptions, but every Azure customer's resources as well, and therefore this is not recommended.

6. For this demonstration, we will add the IP address of our computer. First, we need to find our public address – usually, this will be detected in the portal and the boxes will be prefilled, but if it isn't, use your search engine of choice and search for `What is my IP`. This will return your public IP address.
7. Back in the Azure portal, under **Firewall rules**, click **Add your client Ipv4 address()**. Give the rule a name, such as `My Laptop`, then enter your public IP address in the **Start IPv4 address** and **End IPv4 address** boxes. Click **Save**.

   Your screen should look something like what's shown in *Figure 4.13*:

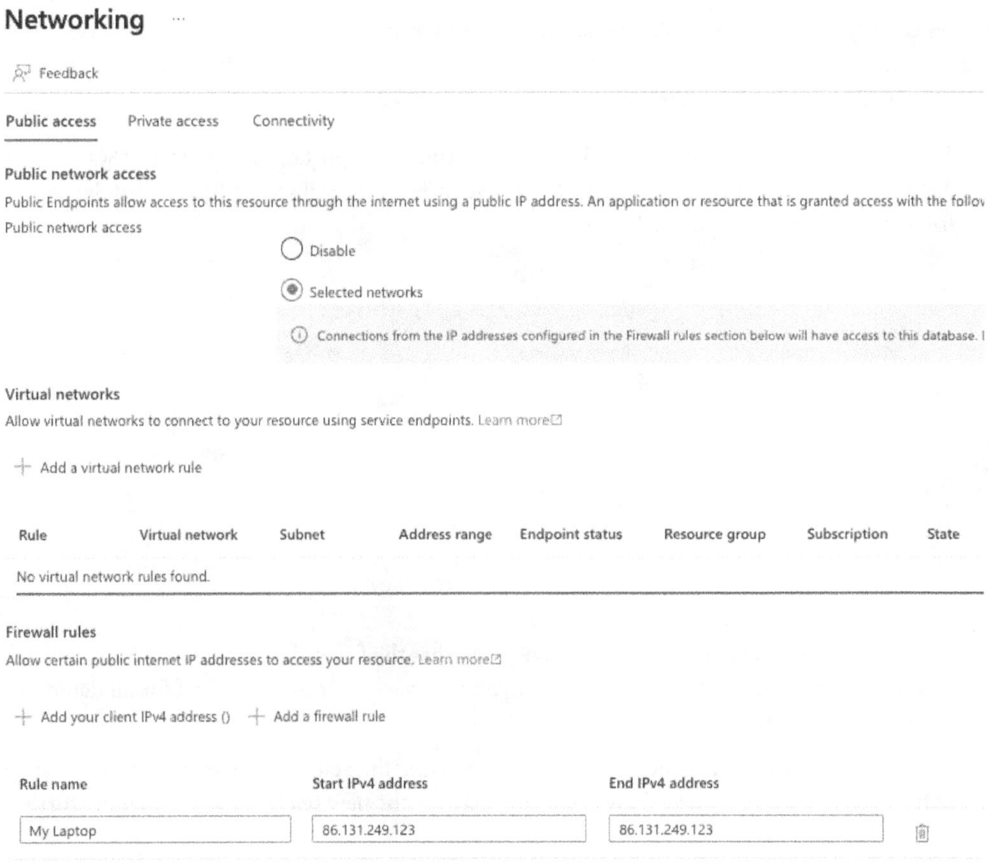

Networking
Feedback
Public access
Private access
Connectivity
Public network access
Public Endpoints allow access to this resource through the internet using a public IP address. An application or resource that is granted access with the follo
Public network access
Disable
Selected networks
Connections from the IP addresses configured in the Firewall rules section below will have access to this database.
Virtual networks
Allow virtual networks to connect to your resource using service endpoints. Learn more
Add a virtual network rule
Rule
Virtual network
Subnet
Address range
Endpoint status
Resource group
Subscription
State
No virtual network rules found.
Firewall rules
Allow certain public internet IP addresses to access your resource. Learn more
Add your client IPv4 address ()
Add a firewall rule
Rule name
Start IPv4 address
End IPv4 address
My Laptop
86.131.249.123
86.131.249.123

Figure 4.13 – Opening the SQL server firewall

8. Next, back on the SQL database overview page, at the top, copy the server name; again, refer to *Figure 4.12*. We can now run Azure Data Studio.

9. From the main page, click **New -> New connection** and enter the following details:
   - **Server**: Paste in the server name you just copied
   - **Authentication type**: **Azure Active Directory – Universal with MFA support**
   - **Trust server certificate**: **True**
10. Leave the rest of the options as-is and click **Connect**.
11. You will follow the standard Azure authentication process, but then it should connect you. From here, we can start to issue T-SQL commands. To try this out, create a new table with some sample data and then query it.
12. Click **New Query**, as highlighted in *Figure 4.14*:

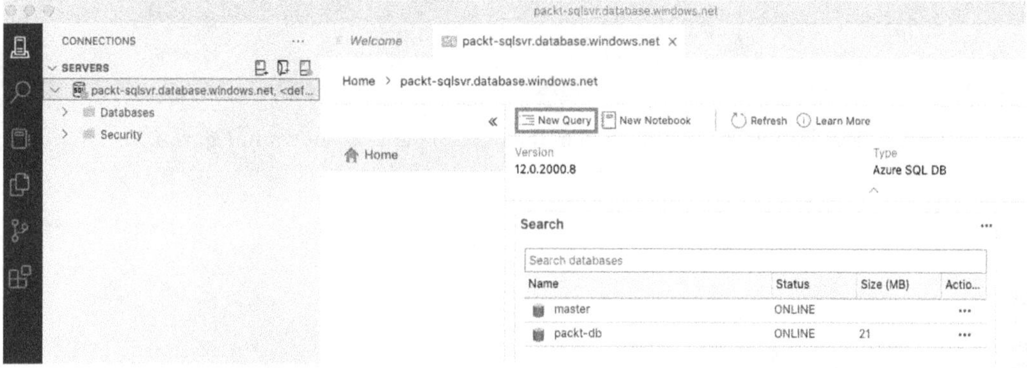
packt-sqlsvr.database.windows.net
CONNECTIONS
SERVERS
packt-sqlsvr.database.windows.net, <def...
Databases
Security
Welcome
packt-sqlsvr.database.windows.net
Home > packt-sqlsvr.database.windows.net
New Query
New Notebook
Refresh
Learn More
Home
Version
12.0.2000.8
Type
Azure SQL DB
Search
Search databases
Name
Status
Size (MB)
Actio...
master
ONLINE
packt-db
ONLINE
21

Figure 4.14 – Connecting to the database from your computer

13. On the new screen, change the **Change connection** dropdown so that it specifies your database (it defaults to master) and enter the following code:

```
CREATE TABLE Customers
(
CustomerId int NOT NULL,
FirstName varchar(20) NOT NULL,
Surname varchar(20) NOT NULL
)
INSERT INTO Customers VALUES (1, 'Albert','Einstein')
INSERT INTO Customers VALUES (1, 'Thomas','Edison')
```

Your screen should look similar to *Figure 4.15*. Click **Run**:

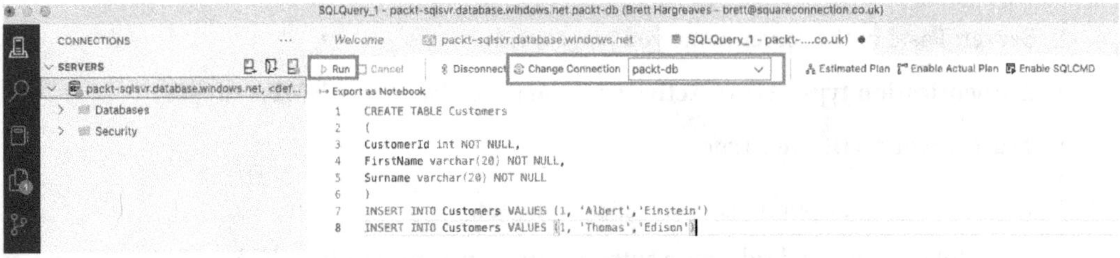
SQLQuery_1 - packt-sqlsvr.database.windows.net.packt-db (Brett Hargreaves - brett@squareconnection.co.uk)
CONNECTIONS
SERVERS
packt-sqlsvr.database.windows.net, <def...
Databases
Security
Welcome
packt-sqlsvr.database.windows.net
SQLQuery_1 - packt-....co.uk)
Run
Cancel
Disconnect
Change Connection
packt-db
Estimated Plan
Enable Actual Plan
Enable SQLCMD
Export as Notebook
CREATE TABLE Customers
(
CustomerId int NOT NULL,
FirstName varchar(20) NOT NULL,
Surname varchar(20) NOT NULL
)
INSERT INTO Customers VALUES (1, 'Albert','Einstein')
INSERT INTO Customers VALUES (1, 'Thomas','Edison')

Figure 4.15 – Creating a SQL table and inserting data

14. Finally, delete the code you just wrote and replace it with the following:

```
SELECT * FROM Customers
```

15. Again, click **Run**.

Your new data should now be returned in the bottom pane, as shown in *Figure 4.16*:

CONNECTIONS
SERVERS
packt-sqlsvr.database.windows.net, <def...
Databases
Security
Welcome
packt-sqlsvr.database.windows.net
SQLQuery_1 - packt-....co.uk)
Run
Cancel
Disconnect
Change Connection
packt-db
Estimated Plan
Enable Actual Plan
Enable SQLCMD
Export as Notebook
SELECT * FROM Customers
Results
Messages
CustomerId
FirstName
Surname
1
Albert
Einstein
1
Thomas
Edison

Figure 4.16 – Querying your new table

This example just shows how quick and easy it is to get up and running with Azure SQL – no need to build VMs, install software, or configure lots of settings.

From a security perspective, you should get in the habit of removing your public IP address when you're finished. Simply go back to the SQL Server firewall view and click the trashcan icon next to your connection.

Azure SQL is the flagship SQL offering in Azure, and for new, modern solutions, it should be the first choice in the majority of cases.

In recent years, however, a new technology has appeared called NoSQL. We discussed NoSQL at the start of this chapter when we looked at the differences between structured and semi-structured data since it is a database mechanism that stores semi-structured data.

Several NoSQL databases have arisen, with one of the most well-known being MongoDB. However, Microsoft has created an offering called Cosmos DB. We will discuss this next.

Azure Cosmos DB

Microsoft introduced a new NoSQL database engine many years ago called DocumentDB. The technology has matured since it was originally introduced and was rebranded Cosmos DB. This is useful to know as there are still sometimes references to DocumentDB!

Cosmos DB is a globally distributed database that is designed to scale horizontally on demand with the ability to have read/write replicas in any Azure region. It has low latency and fast IOPS, and as a NoSQL database, it is ideal for modern applications.

As a NoSQL database, you don't need to predefine schemas before you use it – you can chop and change your models as required.

Cosmos DB was built from the ground up as a fully managed cloud service, and as such was designed to have deep integration with other Azure services, such as IoT Hub, AKS, and Azure Functions.

From a programmer's point of view, it offers multi-APIs – this makes it easy for developers who are used to other database types to get up to speed quickly or adapt existing applications to swap over to it. These APIs include a native NoSQL API, MongoDB, PostgreSQL, Apache Cassandra, Gremlin, and Azure Table. It's worth noting that this can only be selected at the time of account creation as it affects how data is stored.

Cosmos DB uses a different structure compared to traditional SQL databases. Instead of a server, database, tables, and rows or records, we have an account as the top level, then a database. Then, within that database, we have containers, and in those containers, we store items.

Figure 4.17 shows the hierarchical differences:

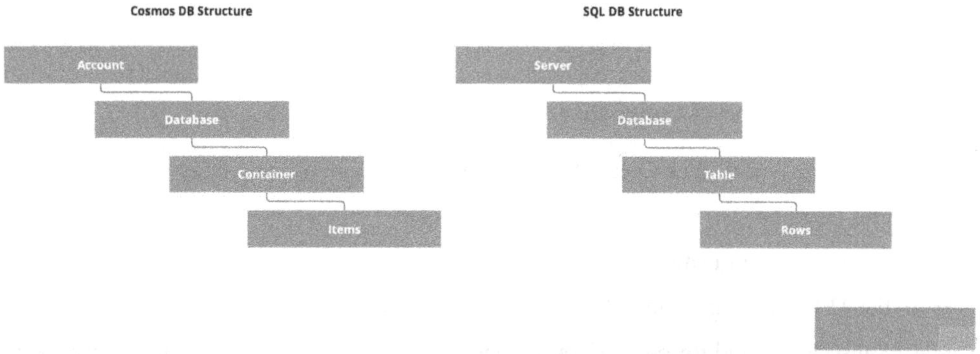
Cosmos DB Structure
Account
Database
Container
Items
SQL DB Structure
Server
Database
Table
Rows

Figure 4.17 – Comparing Cosmos DB and SQL

Containers are like tables, and items are synonymous with records. However, whereas a table in a SQL database must always store the same records, in a NoSQL database, we can store completely different records!

This, of course, can make things confusing when we're trying to query and sort data, so each record must use a partition as well as an ID key. The ID uniquely defines each record, and the partition can be used to define record types such as Customer, Supplier, and Product. It doesn't have to be used in this way, though. If you are storing the same records in a container, the partition key can be used to separate data based on location, year, or any other natural grouping that makes sense.

Pricing models

Cosmos DB, like Azure SQL, has two pricing models – **Provisioned** and **Serverless**. Serverless, as you would expect, means you simply set upper limits for performance, and the platform scales accordingly. As with Azure SQL Serverless, you only pay for compute as it's being used, and you are only charged for storage when the service is idle.

With Provisioned billing, you define how many **request units** (**RUs**) you want. An RU is a unit of action, such as reading or writing data of 1 KB in size. More complex operations such as indexing and querying cost more RUs and how many will be required depends on the complexity or size of the operation.

It can sometimes be difficult to estimate how many RUs to provision until you have experimented with and load-tested your service. It is also worth noting that you will always be billed for at least 10% of provisioned RUs, whether you use them or not.

RUs can be set at the account or container level. When you set them at the account level, resources are shared by all containers in that account. Setting RUs at the container level dedicates those resources to each container, but it can potentially cost more.

Before we look at some more features, let's go ahead and create a Cosmos DB account.

Creating a Cosmos DB account

Let's go ahead and create a new Cosmos DB account:

1. From the Azure portal, click **Create Resource** from the menu or main page.
2. Search for `Cosmos DB` and select **Azure Cosmos DB**.
3. On the Cosmos DB information page, click **Create**.
4. From the API select page, under **Azure Cosmos DB for NoSQL**, click **Create**.
5. Select **Subscription** and **Resource Group** properties or create a new resource group by clicking **Create new**.
6. Name your account – for example, `packt-cosmosdb`.
7. Select a **Location** type, such as **East US**.

8. Leave the capacity mode set to **Provisioned**.
9. For **Apply Free Tier Discount**, select **Apply** (if you haven't already used it before).
10. We will leave the other settings as-is, so click **Review and Create**.
11. Click **Create**.

It can take a few minutes to provision, but once completed, we will have a database. The next step is to create a container that can hold records.

In the Azure portal, find your new Cosmos DB account and select it, then choose **Data Explorer** in the left-hand menu. Follow these steps:

1. Click **New Container**.
2. Under **Database id**, select **Create new** and provide `sampledb` as the name.
3. Set **Database Max RU/s** to `1000`.
4. Under **Container id**, enter `samplecontainer`.
5. Under **Partition key**, enter `/City`.
6. Click **OK**.

This will create a new database and container and set the maximum RUs to 1,000. However, it will automatically scale from 10% of that up to the maximum. *Figure 4.18* shows what your screen should look like before you click **OK**:

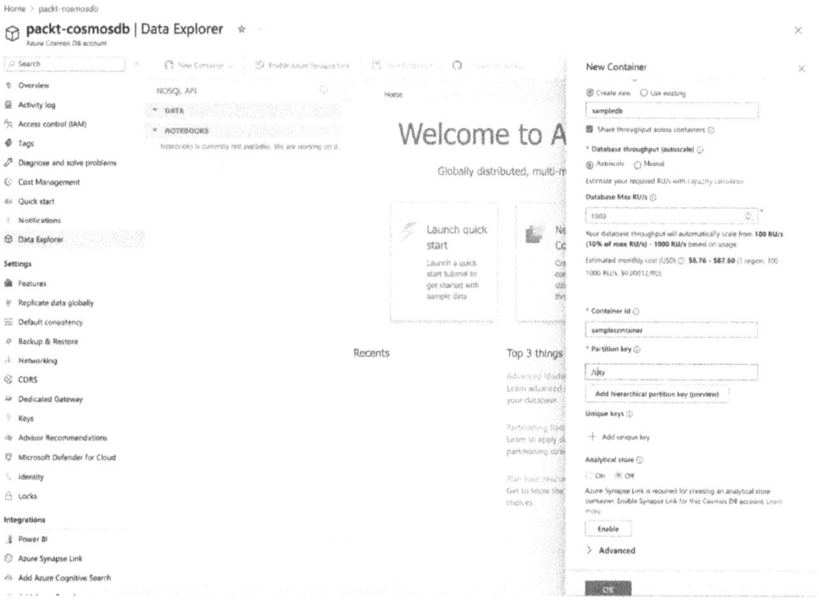
Home > packt-cosmosdb
packt-cosmosdb | Data Explorer
Azure Cosmos DB account
Search
Overview
Activity log
Access control (IAM)
Tags
Diagnose and solve problems
Cost Management
Quick start
Notifications
Data Explorer
Settings
Features
Replicate data globally
Default consistency
Backup & Restore
Networking
CORS
Dedicated Gateway
Keys
Advisor Recommendations
Microsoft Defender for Cloud
Identity
Locks
Integrations
Power BI
Azure Synapse Link
Add Azure Cognitive Search
NOSQL API
DATA
NOTEBOOKS
Welcome to A
Globally distributed, multi-m
Launch quick start
Recents
Top 3 things
New Container
Create new
Use existing
sampledb
Share throughput across containers
Database throughput (autoscale)
Autoscale
Manual
Database Max RU/s
1000
Container id
samplecontainer
Partition key
/city
Add hierarchical partition key (preview)
Unique keys
Add unique key
Analytical store
On
Off
Enable
Advanced
OK

Figure 4.18 – Creating a Cosmos DB container

Once our database has been created, we can test it by creating and viewing records directly within the Azure portal. From the left-hand menu of the Cosmos DB screen, click **Data Explorer**.

From here, we can see our databases and query them, much like in a SQL tool such as Azure Data Explorer on a SQL database.

Under **Data**, select the **sampledb** database we created and then expand the sample container. Now, click **New Item** at the top of the page and enter the following:

```
{
  "id": "001",
  "firstName": "Sherlock",
  "surname": "Holmes",
  "city": "London"
}
```

Your screen should look like what's shown in *Figure 4.19*:

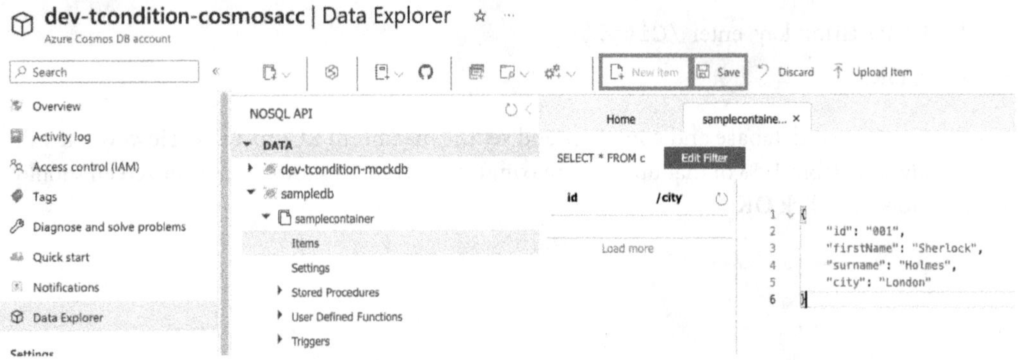

Figure 4.19 – Creating a Cosmos DB data item

Now, click **Save**. The data you typed will be replaced with something like the following:

```
{
  "id": "001",
  "firstName": "Sherlock",
  "surname": "Holmes",
  "city": "London",
  "_rid": "hB9eAJCJiu0BAAAAAAAAAA==",
  "_self": "dbs/hB9eAA==/colls/hB9eAJCJiu0=/docs/
hB9eAJCJiu0BAAAAAAAAAA==/",
  "_etag": "\"c9019b1e-0000-1100-0000-641755fe0000\"",
  "_attachments": "attachments/",
  "_ts": 1679250942
}
```

Now, click **New Item** again and enter the following:

```
{
  "id": "002",
  "firstName": "Albert",
  "surname": "Einstien",
  "city": "Princeton",
  "birthYear": 1879
}
```

Again, you will see something similar to before, but an additional record will appear in the record list on the left.

Now, click the ellipses next to **samplecontainer** and select **New SQL Query**. The default text of `SELECT * FROM c` will be displayed. Upon clicking **Execute Query**, you will see both records you have entered – see *Figure 4.20* for an example:

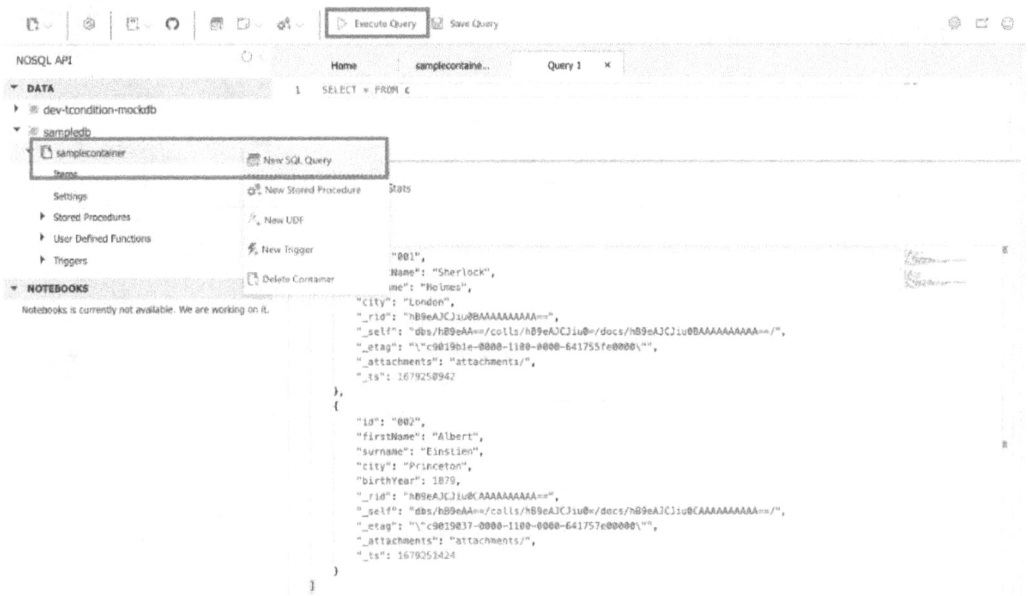

Figure 4.20 – Querying a Cosmos DB container

Normally, operations would be processed within an application, but the built-in Data Explorer in the Azure portal is a great tool to help confirm and investigate issues when building solutions.

Global distribution

Earlier, we mentioned that Cosmos DB is globally distributed and that each replica can either be read-only or read-write. This technology is about more than just availability and redundancy. The ability to have a globally distributed database provides the foundation for building truly global solutions.

As a simple example, imagine an e-commerce site that is hosted in the US. You could simply have all users access the US site, wherever they are in the world, but you would experience different performance levels for those outside the US than those in it.

Using Cosmos DB, you could deploy a frontend application on multiple continents with a replica of a Cosmos DB database behind them. In this configuration, the user experience is far superior as the local instance provides better localized performance, but the database is kept in sync across replicas.

As well as a better user experience, it also allows you to scale individual instances based on the usage in that location, and of course, you also get full redundancy since if one region fails, you can redirect traffic to the next closest region.

Figure 4.21 shows an example of this setup:

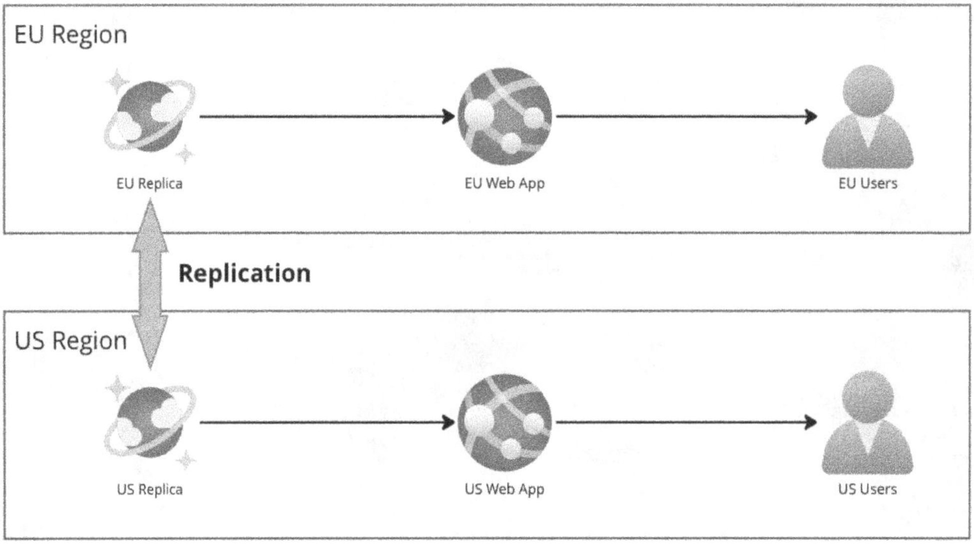
EU Region
EU Replica
EU Web App
EU Users
Replication
US Region
US Replica
US Web App
US Users

Figure 4.21 – Building globally distributed solutions

We can enable and disable where our replicas are stored when we create our Cosmos DB account or afterward – in this way, we can add new replicas as required.

In the walkthrough earlier, we didn't enable any replicas, so we will go ahead and create one. In the Azure portal, find your new Cosmos DB account and select it, and go through the following steps. Refer to *Figure 4.22* to see where the different options are:

1. Under **Settings**, click **Replicate data globally**.
2. Click + **Add region**.
3. Select the region where you want to replicate to – for example, **UK South**.
4. Click **OK**.
5. Change **Multi-region writes** to **Enable**.
6. Click **Save**.

This will replicate any data you have to the chosen region. Note that you will be billed per replica, depending on the RUs we set when we created our first container:

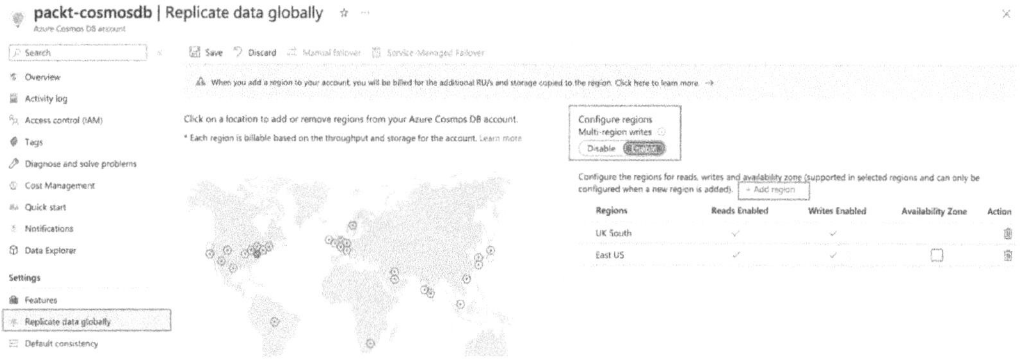

packt-cosmosdb | Replicate data globally
Azure Cosmos DB account
Search
Save
Discard
Manual failover
Service-Managed Failover
Overview
Activity log
Access control (IAM)
Tags
Diagnose and solve problems
Cost Management
Quick start
Notifications
Data Explorer
Settings
Features
Replicate data globally
Default consistency
Click on a location to add or remove regions from your Azure Cosmos DB account.
Configure regions
Multi-region writes
Disable
Enable
+ Add region
Regions
Reads Enabled
Writes Enabled
Availability Zone
Action
UK South
East US

Figure 4.22 – Setting up global replicas

It can take some time to replicate data, depending on how much data you have. A globally distributed database that can be read and written to from multiple locations is a great way to build solutions that cross geographical boundaries. However, it comes at a price: you must decide which is the most important – performance or keeping all the copies of that database in sync. This is known as consistency.

Replica consistency

If we have a database that has multiple copies of itself, we must choose between how fast we can write data to it and how quickly those updates reflect among the various copies. How quickly updates reflect amongst replicas is known as consistency.

We can choose optimal performance, but here, any writes are written immediately to the local copy of the database and then replicated out to the copies. The downside to this is that those copies may take some time to receive the updates.

The other option is optimal consistency. With this option, writes are made to all copies at the same time, but this impacts performance.

Azure Cosmos DB allows us to choose from one of five levels of consistency, from best performance, which is eventual consistency, or having all replicas up to date, known as strong consistency.

The default is *session consistency*, whereby the balance between consistency and performance is equal, or we can choose two levels on either side. We can set the desired consistency in the Azure portal. From the Cosmos DB page, on the left-hand menu, select **Default consistency**.

You will see a page similar to the one shown in *Figure 4.23*:

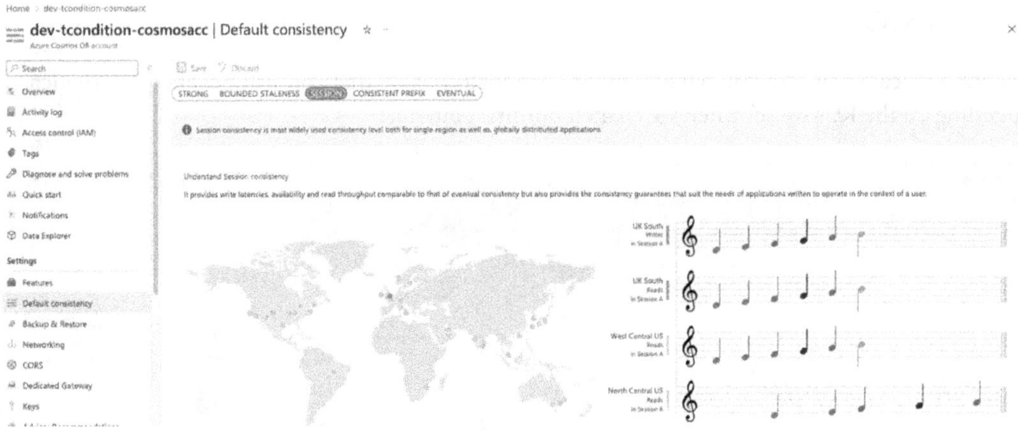
dev-tcondition-cosmosacc | Default consistency
Search
Overview
Activity log
Access control (IAM)
Tags
Diagnose and solve problems
Quick start
Notifications
Data Explorer
Settings
Features
Default consistency
Backup & Restore
Networking
CORS
Dedicated Gateway
Keys
STRONG
BOUNDED STALENESS
CONSISTENT PREFIX
EVENTUAL
UK South
West Central US
North Central US

Figure 4.23 – Cosmos DB – Default consistency

The default consistency can be changed at any time, but careful consideration needs to be given to the effect of keeping data across replicas in that it can impact performance and vice versa.

Cosmos DB is a great database for modern solutions; however, it is not always the right choice. If we need to build applications that need to replicate with traditional SQL servers, for example, Azure SQL or Azure SQL Managed Instance may be better options.

Summary

In this chapter, we have delved into the various options for storing data in Azure. First, we examined how we need to consider what type of data we are storing – structured, semi-structured, and unstructured data.

We then looked at how we can use and configure Azure storage accounts, before diving into the various SQL options we have available to us.

Finally, we looked at Cosmos DB, a modern, cloud-native, NoSQL database solution that enables us to build truly global solutions.

In the next chapter, we will look at how we might migrate existing applications to Azure. We will consider the options from a compute perspective, as well as how Azure data solutions map to on-premises data storage mechanisms.

5

Migrating to the Cloud

In the previous chapter, we looked at the different options for storing data in Azure— specifically, Azure Storage accounts, Azure SQL, and Azure Cosmos DB.

In this chapter, we will look at how we can best leverage Azure for our existing applications that are currently hosted on traditional infrastructure.

When migrating we sometimes need to consider servers, data, and applications separately and think about the trade-off between more work upfront versus less work and more flexibility later on. Therefore, in this chapter, we will address what options are available to us and how they differ by covering the following topics:

- Understanding migration options
- Managing servers
- Modernizing applications
- Migrating data

Technical requirements

This chapter will use the Azure portal (`https://portal.azure.com`) throughout for examples.

Understanding migration options

When migrating applications, we have different options for **virtual machines** (**VMs**) and for data. We must also choose how much work we want to perform modernizing the application before we move it. At a high level, our choices include the following:

- **Lift and shift/Rehost**—Leave the application architecture as-is by simply replicating the VM into Azure. This takes a complete copy of the VM—OS, application, configurations, and data.

- **Refactor**—Refactoring involves re-deploying your solution with minimal (if any) code changes but to **Platform-as-a-Service** (**PaaS**) components—for example, deploying a .NET app onto an Azure web app instead of a VM or a database to Azure SQL instead of SQL on a VM (if it supports it).
- **Re-architect**—Take refactoring to the next level by making code changes that will enable you to leverage other cloud services. This might involve breaking up a monolithic application to use microservices, building new messaging systems, and basically redesigning the underlying architecture, but stopping short of a complete rebuild—for example, rebuilding a .NET app to take advantage of Azure Functions, Service Bus, or containerization. From a database perspective, this may include resolving any database design issues that prevent the use of Azure SQL, or even moving to a NoSQL option such as Cosmos DB.
- **Rebuild**—The ultimate decision may be to completely rebuild and replace your solution with a new one built from the ground up.

Each has its own pros and cons. Lift and shift is the fastest migration path, and ideal when you have a deadline to meet; however, you can't take full advantage of the cost savings and efficiencies provided by Azure.

Rebuilding means you can create modern, dynamically scalable solutions but the process will take far longer.

In reality, you may use a combination depending on your application landscape; however, in the next section, we'll look at the first of these options—lift and shift.

Managing servers

The simplest way to migrate services into Azure is via the lift-and-shift methodology; this means that we simply copy our existing VMs into Azure as-is.

Although you don't leverage Azure's best features, the lift-and-shift option enables you to move into Azure in the quickest time and with the least amount of effort.

Before we perform a migration, even one just using VMs, we need to consider how we will manage and secure those VMs. Although Azure provides a range of tools to help with this, you must build and configure those services.

In this section, we'll consider two of these features—update management and backups. Logging, security, and networking are also important; however, we cover these separately throughout this book.

Update management

One of the key aspects of maintaining servers in the cloud is keeping them up to date with the latest software updates and security patches. **Azure Update Management** is a service that allows you to manage updates and patches for your servers in a centralized manner, regardless of whether they are running in Azure, on-premises, or in other cloud environments.

Azure Update Management is part of a Log Analytics workspace, and therefore our first task would be to create and configure one.

Although the goal is to migrate existing servers, we will first create a Log Analytics workspace, then onboard an existing VM. We'll then see how to configure the workspace to onboard all future VMs automatically.

To verify and walk through the following examples, we will need a VM up and running; therefore, we will first create a basic VM.

Creating a VM

Navigate to the Azure portal and perform the following steps:

1. Click + **Create a Resource.**
2. Select **Virtual Machine** from the **Popular resources** option.
3. Complete the details as follows:
   - **Subscription**: Select your subscription
   - **Resource group**: Click **Create new** and call your resource group `rsg-VirtualMachines`
   - **Virtual machine name**: `my-test-vm`
   - **Region**: Select your closest region
   - **Availability options**: Leave as **No infrastructure redundancy required**
   - **Security type**: **Standard**
   - **Image**: **Windows Server 2022 Datacenter: Azure Edition - x64 Gen2**
   - **VM architecture**: **x64**
   - **Size**: **Standard_B2s - 2 vcpus, 4 GiB memory**
   - **Username**: Enter a username
   - **Password**: Enter a complex password
   - **Confirm password**: Re-enter the same password
   - **Public inbound ports**: **Allow selected ports**
   - **Select inbound ports**: **RDP (3389)**

Your screen should look similar to the one shown in *Figure 5.1*:

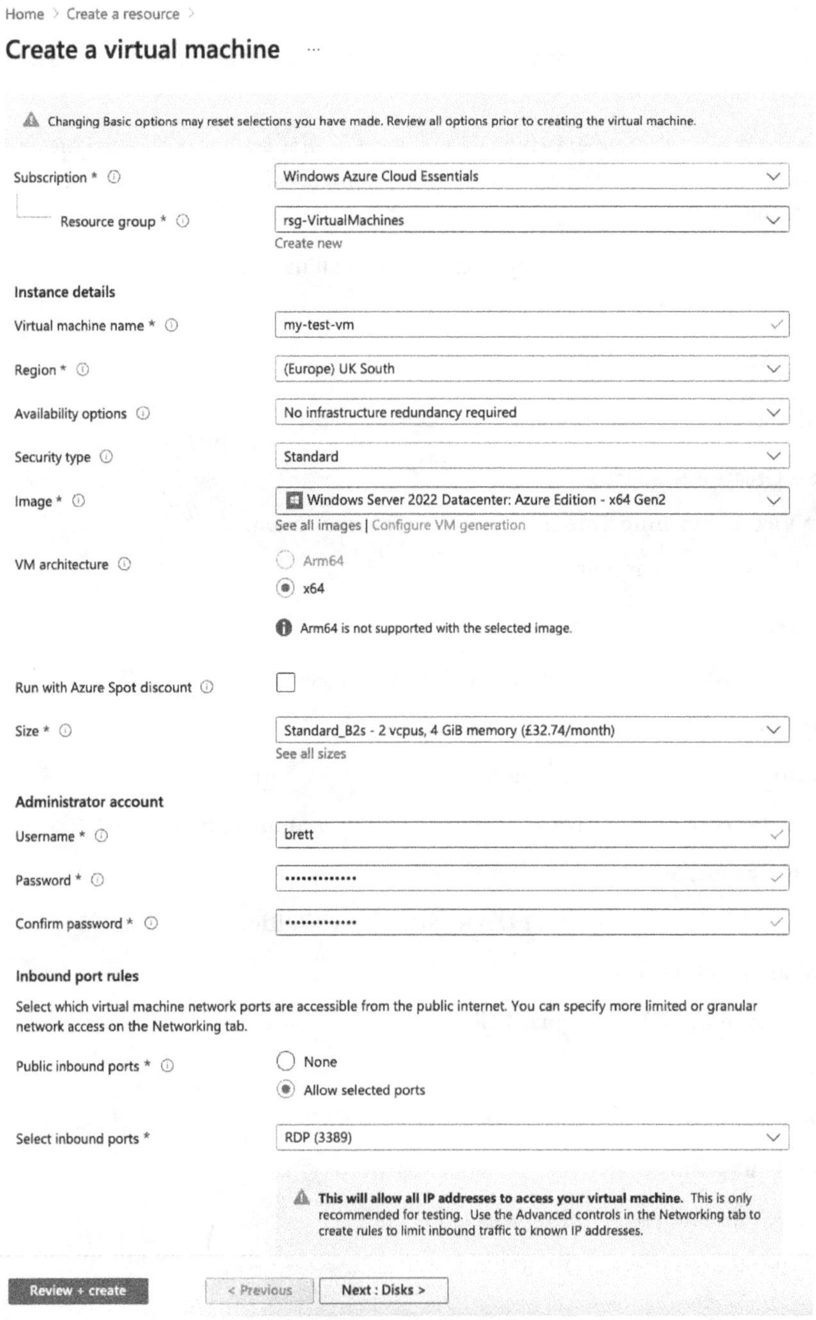
Home > Create a resource >
Create a virtual machine
Changing Basic options may reset selections you have made. Review all options prior to creating the virtual machine.
Subscription *
Windows Azure Cloud Essentials
Resource group *
rsg-VirtualMachines
Create new
Instance details
Virtual machine name *
my-test-vm
Region *
(Europe) UK South
Availability options
No infrastructure redundancy required
Security type
Standard
Image *
Windows Server 2022 Datacenter: Azure Edition - x64 Gen2
See all images | Configure VM generation
VM architecture
Arm64
x64
Arm64 is not supported with the selected image.
Run with Azure Spot discount
Size *
Standard_B2s - 2 vcpus, 4 GiB memory (£32.74/month)
See all sizes
Administrator account
Username *
brett
Password *
Confirm password *
Inbound port rules
Select which virtual machine network ports are accessible from the public internet. You can specify more limited or granular network access on the Networking tab.
Public inbound ports *
None
Allow selected ports
Select inbound ports *
RDP (3389)
This will allow all IP addresses to access your virtual machine. This is only recommended for testing. Use the Advanced controls in the Networking tab to create rules to limit inbound traffic to known IP addresses.
Review + create
< Previous
Next : Disks >

Figure 5.1 – Creating a basic VM

Finally, click **Review + create**, followed by **Create**. This will deploy a basic VM. We have skipped a lot of options—disks, networking, monitoring, and so on. This is fine for our demonstration purposes as the default options are adequate.

Once deployed, we can go ahead and create a workspace.

Creating a new workspace

From the Azure portal, perform the following steps:

1. Click + **Create a Resource**.
2. Type **Log Analytics Workspace**.
3. From the list of services, click **Log Analytics Workspace**.
4. Leave the plan as **Log Analytics Workspace** and click **Create**.
5. Select your subscription, then select or create a resource group.
6. Name your workspace—for example, `my-la-workspace`.
7. Select your closest region.
8. Click **Review + Create**.
9. Click **Create**.

This will provision a new workspace.

Creating an Automation account

Next, we need to create an Automation account. Again, from the Azure portal, perform the following steps:

1. Click + **Create a Resource**.
2. Type, then select **Automation Account**.
3. From the list of services, click **Automation**.
4. Leave the plan as **Automation** and click **Create**.
5. Select your subscription, then select or create a resource group.
6. Name your workspace—for example, `my-automation`.
7. Select your closest region.
8. Click **Review + Create**.
9. Click **Create**.

Again, this will create your Automation account. Once finished, click **Go to Resource**.

> **Important note**
>
> Not all regions support Automation accounts—go to this page to check available regions: `https://learn.microsoft.com/en-gb/azure/automation/how-to/region-mappings`.

Linking the Automation account

The next step is to link the Automation account to your Log Analytics workspace. Follow these steps to do so:

1. On your Automation account, on the left-hand menu, select **Update management**, then select the workspace you just created in the **Log Analytics workspace** drop-down box. Your screen should look like the one shown in *Figure 5.2*:

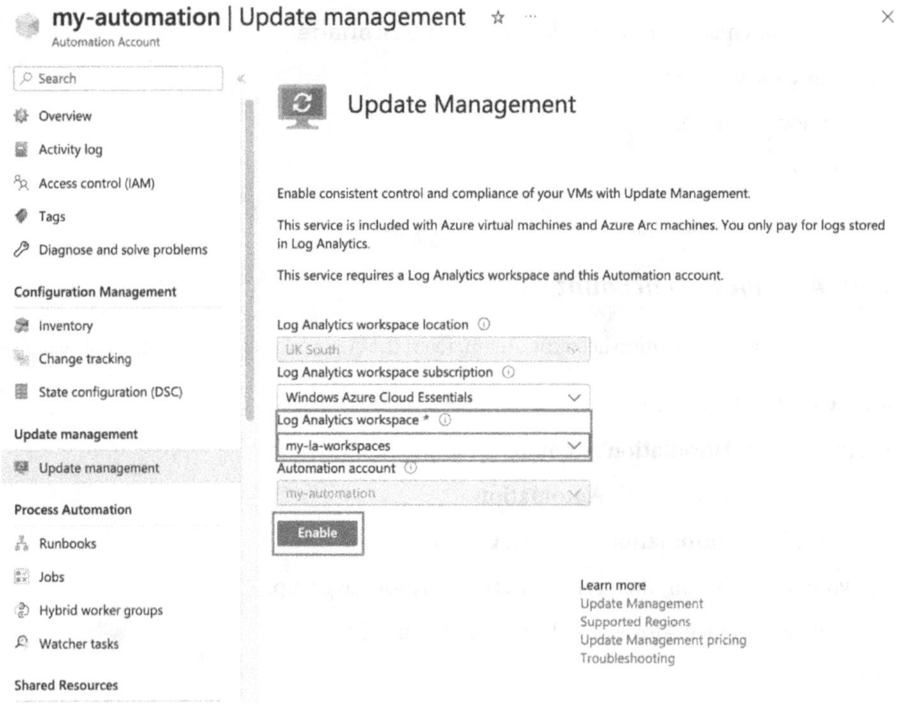
my-automation | Update management
Automation Account
Search
Overview
Activity log
Access control (IAM)
Tags
Diagnose and solve problems
Configuration Management
Inventory
Change tracking
State configuration (DSC)
Update management
Update management
Process Automation
Runbooks
Jobs
Hybrid worker groups
Watcher tasks
Shared Resources
Update Management
Enable consistent control and compliance of your VMs with Update Management.
This service is included with Azure virtual machines and Azure Arc machines. You only pay for logs stored in Log Analytics.
This service requires a Log Analytics workspace and this Automation account.
Log Analytics workspace location
UK South
Log Analytics workspace subscription
Windows Azure Cloud Essentials
Log Analytics workspace *
my-la-workspaces
Automation account
my-automation
Enable
Learn more
Update Management
Supported Regions
Update Management pricing
Troubleshooting

Figure 5.2 – Setting up the Automation account

2. Once you have selected the workspace, click **Enable**.
3. Finally, we can tell the Automation account to onboard all VMs to the connected workspace to the **Update management** feature. Still on the **Automation Account** page, click **Overview**, then click back on **Update management**.

4. The page now shows details of the **Update management** feature, such as any schedules and the status of onboarded VMs—at present, this will be blank. Click **Manage machines**, then select **Enable on all available and future machines**, as in *Figure 5.3*:

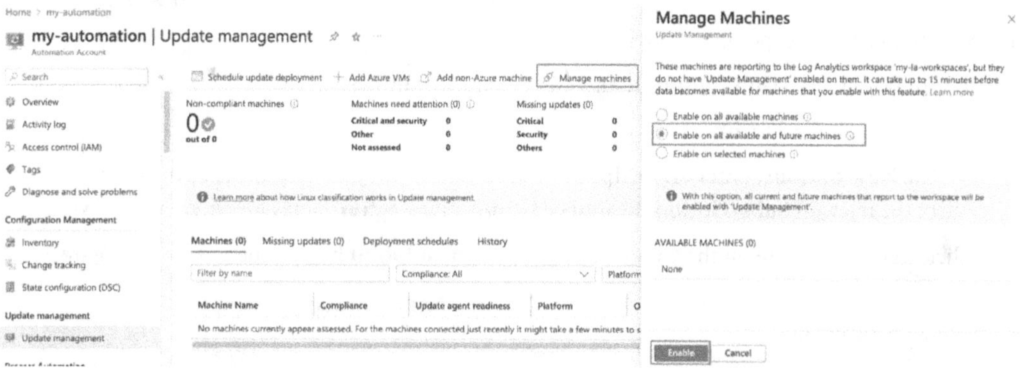
my-automation | Update management
Manage Machines
Schedule update deployment
Add Azure VMs
Add non-Azure machine
Manage machines
Overview
Activity log
Access control (IAM)
Tags
Diagnose and solve problems
Configuration Management
Inventory
Change tracking
State configuration (DSC)
Update management
Update management
Non-compliant machines
Machines need attention (0)
Missing updates (0)
Machines (0)
Missing updates (0)
Deployment schedules
History
Machine Name
Compliance
Update agent readiness
Platform
Enable on all available machines
Enable on all available and future machines
Enable on selected machines
AVAILABLE MACHINES (0)
None
Enable
Cancel

Figure 5.3 – Enabling Update management

To test the onboarding, we will connect a VM to the new Log Analytics workspace.

Connecting to the VM

To do this, we must install the Azure Monitor agent on a VM, and we do that by creating a **data collection rule** (**DCR**) in the Log Analytics workspace. Proceed as follows:

1. In the Azure portal, search for and select the workspace you created earlier.
2. On the left-hand menu, click **Agents**, then **Data Collection Rules**, as per *Figure 5.4*:

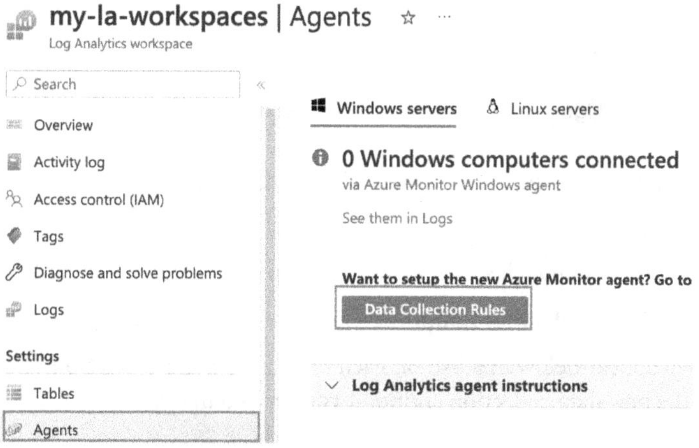
my-la-workspaces | Agents
Log Analytics workspace
Search
Overview
Activity log
Access control (IAM)
Tags
Diagnose and solve problems
Logs
Settings
Tables
Agents
Windows servers
Linux servers
0 Windows computers connected
via Azure Monitor Windows agent
See them in Logs
Want to setup the new Azure Monitor agent? Go to
Data Collection Rules
Log Analytics agent instructions

Figure 5.4 – Installing the Azure Monitor agent

3. On the next page, click **Create data collection rule.**
4. Enter a **Rule name** value—for example, `WindowsCollector`.
5. Select the relevant **Subscription**, **Resource Group**, and **Location** values for where your VMs are.
6. Set the **Platform Type** option to **Windows.**
7. Ignore **Data Collection Endpoint** and click **Next: Resources.**
8. Click **+ Create Endpoint.**
9. On the **New options**, enter an **Endpoint Name** value—for example, `WindowsEndpoint`. Select the relevant **Subscription**, **Resource Group**, and **Region** values for where your VMs are.
10. Click **Review + create** on the **Create data collection endpoint** box on the *right*—see *Figure 5.5*:

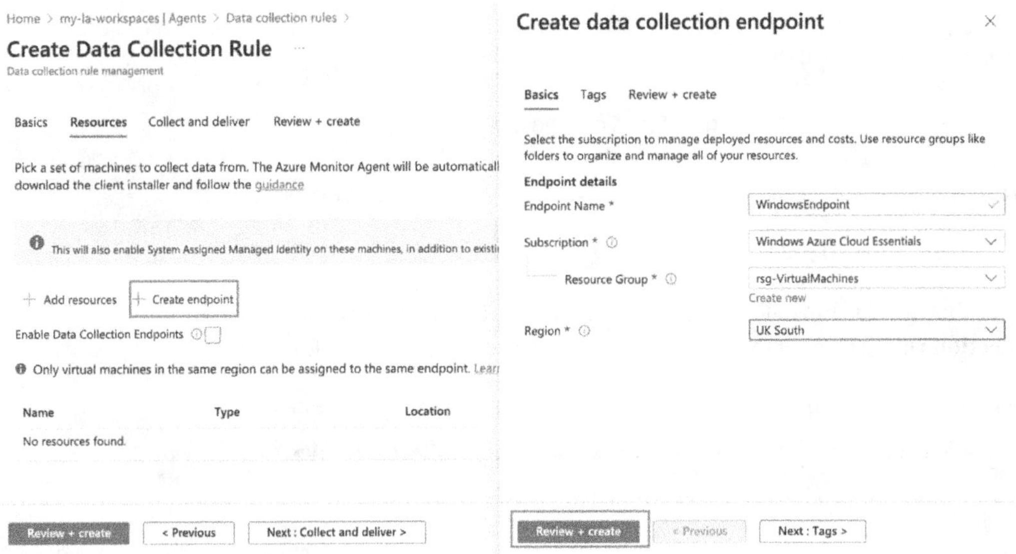
Home > my-la-workspaces | Agents > Data collection rules >
Create Data Collection Rule
Data collection rule management
Basics
Resources
Collect and deliver
Review + create
Pick a set of machines to collect data from. The Azure Monitor Agent will be automaticall
download the client installer and follow the guidance
This will also enable System Assigned Managed Identity on these machines, in addition to existi
+ Add resources
+ Create endpoint
Enable Data Collection Endpoints
Only virtual machines in the same region can be assigned to the same endpoint.
Name
Type
Location
No resources found.
Review + create
< Previous
Next : Collect and deliver >
Create data collection endpoint
Basics
Tags
Review + create
Select the subscription to manage deployed resources and costs. Use resource groups like folders to organize and manage all of your resources.
Endpoint details
Endpoint Name *
WindowsEndpoint
Subscription *
Windows Azure Cloud Essentials
Resource Group *
rsg-VirtualMachines
Create new
Region *
UK South
Review + create
< Previous
Next : Tags >

Figure 5.5 – Creating a data collection endpoint

11. It will say **Validation passed**; again in the right-hand box, click **Create.**
12. Click **Add resources.**
13. In the **Select Scope** window, check the resource group that contains your VMs and click **Apply**.
14. On the **Enable Data Collection Endpoints** option, check the box.
15. Your VMs will appear below in a list; on each one in the **Data collection endpoint** column, change it to the new data collection endpoint you created in *step 12.*

Your screen should look similar to the one shown in *Figure 5.6*:

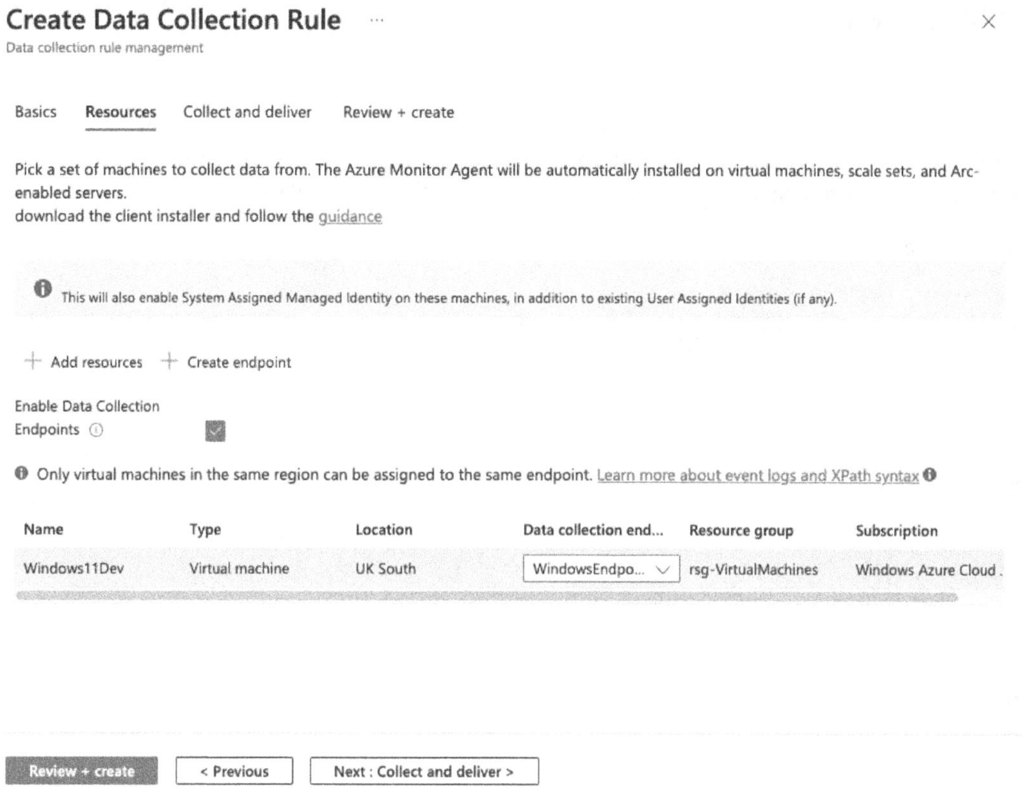
Create Data Collection Rule
Data collection rule management
Basics
Resources
Collect and deliver
Review + create
Pick a set of machines to collect data from. The Azure Monitor Agent will be automatically installed on virtual machines, scale sets, and Arc-enabled servers.
download the client installer and follow the guidance
This will also enable System Assigned Managed Identity on these machines, in addition to existing User Assigned Identities (if any).
Add resources
Create endpoint
Enable Data Collection Endpoints
Only virtual machines in the same region can be assigned to the same endpoint. Learn more about event logs and XPath syntax
Name
Type
Location
Data collection end...
Resource group
Subscription
Windows11Dev
Virtual machine
UK South
WindowsEndpo...
rsg-VirtualMachines
Windows Azure Cloud .
Review + create
< Previous
Next : Collect and deliver >

Figure 5.6 – Creating a DCR

16. Click **Next: Collect and deliver**.
17. On the next screen, click + **Add data source**.
18. Set the **Data source type** option to **Windows Event Logs**.
19. In the logs to collect, select **Critical** for **Application** and **System**.
20. Click **Next: Destination**.
21. You should have a destination already set; if not, click + **Add Destination** and set the following options:

    A. **Destination type**: **Azure Monitor Logs**

    B. **Subscription**: Your subscription

    C. **Account or namespace**: The Log Analytics workspace you created earlier

22. Click **Add data source**.
23. Click **Next: Review + create**.
24. Finally, click **Create**.

This will install **AzureMonitorWindowsAgent** on the VM, which is needed for both collecting logs and using the **Update management** feature. You can confirm this by searching for and selecting the VM in the Azure portal, then clicking **Extensions + applications** in the **Settings** menu.

You should see the agent is installed, as in *Figure 5.7*:

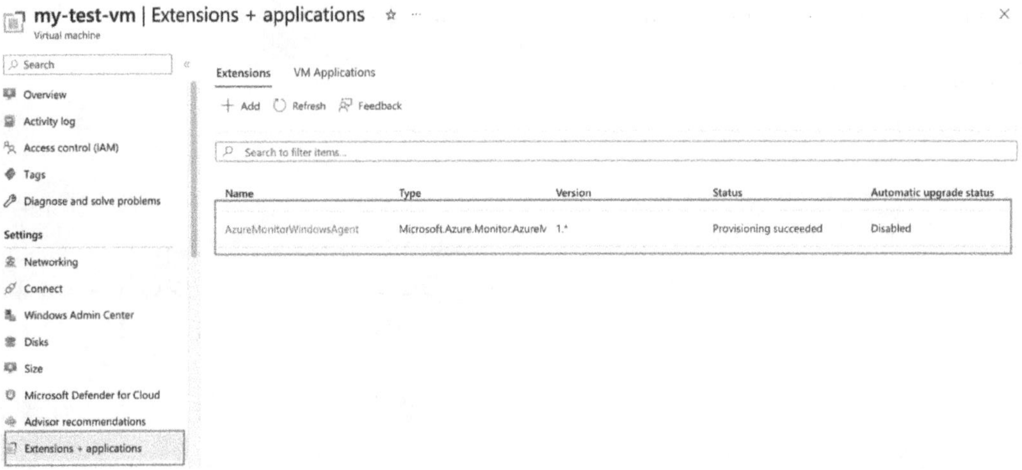
my-test-vm | Extensions + applications
Virtual machine
Search
Overview
Activity log
Access control (IAM)
Tags
Diagnose and solve problems
Settings
Networking
Connect
Windows Admin Center
Disks
Size
Microsoft Defender for Cloud
Advisor recommendations
Extensions + applications
Extensions
VM Applications
Add
Refresh
Feedback
Search to filter items...
Name
Type
Version
Status
Automatic upgrade status
AzureMonitorWindowsAgent
Microsoft.Azure.Monitor.AzureM
1.*
Provisioning succeeded
Disabled

Figure 5.7 – Confirming the monitoring agent is installed

Finally, we can check that updates are being managed, and start creating an update schedule by going to the Automation account we created.

> **Note**
>
> It can take about 15-20 minutes for a newly onboarded VM to be reported.

Creating an update schedule

To create a schedule, which we can do before any VMs are fully onboarded, go to the Automation account, and select **Update management**. Then, follow these steps:

1. From the top menu, click **Schedule update deployment**.
2. Enter a name—for example, `Weekly Windows Updates`.
3. Set **Operating system** to **Windows**.

4. Under **Items to update** -> **Groups to update**, click **Click to Configure**.
5. On the next screen, select your subscription from the **Subscription** dropdown, then click **Add**.
6. Click **OK**.
7. You will be returned to the **New update deployment** screen.
8. Under **Schedule settings**, click **Click to Configure**.
9. Under **Starts**, change the date and time to a weekend at `11.30` PM.
10. Change **Recurrence** to **Recurring**.
11. Change **Recur every** to **1 Week**.
12. Select either **Saturday** or **Sunday**.
13. Click **OK**.
14. You will be returned to the **New update deployment** screen.
15. Click **Create**.

Once that is done, you will need to wait a while for the VMs to be discovered and the patching status known. So, leave it an hour or so, then revisit the **Update management** view on your Automation account—your VM and its patch status should now be displayed, as in *Figure 5.8*:

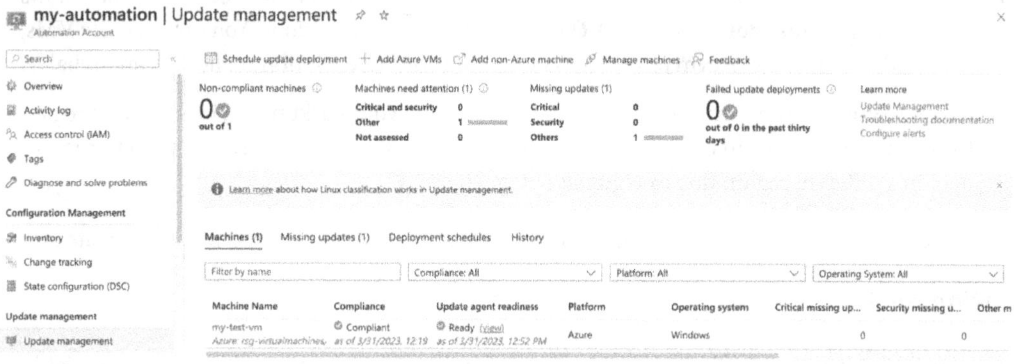
my-automation | Update management
Automation Account
Search
Schedule update deployment
Add Azure VMs
Add non-Azure machine
Manage machines
Feedback
Overview
Activity log
Access control (IAM)
Tags
Diagnose and solve problems
Configuration Management
Inventory
Change tracking
State configuration (DSC)
Update management
Update management
Non-compliant machines
0
out of 1
Machines need attention (1)
Critical and security 0
Other 1
Not assessed 0
Missing updates (1)
Critical 0
Security 0
Others 1
Failed update deployments
0
out of 0 in the past thirty days
Learn more
Update Management
Troubleshooting documentation
Configure alerts
Learn more about how Linux classification works in Update management.
Machines (1)
Missing updates (1)
Deployment schedules
History
Filter by name
Compliance: All
Platform: All
Operating System: All
Machine Name
Compliance
Update agent readiness
Platform
Operating system
Critical missing up...
Security missing u...
my-test-vm
Compliant
Ready
Azure
Windows

Figure 5.8 – VM patch status

Patching is an important aspect of managing your VM estate, so if you are planning on using VMs as part of your solution, it is advisable to set up **Update management**.

Another important aspect of keeping your solutions healthy is backups, which we will discuss next.

VM backups

Backing up your servers is essential to protect your data and ensure **business continuity** (**BC**) in case of server failures or data loss. Azure Backup is a service that provides a simple, secure, and cost-effective solution for backing up your servers in the cloud.

Understanding backup options

An Azure backup and recovery vault, known as a **Recovery Services** vault, is a storage entity within Azure that consolidates and manages your backup and recovery data. When using a **Recovery Services** vault, you have several main options to configure and manage your backup and recovery needs, as follows:

- **Backups**—You can configure backups for different types of items such as Azure VMs, on-premises VMs and physical servers, and Azure File shares. The vault supports backups for various workloads such as SQL Server, Exchange, and SharePoint.
- **Backup policies**—These allow you to define the frequency of backups, retention periods, and other settings. Backup policies also help you automate backup schedules.
- **Monitoring and reporting**—Monitoring and reporting capabilities give you insights into the status of backup jobs, the health of your protected items, and the usage of your vault.
- **Site Recovery**—Azure Site Recovery, which is integrated into the **Recovery Services** vault, allows you to configure **disaster recovery** (**DR**) solutions for your applications. You can replicate VMs, create DR plans, and perform test failovers to ensure your DR solution is effective and reliable.
- **Cross-region-restore**—This feature allows you to restore your backup data in a different region from where it was originally backed up. This is useful in scenarios where you need to recover data in a different region due to regional outages.

Now we understand the different options and components for backups, we can go ahead and create one.

Creating an Azure Recovery Services vault

We will create a **Backup and Recovery Services** vault so that we can look at how we can configure it. Follow these steps:

1. From the Azure portal, click + **Create a Resource**, then search for and select **Backup and Site Recovery**.
2. Click **Backup and Site Recovery**, then click **Create**.
3. Select your subscription and either create a new *resource group* or select an existing one. In my example, I will choose my VM resource group.
4. Give your vault a name—for example, `VMBackupVault`.
5. Select a **Region** value—this is best being the region your VMs are in.

6. Click **Review + Create**.
7. Click **Create**.

Wait for your backup vault to be created, and once it has, select it in the portal. Before we perform any backups, let's look at some of the key options.

From the **Recovery Services vault** page, on the left-hand menu, click **Properties**. Underneath **Backup Configuration**, click **Update**.

Figure 5.9 shows you the configuration page:

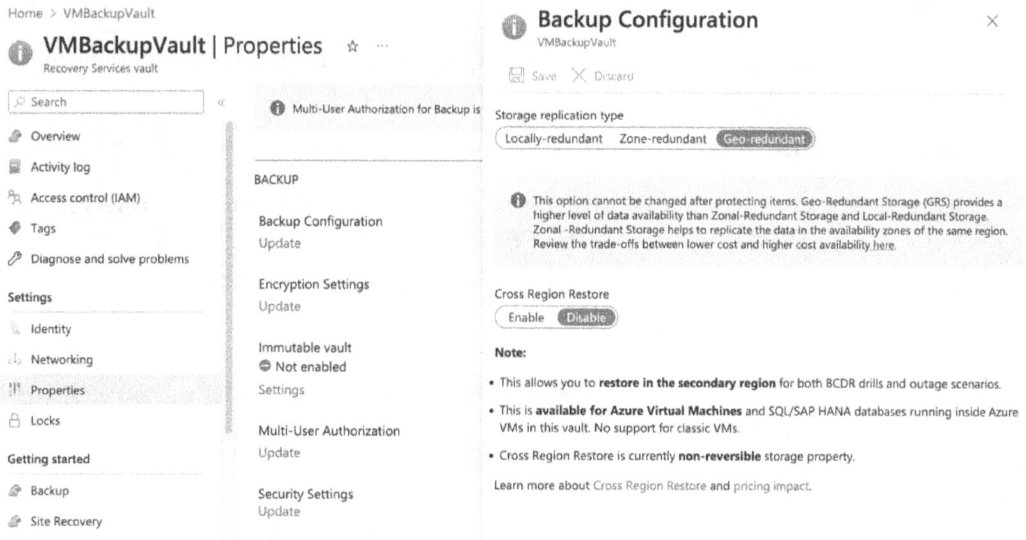

Home > VMBackupVault
VMBackupVault | Properties
Recovery Services vault
Search
Overview
Activity log
Access control (IAM)
Tags
Diagnose and solve problems
Settings
Identity
Networking
Properties
Locks
Getting started
Backup
Site Recovery
Multi-User Authorization for Backup is
BACKUP
Backup Configuration
Update
Encryption Settings
Update
Immutable vault
Not enabled
Settings
Multi-User Authorization
Update
Security Settings
Update
Backup Configuration
VMBackupVault
Save
Discard
Storage replication type
Locally-redundant
Zone-redundant
Geo-redundant
This option cannot be changed after protecting items. Geo-Redundant Storage (GRS) provides a higher level of data availability than Zonal-Redundant Storage and Local-Redundant Storage. Zonal -Redundant Storage helps to replicate the data in the availability zones of the same region. Review the trade-offs between lower cost and higher cost availability here.
Cross Region Restore
Enable
Disable
Note:
This allows you to restore in the secondary region for both BCDR drills and outage scenarios.
This is available for Azure Virtual Machines and SQL/SAP HANA databases running inside Azure VMs in this vault. No support for classic VMs.
Cross Region Restore is currently non-reversible storage property.
Learn more about Cross Region Restore and pricing impact.

Figure 5.9 – Backup configuration

This is an important page to set up first, as you cannot change some of those details once you have created any backups.

The storage replication type sets how your backups are stored and replicated in Azure. Under the hood, backups are stored in Azure Storage accounts—you just can't manage them directly. Because they use Azure Storage accounts, you can define how that data —in this case, our backups—is replicated.

The options are the same as we have on our own storage accounts—namely, the following:

- **Locally-redundant**—Copies are made within a region within an Availability Zone. This means if a single data center goes down, you may not have access to your backups.
- **Zone-redundant**—Copies are made across Availability Zones so that you are protected against a single data center failure, but not against a whole region outage.

- **Geo-redundant**—Copies are made across two regions. This represents high protection; however, if you want to be protected against a region failure, you need to enable the other option on this page: **Cross Region Restore**.

Care needs to be taken to choose the correct option—**Geo-redundant** with **Cross Region Restore** gives the most protection; however, it uses more storage and therefore increases your costs. You can view the current pricing for your location and redundancy options at the following location: `https://azure.microsoft.com/en-gb/pricing/details/storage/blobs/#pricing`.

The next option to look at it is backup policies. On the left-hand menu, under **Manage**, click **Backup policies**.

As already stated, policies enable you to define when backups happen, how often, and how long they are kept.

Azure provides three default policies you can use or amend, as follows:

- **HourlyLogBackup**—Performs a full daily backup of a VM, and SQL Log backups every hour. Obviously, this is built specifically for SQL Server running on a VM.
- **DefaultPolicy**—Performs a full daily backup with a 30-day retention period.
- **EnhancedPolicy**—Performs a backup of the entire VM *every* hour.

You can, of course, create your own policies or modify existing ones, but for now, we will just go ahead and set up a backup job on a VM, as follows:

1. In the **Recovery Service** vault, on the left-hand menu, click **Backup**.
2. Choose your VM location; the options are **Azure**, **Azure Stack Hub**, **Azure Stack HCI**, and **on-premises**—we will choose **Azure**.
3. Choose what to back up—you will see the options are **Virtual Machine**, **File Shares**, **SQL on Virtual Machine**, and **SAP on Virtual Machine**. Choose **Virtual Machine**.
4. Click **Backup**.
5. On the next screen, you choose your **Backup policy** type—**Standard** or **Enhanced** allow the selection of one of the backup policies we mentioned earlier. The **HourlyLogBackup** type is now presented because, in the previous step, we chose a standard VM rather than an SQL VM.
6. Next, we need to add a VM, so click **Add** under **Virtual Machines**.
7. A screen will appear, like the one shown in *Figure 5.10*. Select a VM and click **Add**.
8. Then, click **Enable backup**.

This performs a deployment that can take a few minutes:

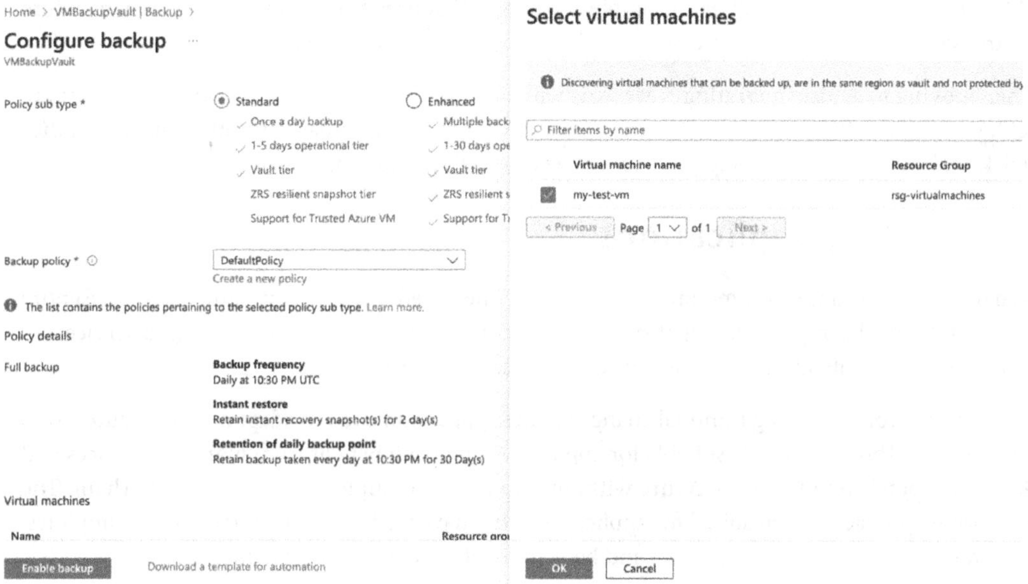
Home > VMBackupVault | Backup >
Configure backup
VMBackupVault
Policy sub type *
Standard
Once a day backup
1-5 days operational tier
Vault tier
ZRS resilient snapshot tier
Support for Trusted Azure VM
Enhanced
Backup policy *
DefaultPolicy
Create a new policy
The list contains the policies pertaining to the selected policy sub type. Learn more.
Policy details
Full backup
Backup frequency
Daily at 10:30 PM UTC
Instant restore
Retain instant recovery snapshot(s) for 2 day(s)
Retention of daily backup point
Retain backup taken every day at 10:30 PM for 30 Day(s)
Virtual machines
Name
Enable backup
Download a template for automation
Select virtual machines
Filter items by name
Virtual machine name
Resource Group
my-test-vm
rsg-virtualmachines
< Previous
Page 1 of 1
Next >
OK
Cancel

Figure 5.10 – Selecting VMs for backup

You can view the status of a backup job back in the vault. Go to your **Recovery Services** vault in the portal, and on the left-hand menu, click **Backup items**.

You will see a list of backup types; select **Azure Virtual Machine**.

You will now see a screen like the one shown in *Figure 5.11*. This shows you all the VMs being backed up, along with the current status:

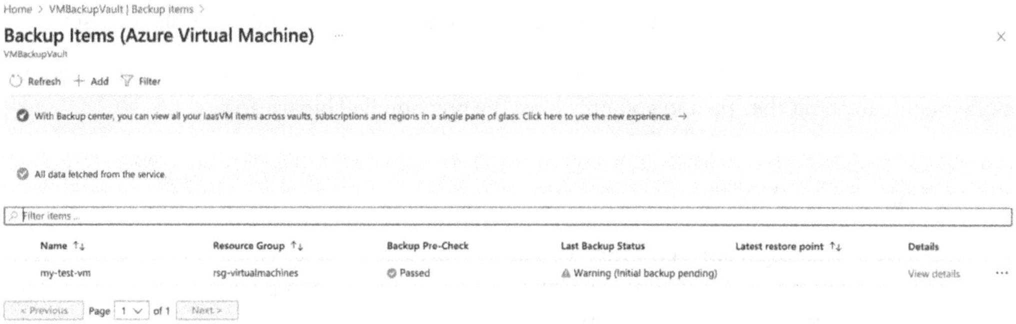
Home > VMBackupVault | Backup items >
Backup Items (Azure Virtual Machine)
VMBackupVault
Refresh
Add
Filter
With Backup center, you can view all your IaasVM items across vaults, subscriptions and regions in a single pane of glass. Click here to use the new experience. →
All data fetched from the service.
Filter items...
Name
Resource Group
Backup Pre-Check
Last Backup Status
Latest restore point
Details
my-test-vm
rsg-virtualmachines
Passed
Warning (Initial backup pending)
View details
< Previous
Page 1 of 1
Next >

Figure 5.11 – VM backup status

When defining a backup strategy for VMs, you need to consider how often data might change on a VM, if it even stores data at all. For example, a web app that uses a backend database may not store any transient files at all, in which case a weekly or even monthly backup may suffice.

When moving to Azure, migrating VMs as-is will often be the first choice, but when timelines are tight, you should always consider, on an application-by-application basis, whether there are other "quick win" options that can help you modernize an existing solution.

Modernizing applications

Even older applications can sometimes be partially modernized—even if you cannot move the entire system, individual components might still be able to make use of better Azure managed services. In these cases, we should always at least consider if we can **refactor**.

Refactoring involves making minimal changes to the application code to adapt it to the Azure cloud environment. This approach is suitable for applications that can benefit from managed services and platform capabilities offered by Azure without requiring a complete architectural overhaul. The refactoring approach is well suited for applications that have well-defined components or modules, use common programming languages, and have relatively straightforward dependencies.

Of course, a lot depends on the age of the application you are considering and its current platform. Solutions built in the past 10 years or so have a high chance of being built as *n-tier* architectures, whereby it is split into different layers or tiers—for example, a frontend user interface, a middle-tier business processing component, and a backend database.

Older applications might not work like this. Many years ago, applications were built to be monolithic, meaning they were basically one big intertwined system. Some applications even used their own proprietary databases, and others were built using languages that are no longer in use.

Another aspect to consider is whether you own the source code or not. In other words, is this an application you or your company built, or was it an off-the-shelf package built by someone else?

Obviously, if you own the source code, refactoring will be far easier; however, there are still options if another company built the app and you only have the precompiled binaries.

Assuming an application has at least some components we can modernize, here are some examples of how we might change them during refactoring:

- **Databases**—Moving from a traditional relational database to Azure SQL Database or Azure Database for PostgreSQL/MySQL/MariaDB. This may involve simply updating connection strings, or it may require modifying the data access code. Azure provides managed database services from several different vendors; however, you need to watch for version differences.

- **Web applications**—Migrating from on-premises web servers to Azure App Service might need an update to configurations, adapting deployment scripts, or ensuring that your application is compatible with Azure managed offerings.
- **Business logic**—Containerizing the application using Docker and running it on Azure container services such as Web App for Containers, App Containers, or **Azure Kubernetes Service** (**AKS**). Depending on the application, you may be able to take a standard OS image (such as Windows or Linux) and programmatically deploy the application code to it. Containerization is actually great for all layers of your application, including web apps and even databases.
- **Other improvements**—Leveraging Azure services such as Azure Cache for Redis, Azure Storage, or Azure Event Hubs. You might need to update your application code to use Azure SDKs, adapt configuration settings, and make necessary changes to accommodate the new services; however, small changes could make big differences in flexibility.

These are some high-level examples, but let's dive deeper into a few simple examples.

Scale sets

Before we look at options that may require some code changes, let's examine an option that actually sits somewhere between a traditional VM or a managed service such as Web Apps.

Scale sets are basically VMs; however, rather than simply building a VM and deploying your application to it, you create an image of the VM with your application already installed.

That image is then used to create multiple instances of your application on demand—however, all those instances sit behind a load balancer that acts as the single entry point and distributes traffic between all running nodes.

The Scale Sets service automatically adds and removes additional images as the system requires, which you define by creating scaling rules. For example, you can create a scaling rule that adds new instances when the CPU of all existing nodes reaches 80% and then removes instances when they all drop below 20%.

Figure 5.12 depicts how this looks:

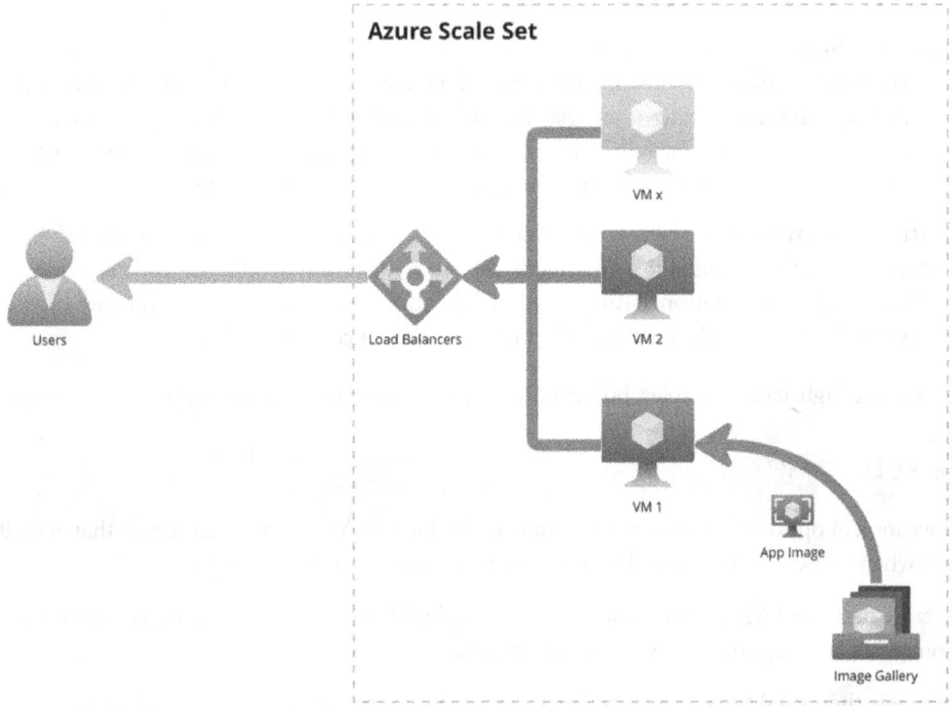
Azure Scale Set
VM x
Users
Load Balancers
VM 2
VM 1
App Image
Image Gallery

Figure 5.12 – Scale set architecture

Scale sets are a great first step if you want to add resiliency and scalability to your application, but you have no choice but to run it as a VM.

Azure App Service/Web Apps

Azure Web Apps is part of a more general service called Azure App Service. App Service essentially provides a managed service for running applications that normally run as a website. In the Windows world, this would be a .NET application that runs on **Internet Information Services** (**IIS**), or if it's a Linux website, it could be a PHP app running on Apache.

App Service has a few different versions, but for simplicity, we will consider web apps, which are running websites that contain a user interface, and API apps, which are still websites, but only expose an API and don't serve up a user interface.

Either way, the premise is the same. If your application is a web-based application, there is a good chance you could simply redeploy it to an app service.

There are occasions when this won't work; for example, some web applications contain background services or require more direct access to the underlying OS. However, this tends to be older applications, and if your application doesn't have any complex components such as this, you could use App Service without any modification to your code.

App Service also supports a range of languages, including ASP.NET, ASP.NET Core, Java, Ruby, Node.js, PHP, and Python.

Creating a web app

The best way to understand the benefits and flexibility of web apps is to create one, so let's walk through a simple setup and then we can look at the different configuration options:

1. From the Azure portal, click + **Create a Resource**, then select **Web App** from the **Popular Azure Services** list.
2. Complete the details as follows:

    - **Subscription**: Select your subscription
    - **Resource Group**: Click **Create new** then enter a name—in my example, I chose `rsg-Dev`
    - **Name**: A unique name—in my example, `brettswebapp`
    - **Publish**: Select **Code**
    - **Runtime stack**: **.NET 6 (LTS)**
    - **Operating System**: **Linux**
    - **Region**: Your closest region
    - **Linux Plan**: Leave as default
    - **Pricing plan**: Change to **Basic B1**

Figure 5.13 shows an example of what your screen should look like:

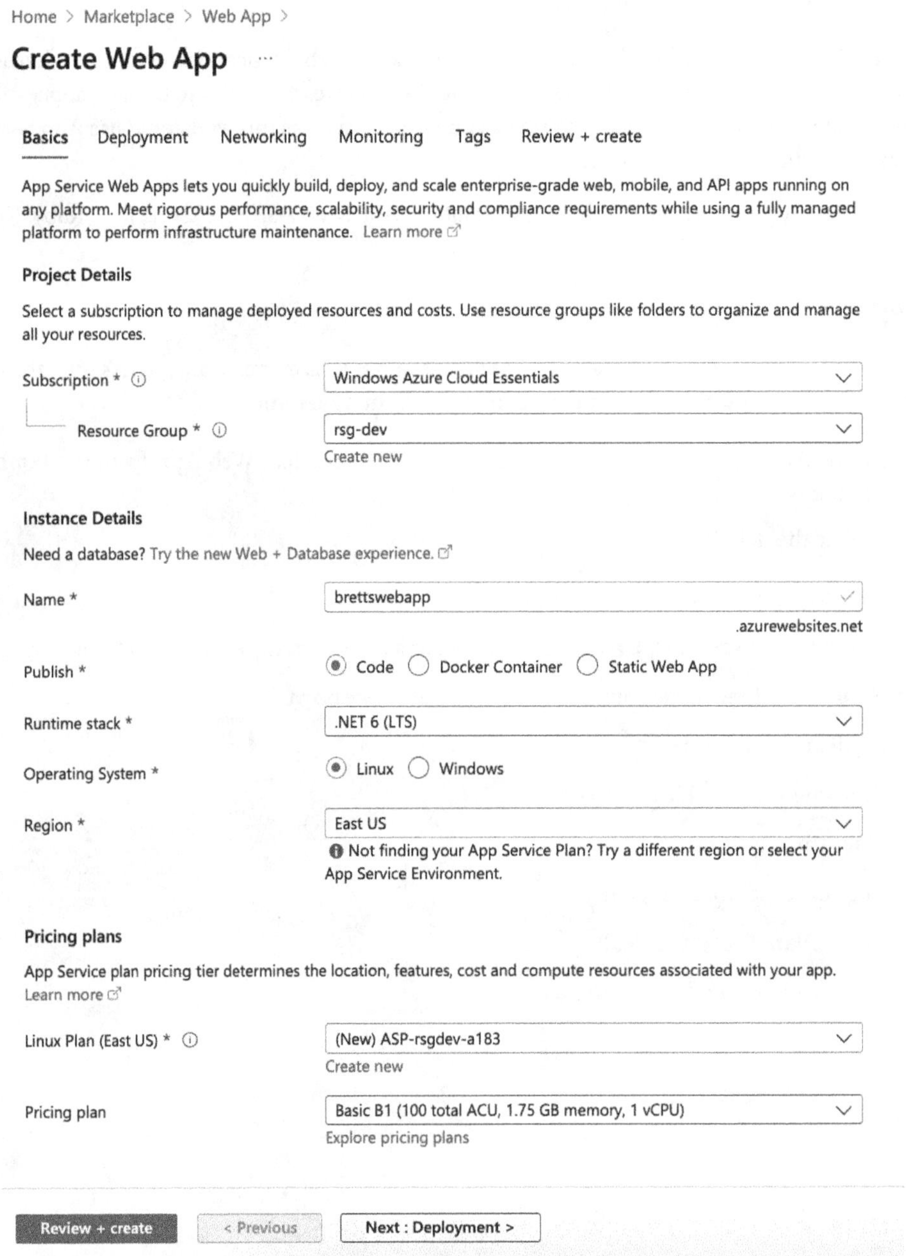

Home > Marketplace > Web App >

Create Web App   ...

Basics   Deployment   Networking   Monitoring   Tags   Review + create

App Service Web Apps lets you quickly build, deploy, and scale enterprise-grade web, mobile, and API apps running on any platform. Meet rigorous performance, scalability, security and compliance requirements while using a fully managed platform to perform infrastructure maintenance. Learn more ☐

Project Details

Select a subscription to manage deployed resources and costs. Use resource groups like folders to organize and manage all your resources.

Subscription * ⓘ	Windows Azure Cloud Essentials ⌄
└── Resource Group * ⓘ	rsg-dev ⌄
	Create new

Instance Details

Need a database? Try the new Web + Database experience. ☐

Name *	brettswebapp ✓
	.azurewebsites.net
Publish *	⦿ Code ◯ Docker Container ◯ Static Web App
Runtime stack *	.NET 6 (LTS) ⌄
Operating System *	⦿ Linux ◯ Windows
Region *	East US ⌄
	ⓘ Not finding your App Service Plan? Try a different region or select your App Service Environment.

Pricing plans

App Service plan pricing tier determines the location, features, cost and compute resources associated with your app. Learn more ☐

Linux Plan (East US) * ⓘ	(New) ASP-rsgdev-a183 ⌄
	Create new
Pricing plan	Basic B1 (100 total ACU, 1.75 GB memory, 1 vCPU) ⌄
	Explore pricing plans

[Review + create]   [< Previous]   [Next : Deployment >]

Figure 5.13 – Creating a web app

Once you're happy with your options, click **Review + create** then **Create**.

A new web app with default settings will be built. Once deployed, click **Go to resource** or search for it in the Azure portal and select it so that we can examine some options.

Understanding App Service plans and scalability

Our web app was built on a **Basic** App Service plan. This is one of the cheapest; however, it does not provide all the different capabilities the more expensive ones have.

The App Service plan determines the amount of RAM and CPU that is allocated. The more RAM and CPU, the more resources you have, but the more it costs. App Service plans allow you to run multiple web apps in the same plan. In other words, you can create a single pricing plan, then create two or more apps on that plan—each app will compete for those resources, however, but it is a great way for apps that may have different usage patterns to use those resources.

But of course, if one app becomes busy and uses lots of resources, then both apps will slow down.

One of the greatest capabilities of Azure is its scalability, and web apps can use this scalability dynamically. Just as we can use VM scale sets to add additional instances of our application, with web apps we can also set scale rules to create additional instances of our application.

This can be a great way to bring costs down. Most applications have peaks and troughs in their usage, but with a traditional on-premises application that runs on a VM or physical hardware, the infrastructure must be built to meet peak demand. The rest of the time, the hardware is sitting idle but still costing money!

With web apps, when demand increases, we could change the pricing plan to one that has more CPU and RAM—known as scaling up. A more efficient option would be to use a lower-priced plan that has far fewer resources but to add more instances as demand increases—known as scaling out.

You can scale out manually whenever you want, or you can define rules or schedules to do this for you.

Rules-based scaling is only available on Production-level plans—**Standard** and **Premium**—but manual scaling is still possible on Dev/Test plans.

You can check the specs and scaling options in your web app. On the **Web App** view in Azure, on the left-hand menu there is a **Scale up** option—clicking this allows you to view the different capabilities, as we can see in *Figure 5.14*, which shows the hardware configuration:

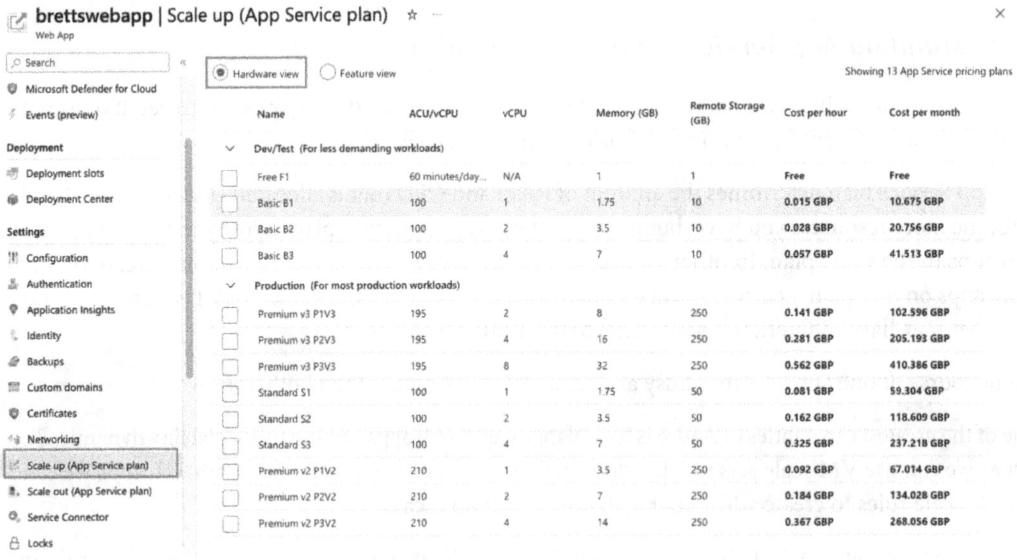
brettswebapp | Scale up (App Service plan)
Web App
Search
Microsoft Defender for Cloud
Events (preview)
Deployment
Deployment slots
Deployment Center
Settings
Configuration
Authentication
Application Insights
Identity
Backups
Custom domains
Certificates
Networking
Scale up (App Service plan)
Scale out (App Service plan)
Service Connector
Locks
Hardware view
Feature view
Showing 13 App Service pricing plans
Name | ACU/vCPU | vCPU | Memory (GB) | Remote Storage (GB) | Cost per hour | Cost per month
Dev/Test (For less demanding workloads)
Free F1 | 60 minutes/day... | N/A | 1 | 1 | Free | Free
Basic B1 | 100 | 1 | 1.75 | 10 | 0.015 GBP | 10.675 GBP
Basic B2 | 100 | 2 | 3.5 | 10 | 0.028 GBP | 20.756 GBP
Basic B3 | 100 | 4 | 7 | 10 | 0.057 GBP | 41.513 GBP
Production (For most production workloads)
Premium v3 P1V3 | 195 | 2 | 8 | 250 | 0.141 GBP | 102.596 GBP
Premium v3 P2V3 | 195 | 4 | 16 | 250 | 0.281 GBP | 205.193 GBP
Premium v3 P3V3 | 195 | 8 | 32 | 250 | 0.562 GBP | 410.386 GBP
Standard S1 | 100 | 1 | 1.75 | 50 | 0.081 GBP | 59.304 GBP
Standard S2 | 100 | 2 | 3.5 | 50 | 0.162 GBP | 118.609 GBP
Standard S3 | 100 | 4 | 7 | 50 | 0.325 GBP | 237.218 GBP
Premium v2 P1V2 | 210 | 1 | 3.5 | 250 | 0.092 GBP | 67.014 GBP
Premium v2 P2V2 | 210 | 2 | 7 | 250 | 0.184 GBP | 134.028 GBP
Premium v2 P3V2 | 210 | 4 | 14 | 250 | 0.367 GBP | 268.056 GBP

Figure 5.14 – Pricing plans in Web Apps

Changing the view option at the top from **Hardware view** to **Feature view** shows us which plans support rules-based scaling, as we can see in *Figure 5.15*:

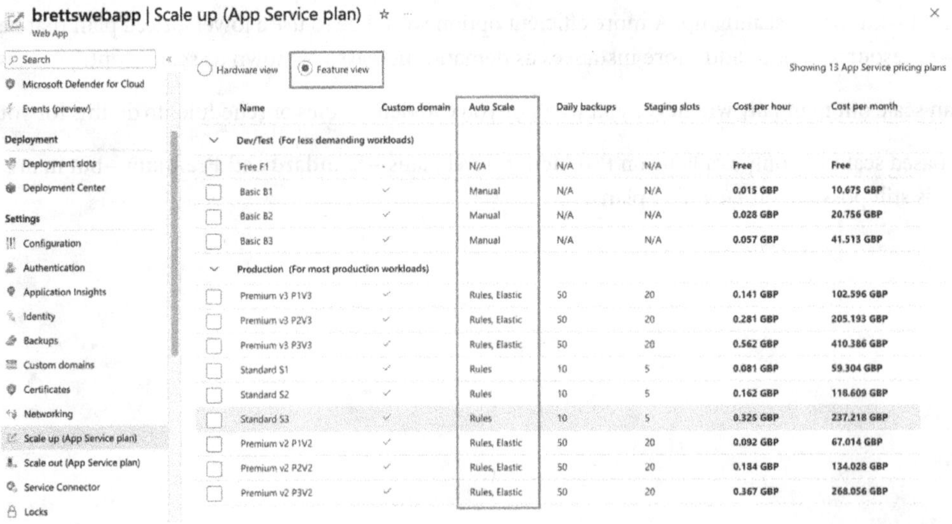
brettswebapp | Scale up (App Service plan)
Web App
Search
Microsoft Defender for Cloud
Events (preview)
Deployment
Deployment slots
Deployment Center
Settings
Configuration
Authentication
Application Insights
Identity
Backups
Custom domains
Certificates
Networking
Scale up (App Service plan)
Scale out (App Service plan)
Service Connector
Locks
Hardware view
Feature view
Showing 13 App Service pricing plans
Name | Custom domain | Auto Scale | Daily backups | Staging slots | Cost per hour | Cost per month
Dev/Test (For less demanding workloads)
Free F1 | | N/A | N/A | N/A | Free | Free
Basic B1 | ✓ | Manual | N/A | N/A | 0.015 GBP | 10.675 GBP
Basic B2 | ✓ | Manual | N/A | N/A | 0.028 GBP | 20.756 GBP
Basic B3 | ✓ | Manual | N/A | N/A | 0.057 GBP | 41.513 GBP
Production (For most production workloads)
Premium v3 P1V3 | ✓ | Rules, Elastic | 50 | 20 | 0.141 GBP | 102.596 GBP
Premium v3 P2V3 | ✓ | Rules, Elastic | 50 | 20 | 0.281 GBP | 205.193 GBP
Premium v3 P3V3 | ✓ | Rules, Elastic | 50 | 20 | 0.562 GBP | 410.386 GBP
Standard S1 | ✓ | Rules | 10 | 5 | 0.081 GBP | 59.304 GBP
Standard S2 | ✓ | Rules | 10 | 5 | 0.162 GBP | 118.609 GBP
Standard S3 | ✓ | Rules | 10 | 5 | 0.325 GBP | 237.218 GBP
Premium v2 P1V2 | ✓ | Rules, Elastic | 50 | 20 | 0.092 GBP | 67.014 GBP
Premium v2 P2V2 | ✓ | Rules, Elastic | 50 | 20 | 0.184 GBP | 134.028 GBP
Premium v2 P3V2 | ✓ | Rules, Elastic | 50 | 20 | 0.367 GBP | 268.056 GBP

Figure 5.15 – Scaling options in Web Apps

In order to create scaling rules, we will need to change our current **B1** plan to a **Standard** plan. In the **Scale up** view, check the **Standard S1** box then click **Select**.

Wait for a few seconds until you see a message notifying you that the plan has updated; then, still in the **Web App** view, click the **Scale out** left-hand menu option, which allows us to create rules for autoscaling our app.

We create a new scaling rule by clicking the **Custom autoscale** option, then under the **Default** scale condition, change the **Scale mode** value to **Scale based on a metric**, then click **Add a rule**. You can now define metrics on which to scale out or back in. *Figure 5.16* and *Figure 5.17* show an example of configuring a scaling rule:

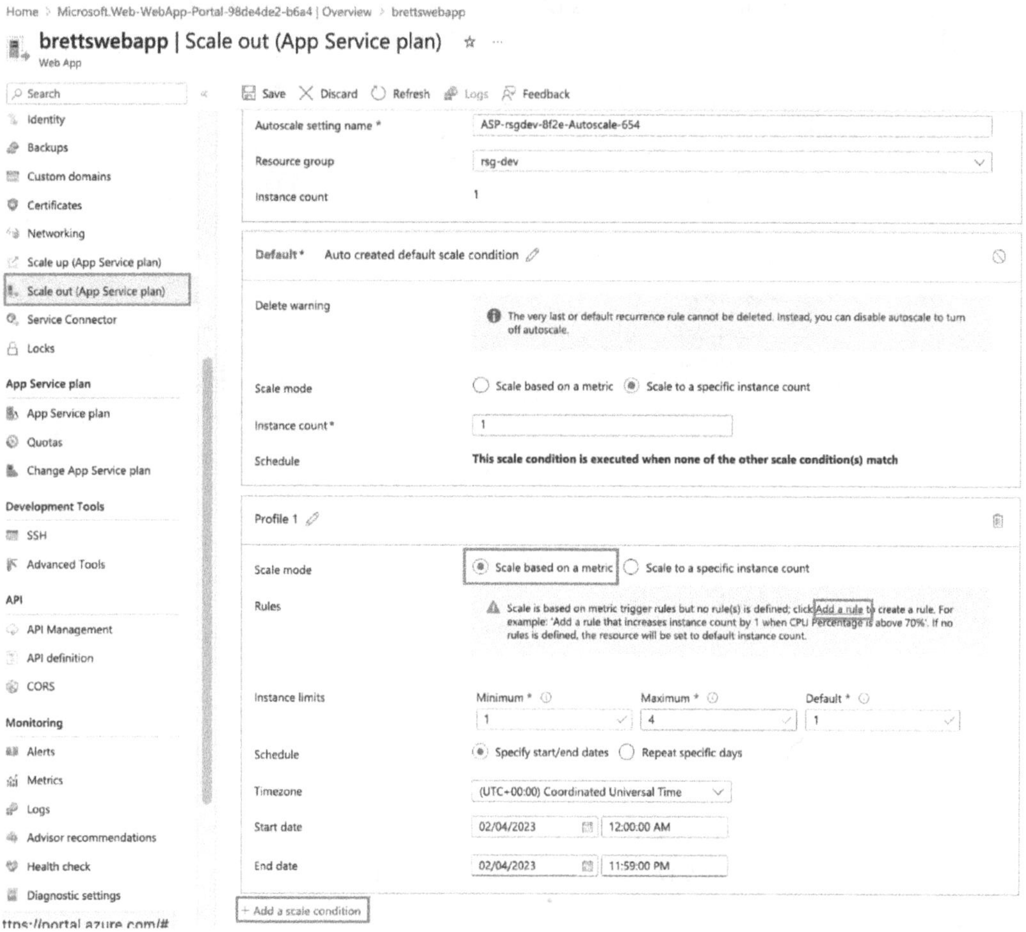
Home > Microsoft.Web-WebApp-Portal-98de4de2-b6a4 | Overview > brettswebapp
brettswebapp | Scale out (App Service plan)
Web App
Search
Identity
Backups
Custom domains
Certificates
Networking
Scale up (App Service plan)
Scale out (App Service plan)
Service Connector
Locks
App Service plan
App Service plan
Quotas
Change App Service plan
Development Tools
SSH
Advanced Tools
API
API Management
API definition
CORS
Monitoring
Alerts
Metrics
Logs
Advisor recommendations
Health check
Diagnostic settings
Save
Discard
Refresh
Logs
Feedback
Autoscale setting name *
ASP-rsgdev-8f2e-Autoscale-654
Resource group
rsg-dev
Instance count
1
Default *
Auto created default scale condition
Delete warning
The very last or default recurrence rule cannot be deleted. Instead, you can disable autoscale to turn off autoscale.
Scale mode
Scale based on a metric
Scale to a specific instance count
Instance count *
1
Schedule
This scale condition is executed when none of the other scale condition(s) match
Profile 1
Scale mode
Scale based on a metric
Scale to a specific instance count
Rules
Scale is based on metric trigger rules but no rule(s) is defined; click Add a rule to create a rule. For example: 'Add a rule that increases instance count by 1 when CPU Percentage is above 70%'. If no rules is defined, the resource will be set to default instance count.
Instance limits
Minimum *
1
Maximum *
4
Default *
1
Schedule
Specify start/end dates
Repeat specific days
Timezone
(UTC+00:00) Coordinated Universal Time
Start date
02/04/2023
12:00:00 AM
End date
02/04/2023
11:59:00 PM
+ Add a scale condition

Figure 5.16 – Scaling a web app

In this example, we can see that we are setting up a new scaling condition:

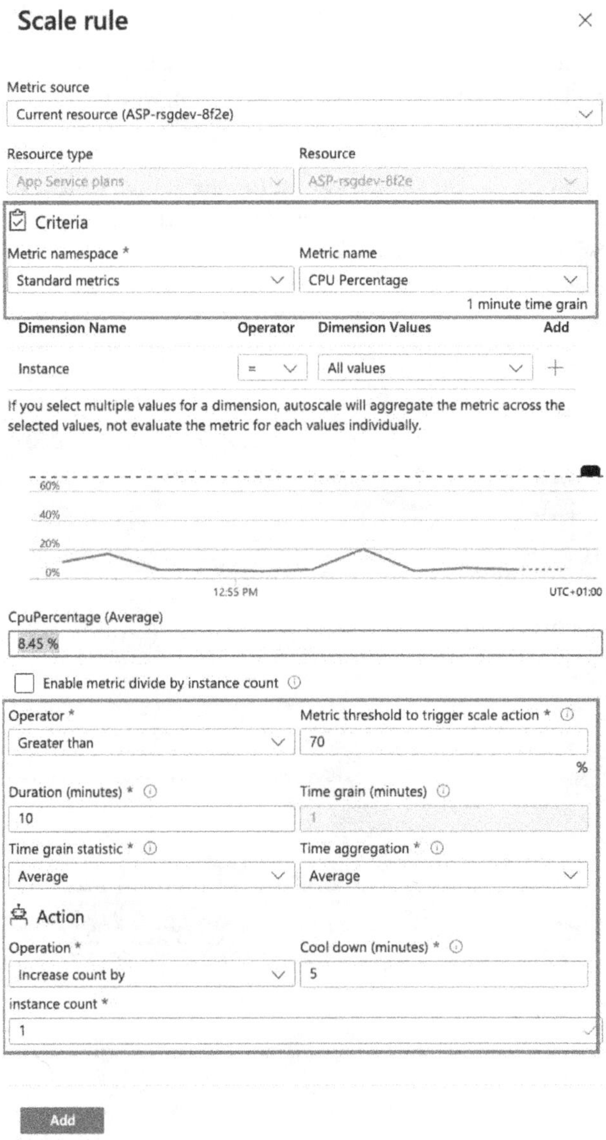
Scale rule
Metric source
Current resource (ASP-rsgdev-8f2e)
Resource type
App Service plans
Resource
ASP-rsgdev-8f2e
Criteria
Metric namespace *
Standard metrics
Metric name
CPU Percentage
1 minute time grain
Dimension Name
Operator
Dimension Values
Add
Instance
=
All values
If you select multiple values for a dimension, autoscale will aggregate the metric across the selected values, not evaluate the metric for each values individually.
60%
40%
20%
0%
12:55 PM
UTC+01:00
CpuPercentage (Average)
8.45 %
Enable metric divide by instance count
Operator *
Greater than
Metric threshold to trigger scale action *
70
%
Duration (minutes) *
10
Time grain (minutes)
1
Time grain statistic *
Average
Time aggregation *
Average
Action
Operation *
Increase count by
Cool down (minutes) *
5
instance count *
1
Add

Figure 5.17 – Configuring the scaling condition

Next, we create a scaling rule based on CPU—we have a range of metrics we can use here, including memory usage, amount of traffic, socket counts, and other advanced options for identifying an app that is struggling to cope.

Once we have a web app provisioned, the final step is to deploy our code to it.

Deploying to web apps

We can deploy our code to our app in several ways, as outlined here:

- **CI/CD pipelines**—Web apps support pipeline deployments from popular tools, including Azure DevOps, GitHub, Bitbucket, or local Git. CI/CD pipelines allow you to deploy code in a repeatable manner and are often the recommended option.
- **Visual Studio**—Being a Microsoft product, Visual Studio and Visual Studio Code (with extensions) integrate deeply into Azure, and as such, offer tools to easily deploy from them directly into your app.
- **FTP**—Failing all else, you can even just upload directly to your web app using traditional **sFTP**, and each web app generates its own set of credentials for you to use.

Opting for services such as Azure Web Apps is a great way to partially modernize your applications without the need to make code changes. With Web Apps, you can benefit from Azure's dynamic scalability, resilience, and cost efficiencies without needing to rewrite your application. However, it will not work for all scenarios, but it is certainly worth exploring before finalizing your migration strategy.

Further modernization

Not all applications are web based, and we've already mentioned that containerization is a great alternative, not just for migrating workloads but also for creating new, modern, microservice-based solutions as well. We will cover containers in more detail in *Chapter 7, Building a Container-Based Solution Design*.

Of course, Azure offers a range of solutions for modern application design, including many serverless options such as Azure Functions, Service Bus, Event Grid, AKS, and more. These cloud-native solutions can take your application architecture to the next level, enabling you to build fully flexible, efficient, and resilient solutions. However, they require you to make changes to your application code and, in some cases, a complete rebuild.

Often, you need to decide whether it is worth the time and effort to do this, which is often a question of how quickly you must migrate, the importance of what you are migrating, and the expected lifespan of your system.

So far, we have talked about how we can migrate our applications, and we've already mentioned how a big component of your solution will undoubtedly involve a database. Therefore, in the next section, we will look at the options for migrating SQL databases, which is the most common backend for on-premises solutions.

Migrating data

SQL Server is one of the most common database types for older, on-premises-based solutions. In the previous chapter, we also discussed NoSQL-based databases, which include products such as MongoDB or Azure Cosmos DB. However, NoSQL databases work differently, and often, the only way to migrate NoSQL to SQL, or SQL to NoSQL, would be customizing code to read the source data and write to the destination.

Although this is, of course, possible, often your solution will be designed with the underlying technology in mind; therefore, we will instead assume you are looking to migrate from a SQL-based solution to an Azure-hosted SQL solution.

We already discussed some of the different options we have in Azure for hosting SQL databases; as a recap, they are Azure SQL Database, Azure SQL Managed Instance, or SQL Server on a VM.

When choosing our desired source, we need to consider several requirements, such as the following:

- **Compatibility**—Does your on-premises SQL Server database have features that may not be supported or have limited support in Azure SQL Database or Azure SQL Managed Instance?
- **Performance**—What are the performance requirements of your application, and which Azure service can meet those requirements?
- **Scalability**—How much do you expect your database to grow, and which Azure service can best handle that growth?
- **Management**—Are you looking for a fully managed service (PaaS) or more control over the infrastructure (IaaS)?
- **Cost**—What are the cost implications of each migration option, and which one best aligns with your budget?

Some of these questions are, of course, down to your business needs; however, when migrating, the first area—compatibility—is perhaps the most important.

Certain features, data types, or functions in your on-premises SQL Server instance might not be supported or may require modification when migrating to Azure SQL Database or Azure SQL Managed Instance, and therefore we need to first understand what any issues might be.

We can then get an idea of how much effort would be involved in modifying our source database and decide if the upfront effort is worth the longer-term gains we would get from having a fully managed and flexible solution.

Luckily, Azure provides a tool called the **Data Migration Assistant** (**DMA**) that we can install on a server and use to scan our databases.

The first step is to download the DMA tool. Unlike the VM migration tool that comes pre-installed on a VM image, the DMA tool is installed on an existing server that can connect to your SQL databases.

The process takes place on the VM. Once you launch the tool, perform the following steps:

1. Download and install the DMA tool on a server that can connect to your on-premises databases.
2. Run the DMA tool and create a new **Assessment** project.
3. Choose a **Source server type** option from SQL Server or **Amazon Relational Database Service** (**Amazon RDS**) for SQL Server.
4. Choose a **Target server type** option from the following:
   - **Azure SQL Database**
   - **Azure SQL Database Managed Instance**
   - **SQL Server on Azure Virtual Machines**
   - **SQL Server**
5. Depending on the **Target Server type** option chosen, you will see a list of report options; for example, if you choose **Azure SQL Database**, you will see the following options:
   - **Check database compatibility**—This option will report on any features you are currently using that are not supported in the target type that will prevent a migration
   - **Check feature parity**—Report on unsupported features that won't block a migration but could prevent your database from functioning as expected
6. Click **Next**, then click **Add sources**.
7. Enter your source database details and click **Connect**.
8. Click **Start Assessment**.

Depending on the options you chose, the report will show you a list of issues along with potential fixes—as you can see in the following example:

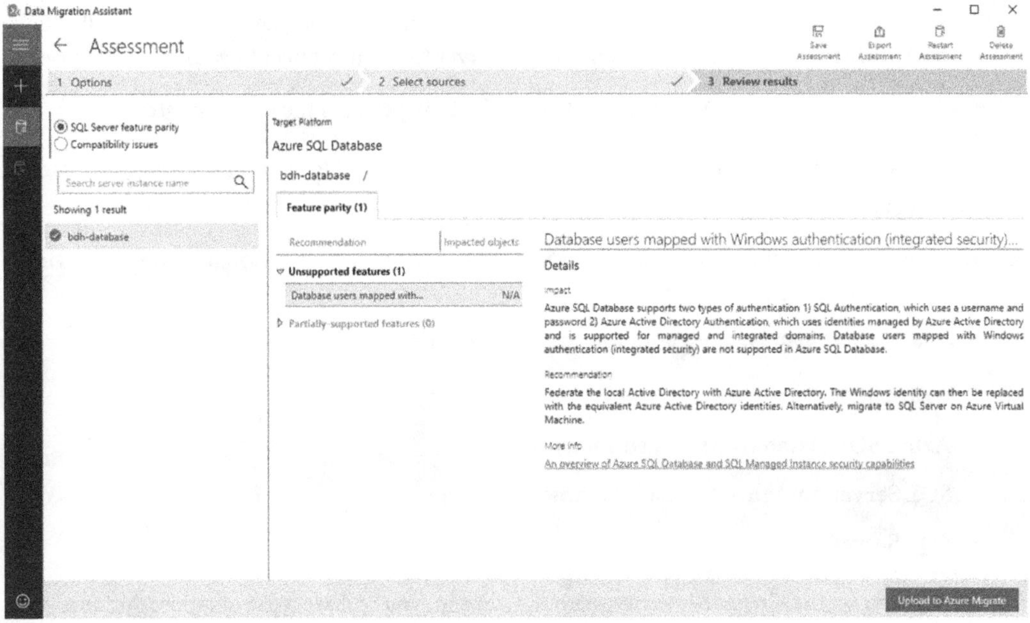
Data Migration Assistant
Assessment
Save Assessment
Export Assessment
Restart Assessment
Delete Assessment
1 Options
2 Select sources
3 Review results
SQL Server feature parity
Compatibility issues
Search server instance name
Showing 1 result
bdh-database
Target Platform
Azure SQL Database
bdh-database /
Feature parity (1)
Recommendation
Impacted objects
Unsupported features (1)
Database users mapped with...
N/A
Partially supported features (0)
Database users mapped with Windows authentication (integrated security)...
Details
Impact
Azure SQL Database supports two types of authentication 1) SQL Authentication, which uses a username and password 2) Azure Active Directory Authentication, which uses identities managed by Azure Active Directory and is supported for managed and integrated domains. Database users mapped with Windows authentication (integrated security) are not supported in Azure SQL Database.
Recommendation
Federate the local Active Directory with Azure Active Directory. The Windows identity can then be replaced with the equivalent Azure Active Directory identities. Alternatively, migrate to SQL Server on Azure Virtual Machine.
More info
An overview of Azure SQL Database and SQL Managed Instance security capabilities
Upload to Azure Migrate

Figure 5.18 – DMA

There are two views for this screen; the first shows **Feature parity**. This will flag any features that aren't compatible with your desired target—in the example, **Azure SQL Database**.

At the top left, change the assessment to **Compatibility issues**, and the screen will change to that shown in *Figure 5.19*:

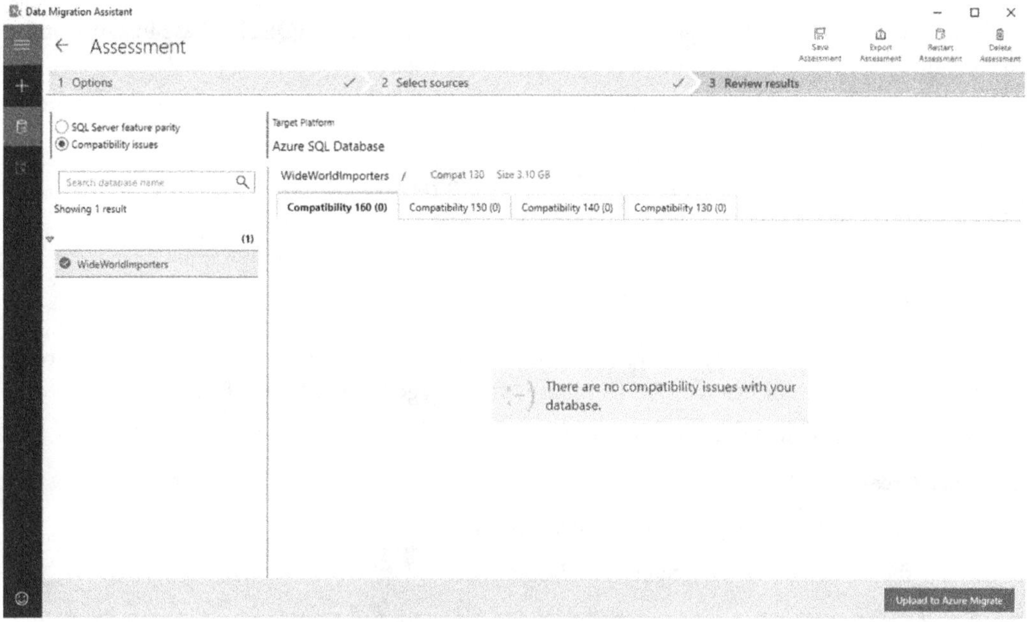
Data Migration Assistant
Assessment
Save Assessment
Export Assessment
Restart Assessment
Delete Assessment
1 Options
2 Select sources
3 Review results
SQL Server feature parity
Compatibility issues
Search database name
Showing 1 result
(1)
WideWorldImporters
Target Platform
Azure SQL Database
WideWorldImporters / Compat 130 Size 3.10 GB
Compatibility 160 (0)
Compatibility 150 (0)
Compatibility 140 (0)
Compatibility 130 (0)
There are no compatibility issues with your database.
Upload to Azure Migrate

Figure 5.19 – Database compatibility issues

In this example, we see no issues. If this were an older or more complex database, you might see different problems.

You can save the assessment report or upload it to Azure for visibility in the portal. To perform the latter, simply click the **Upload to Azure Migrate** button, then connect. You will be prompted to sign in and select a target subscription.

Once you have decided how you want to migrate your database, you can use the DMA tool to copy it.

You can perform either an offline migration, whereby the source server must be shut down for the migration, or an online migration that can be performed while the server is running. Regardless of which option you choose, you need to open access from your source database server to a new database server in Azure.

How you do this depends a lot on the source server—if this is an on-premises database, there will no doubt be firewalls between the server and Azure. The Azure SQL database itself will also need its firewalls to be set to allow access to the server.

In my example, I'm keeping it simple; I'm just going to create an Azure SQL Database instance and then configure the firewalls on it to be fully open to the internet.

> **Note**
> The SQL Database instance must already be created in Azure; the DMA tool will not create it for you.

In the DMA tool, perform the following steps:

1. Create a new **Migration project** type—give it a name, and again select **SQL Server** as the **Source server type** value and **Azure SQL Database** as the **Target server type** value.
2. Set **Migration scope** to **Schema and data**.
3. Click **Create**.
4. Select your source database type and click **Connect**.
5. Select the database you want to copy and click **Next**.
6. Select the target database server, which is the SQL Database instance you created in Azure; in my example, its name is `packtsqlserver.database.windows.net`.
7. Select your credentials—for example, **SQL Server Authentication**—and your Azure SQL Database instance's SQL login details you set when creating the SQL Server instance.
8. Click **Connect**.
9. Select the Azure database you created and click **Next**.
10. The next screen allows us to select individual tables, stored procedures, or views to copy. By default, all are selected, so click **Generate SQL script**.
11. The next screen has all the T-SQL commands required to create your database and tables; this allows you to make amendments if you wish. Click **Deploy schema**.
12. The data schema is now deployed to your Azure database. The final step is to migrate the data by clicking **Migrate data**, then **Start data migration**.

How big your database is will dictate how long this process takes.

The DMA tool is a quick and easy tool to check if your existing databases are compatible, as well as providing the ability to move your data.

Depending on the size and version of your database, it is sometimes better to perform a backup of your database, copy it into an Azure Storage account, then run a database restore.

The best method depends on your source database, but the point is, we have several different avenues for migrating data.

Any migration from on-premises into Azure should look at individual components—that is, compute and data. In many cases, even moving part of your solution to an Azure native option will bring greater efficiency and scalability.

Summary

In this chapter, we looked at how we migrate our existing, on-premises workloads in Azure. We first discussed the main strategies we needed to consider—lift and shift, refactor, re-architect, or rebuild.

We then looked at the options for moving the compute portion of a solution into Azure and discussed the various options for either partially or fully modernizing our solutions. As part of this, we looked at Scale Sets and Web Apps, which are great options when you want to make minimal code changes but take full advantage of Azure's benefits.

Finally, we looked at how we can migrate an existing SQL database into Azure, what questions we need to ask, what the potential issues were, and how we can use the DMA tool to help analyze and migrate those databases.

In the next chapter, we will look at monitoring—why we need it and how to do it. We will examine the specific monitoring and logging requirements for some different components and how we can use them to help us optimize our solutions.

6

End-to-End Observability in Your Cloud and Hybrid Environments

Observability is a crucial aspect of managing cloud-based applications as it enables organizations to gain insights into their systems and troubleshoot issues quickly and efficiently.

Observability pertains to the capacity to comprehend the occurrences or events taking place within your environment by proactively collecting, visualizing, and analyzing all the data it produces. And to achieve observability, a monitoring tool must collect and analyze different kinds of data, known as the pillars of observability. These pillars are typically metrics, logs, and distributed traces.

When an environment is observable, it means that users can quickly identify the root cause of any performance problems by analyzing the data it produces.

When organizations move their workloads to the cloud, it's crucial to create a monitoring strategy that involves input from developers, operations staff, and infrastructure engineers.

We will refer to the case study mentioned in the following section as a practical example to enable observability in your cloud and hybrid environments.

In this chapter, we'll cover the following topics:

- Understanding the importance of a monitoring strategy
- Working on an effective monitoring strategy
- Azure Monitor – a comprehensive solution for observability and efficiency

Understanding the importance of a monitoring strategy

SpringToys is a multi-cloud organization with a distributed and dispersed IT environment. They lack a consistent architectural pattern, which has led to them having various solutions in place, making it challenging to manage security configurations across all environments. Additionally, platform security and configuration management have never been a top priority for the organization, resulting in several environments being built manually by different individuals with different security standards.

SpringToys has multiple Azure subscriptions, and their on-premises data centers run a variety of virtualization platforms managed by different teams with varying security standards. They use an unknown number of Linux distributions and Windows servers with a disjointed, often ad hoc patch management strategy. Furthermore, SpringToys runs container-based solutions both on-premises and in Azure.

To address these challenges, SpringToys has tasked a cloud architect with investigating Azure Monitor and developing an implementation plan using it. SpringToys currently uses a third-party vendor for monitoring. While they are satisfied with this third-party vendor, they are concerned about the high cost of rolling it out to their entire ecosystem. They are looking for a remediation strategy to address various issues, such as the lack of visibility into patch management, the use of monitoring or logging solutions managed by different teams with different remediation strategies, and the need to simplify server log retention and configure all servers for five-year log retention.

Additionally, they currently use Splunk as their **Security Information and Event Management** (**SIEM**) system, but they are considering Sentinel. While they evaluate Sentinel, they would like to use Splunk.

The long-term goals of this implementation include gaining visibility into new and existing shadow IT resources, ensuring compliance and reporting on regulatory standards across the enterprise, and alerting the security team to security events.

Now that we understand the case study, let's learn how SpringToys can leverage Azure Monitor to gain better visibility into its diverse environment.

Working on an effective monitoring strategy

A monitoring strategy should make the organization agile and proactive in monitoring complex distributed applications essential to the business. It should help the organization quickly detect and address any business issues. By regularly monitoring the system, the organization can ensure that the cloud environment is performing as expected and meeting the business's needs.

An effective monitoring strategy will help the organization save money by reducing the cost of downtime, reducing the time it takes to identify and fix problems, and optimizing resources. This strategy will also help improve customer satisfaction by ensuring that the systems are working properly and that the organization is responsive to any issues.

To start designing a monitoring strategy, you can leverage the Cloud Adoption Framework, which recommends including monitoring in any cloud migration project's strategy and planning phases. A sound monitoring strategy is crucial to ensuring the health of cloud applications, measuring end user experience, maintaining security compliance, prioritizing alerts, and integrating with existing IT service management solutions.

A service-based approach to monitoring is recommended, with monitoring seen as an advisory service and provider of expertise to both business and IT consumers. The monitoring plan should be designed to provide visibility to the relevant parties based on their roles and responsibilities.

Centralized governance and strict delegation are also necessary to effectively manage IT services.

As an architect, it is essential to consider the following four outcomes when formulating a monitoring strategy:

- Managing cloud production services when they go live
- Rationalizing existing monitoring tools
- Making monitoring solution processes more efficient
- Planning for and hosting monitoring based on cloud models

To achieve these outcomes, we recommend applying limited resources to rationalize existing monitoring tools, skills, and expertise and use cloud monitoring to reduce complexity. Additionally, it is crucial to make the monitoring solution processes more efficient, work faster and smoother, and be able to change quickly at scale. Organizations should also account for how they plan to host monitoring based on cloud models and work toward reducing requirements as they transition from IaaS to PaaS and SaaS.

The next step in developing a monitoring strategy is collaborating with stakeholders such as architects, strategic planners, and management. Your first step could be to assess your current systems management, including people, partners, tooling, and risks. This assessment will help you prioritize the areas that require improvement and identify key opportunities. It would be best to consider which services, systems, and data will likely remain on-premises and which will move to the cloud.

Ideally, your management team will want a roadmap of initiatives that align with the planning horizon. However, discussing any unknown factors that may impact the monitoring strategy is important. By collaborating with stakeholders and conducting a thorough assessment, you can develop a monitoring strategy that aligns with business needs and supports the organization's goals for cloud adoption.

Determining which users require access to monitoring tools, alerts, and analytics is essential. Azure Monitor and Azure Arc offer flexible monitoring capabilities across all four cloud models, not just resources within Azure. However, it's important to look beyond the traditional cloud models and consider monitoring needs for Microsoft Office applications, which may require additional security and compliance monitoring with Microsoft 365 and Microsoft Defender for Cloud. Additionally, monitoring should extend to identities, endpoint management, and device monitoring beyond the corporate network.

Monitoring plays a critical role in incremental approaches to protect and secure the digital estate. Security monitoring is essential in measuring directory usage and externally sharing sensitive content to achieve the right balance with privacy monitoring. Policies and baselines can also help rationalize the objective of migrating data from on-premises to cloud services and improve confidence in the migration process.

Defining your architecture for monitoring means planning how you will keep track of your systems and services to ensure they are working properly. This involves making decisions on how to use your resources efficiently and effectively and ensuring that your monitoring strategy can support future services that your business may need. You should also consider how you will monitor resources hosted in the cloud and identify any gaps in your current monitoring plan. Finally, you should also consider the financial costs and support needed to implement your monitoring plan and use this information to guide any hybrid decisions you need to make.

To understand how well a service, asset, or component is working, you need to monitor it. There are three sides to the monitoring principle, called service visibility:

- **Monitor in-depth**: This means collecting useful information about how the service works and ensuring you're monitoring the right things
- **Monitor end-to-end**: This involves monitoring every service layer from the bottom to the top so that you can see how everything works together
- **Monitor east to west**: This means focusing on the aspects of health that matter the most, such as ensuring the service is available, performs well, is secure, and can continue running even if something goes wrong

When it comes to monitoring, there are several essential questions to consider:

- How will you secure access to security logs and ensure they are only shared with authorized individuals or teams?
- Which services require global monitoring, and how can you monitor them at the service edge?
- How can you quickly monitor the network points between your infrastructure and the cloud provider's endpoints to identify and troubleshoot connectivity issues?
- What are the boundaries between security operations, health monitoring, and performance monitoring? How can you provide summaries of health and status to all the relevant teams and stakeholders and ensure they have the information they need to do their jobs effectively?

Conversely, it's crucial to consider how to monitor the cloud or hybrid environment at a large scale. One way to do this is to create standard solutions that everyone in your organization can use. To create these solutions, you can use tools such as Azure Resource Manager templates and Azure Policy monitoring initiative definitions and policies. You can also use GitHub to store all the scripts, code, and documentation related to monitoring in one place. By doing this, you can ensure that everyone uses the same monitoring tools and approaches, which can help with consistency and efficiency.

If you're looking to keep a close eye on your applications and services, whether they're hosted in the cloud or on-premises, Azure Monitor can help you do just that. This powerful tool can collect data from various parts of your resources and services, and consolidate it in a single location for easy analysis. With Azure Monitor, you can quickly identify and fix any issues that might arise, ensuring your applications run smoothly.

Azure Monitor – a comprehensive solution for observability and efficiency

Azure Monitor helps you keep track of what's happening in your environment, whether they're in the cloud or on your servers. It can collect data from all parts of your services and resources and put it in one place so that you can analyze it easily. This helps you ensure your applications and services are running smoothly and quickly find and fix any problems.

By collecting data from various pillars and consolidating information from all monitored resources, Azure Monitor can help you accomplish an end-to-end observability strategy.

By providing a common set of tools to correlate and analyze data from multiple Azure subscriptions and tenants, Azure Monitor makes it easier to achieve observability across different services.

With Azure Monitor, all your environment data can be stored together and analyzed using common tools. This means you can easily see how different parts of your environment work together and respond automatically to any system events. Additionally, Azure Monitor includes a feature called SCOM Managed Instance, which lets you move your monitoring tools to the cloud. This can make monitoring your workloads more efficient and help you take advantage of the benefits of cloud computing.

Azure Monitor can help you keep track of different types of resources located in Azure, other clouds, or on your servers. You can use Azure Monitor to monitor various things, such as applications, virtual machines, guest operating systems, containers, databases, security events, networking events, and custom sources that use APIs.

By monitoring these different resources, you can better understand how your computer systems are working and find any issues that need to be fixed. Azure Monitor can help you see how all these different resources are connected, making it easier for you to respond quickly to any problems. Azure Monitor can also work with other tools, such as Azure Sentinel and Network Watcher, to provide even more detailed monitoring and security capabilities.

Azure Monitor allows you to export monitoring data so that you can use it with other tools and systems. For example, you can integrate Azure Monitor with third-party monitoring and visualization tools and ticketing and **Information Technology Service Management** (**ITSM**) solutions. By exporting monitoring data, you can analyze and visualize it in different ways and use it to create reports or alerts that can be sent to other systems. This helps you work more efficiently and use the best tools for you and your team.

Components

Azure Monitor comprises different components that work together to collect, analyze, and respond to data from monitored resources. The data sources refer to the various kinds of information obtained from each monitored resource and transmitted to the data platform.

The *data platform* is made up of different data stores for different types of data, such as *metrics*, *logs*, *traces*, and *changes*:

- **Metrics**: These are numerical values that describe the state of a system at a particular time, such as CPU usage or memory utilization. Azure Monitor Metrics is a time series database that collects and stores metrics data at regular intervals, allowing you to track trends and identify performance issues.
- **Logs**: These are recorded system events, such as errors or warnings, that can contain different types of structured or unstructured data. Azure Monitor Logs stores log data of all types, which can be analyzed and queried using Log Analytics workspaces.
- **Traces**: These are used to measure the operation and performance of your application across multiple components in your system. Traces help you understand the behavior of your application code and the performance of different transactions. Azure Monitor gets distributed trace data from the Application Insights SDK and stores it in a separate workspace in Azure Monitor Logs.
- **Changes**: Azure Monitor also tracks changes in your application and resources, such as updated code that's been deployed, which may have caused issues in your environment. Change Analysis helps you understand which changes may have caused issues in your systems by tracking and storing these events using Azure Resource Graph.

The following figure highlights the data sources, data platform, and consumption components that can be integrated with Azure Monitor:

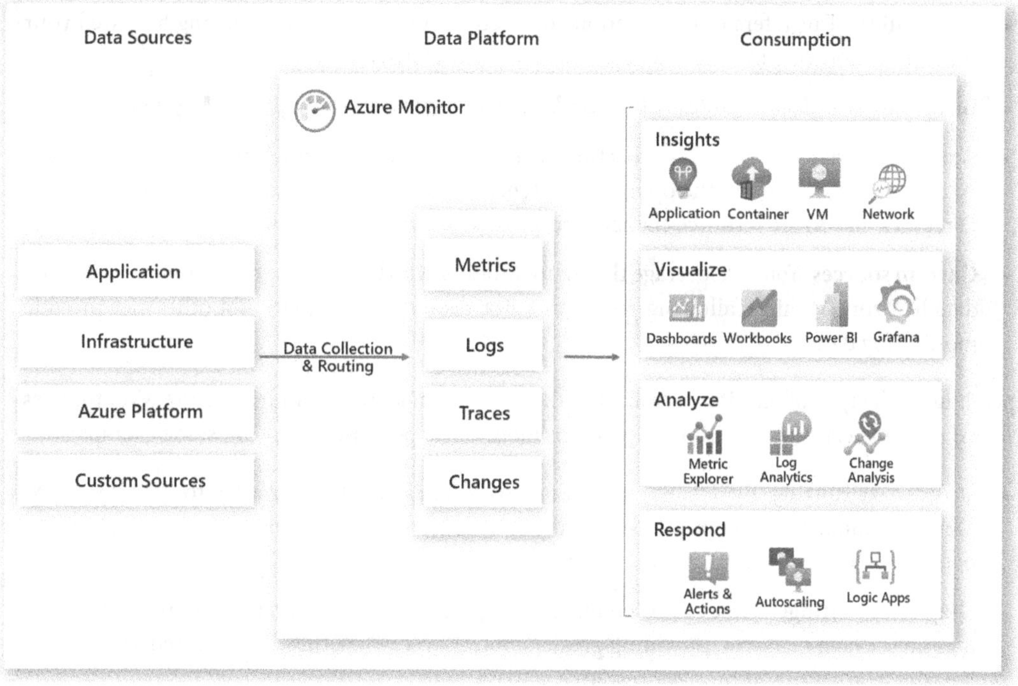
Data Sources
Data Platform
Consumption
Azure Monitor
Application
Infrastructure
Azure Platform
Custom Sources
Data Collection & Routing
Metrics
Logs
Traces
Changes
Insights
Application
Container
VM
Network
Visualize
Dashboards
Workbooks
Power BI
Grafana
Analyze
Metric Explorer
Log Analytics
Change Analysis
Respond
Alerts & Actions
Autoscaling
Logic Apps

Figure 6.1 – Azure Monitor high-level overview

The collected data is then used by various functions and components within Azure Monitor, including analysis, visualizations, insights, and responses. These functions help identify issues, provide insights, and respond to system events automatically.

Azure Monitor also integrates with other services to provide additional functionality. This means that Azure Monitor can work seamlessly with other Azure services to provide a comprehensive monitoring solution for cloud and on-premises environments.

Data sources

Azure Monitor collects data from various sources, including on-premises and non-Microsoft clouds. In addition to collecting data from your applications, operating systems, and services, Azure Monitor allows integration with external sources through the application, infrastructure, and custom data sources. This lets you have a unified view of your entire IT environment and detect issues before they cause significant impact.

Azure Monitor gathers various data types related to your application's and infrastructure's performance and functionality. It collects information from the following data types:

- **Application**: This refers to information collected by Azure Monitor regarding how well your application code is running and how effectively it performs
- **Infrastructure**: This includes collecting data from containers and operating systems.
- **Azure platform**: This includes collecting data from Azure resources, Azure subscriptions, and Azure tenants, including the operation of tenant-level Azure services, such as Azure Active Directory, and Azure resource changes such as data about changes within your Azure resources
- **Custom sources**: You can leverage the Azure Monitor REST API to send custom metrics or log data to Azure Monitor, allowing you to monitor resources that might not otherwise provide monitoring data

Depending on the type of monitoring data and the destination storage platform, Azure Monitor uses various mechanisms to collect and route data for monitoring purposes. Here are some examples:

- **Direct data routing**: By default, platform metrics are automatically routed to Azure Monitor Metrics without any configuration needed, allowing you to easily access and analyze important performance data for your Azure resources.
- **Diagnostic settings**: These can be configured to specify the destination for resource and activity logs on the data platform, providing greater control over where this data is stored and making it easier to analyze and monitor resource activity.
- **Data collection rules**: With data collection rules, you can define the specific data that should be collected, set rules for transforming that data, and specify the destination where the data should be sent.
- **Application SDK**: By integrating the Application Insights SDK into your application code, you can receive, store, and explore monitoring data with the SDK pre-processing telemetry and metrics before sending them to Azure. This data is then further processed by Azure Monitor.
- **Azure Monitor REST API**: With the Logs Ingestion API in Azure Monitor, you can easily send data to a Log Analytics workspace from any REST API client, making collecting and analyzing important monitoring data simpler.
- **Azure Monitor Agent**: **Azure Monitor Agent** (**AMA**) is responsible for gathering monitoring data from the guest operating system of Azure and hybrid virtual machines, with this data then being delivered to Azure Monitor so that it can be used by various features, insights, and services, including Microsoft Sentinel and Microsoft Defender for Cloud.

Aside from data collections, data transformations are a powerful feature in the data collection process, allowing users to modify incoming data using KQL queries. This can be useful in a variety of scenarios, such as filtering out irrelevant data or modifying data to improve query or reporting capabilities.

For example, the following KQL query calculates the average CPU usage percentage per minute for nodes in a Kubernetes cluster:

```
let endDateTime = now();
let startDateTime = ago(1h);
let trendBinSize = 1m;
let capacityCounterName = 'cpuCapacityNanoCores';
let usageCounterName = 'cpuUsageNanoCores';
KubeNodeInventory
| where TimeGenerated < endDateTime
| where TimeGenerated >= startDateTime
| distinct ClusterName, Computer, _ResourceId
| join hint.strategy=shuffle (
  Perf
  | where TimeGenerated < endDateTime
  | where TimeGenerated >= startDateTime
  | where ObjectName == 'K8SNode'
  | where CounterName == capacityCounterName
  | summarize LimitValue = max(CounterValue) by Computer, CounterName,
bin(TimeGenerated, trendBinSize)
  | project Computer, CapacityStartTime = TimeGenerated,
CapacityEndTime = TimeGenerated + trendBinSize, LimitValue
) on Computer
| join kind=inner hint.strategy=shuffle (
  Perf
  | where TimeGenerated < endDateTime + trendBinSize
  | where TimeGenerated >= startDateTime - trendBinSize
  | where ObjectName == 'K8SNode'
  | where CounterName == usageCounterName
  | project Computer, UsageValue = CounterValue, TimeGenerated
) on Computer
| where TimeGenerated >= CapacityStartTime and TimeGenerated <
CapacityEndTime
| project ClusterName, Computer, TimeGenerated, UsagePercent =
UsageValue * 100.0 / LimitValue, _ResourceId
| summarize AggregatedValue = avg(UsagePercent) by bin(TimeGenerated,
trendBinSize), ClusterName, _ResourceId
```

The resulting data provides insights into the average CPU usage percentage per minute for nodes in a Kubernetes cluster over the specified time range. This query can be used to monitor and analyze the CPU performance of the cluster nodes and identify any potential capacity or usage issues.

With data transformations, users can customize the data collection process to meet their specific needs and gain more insights from their data. This feature can ultimately help users make more informed decisions and optimize the performance of their systems.

Transformations allow users to modify incoming data before it is sent to a Log Analytics workspace. Using transformations, users can filter or modify data to better suit their needs.

Once you collect the data, you can start working with the Azure portal, which provides a user-friendly interface that offers seamless access to the monitoring data collected. It also provides easy access to the various tools, insights, and visualizations available in Azure Monitor, enabling users to gain valuable insights into the performance and health of their resources with just a few clicks.

Consumption

To utilize Azure Monitor with the data that's been collected, you must have at least one Log Analytics workspace, which is necessary for various tasks, such as gathering data from Azure resources and the guest operating system of Azure Virtual Machines.

Additionally, Log Analytics enables access to several Azure Monitor insights. For example, if you're using Microsoft Sentinel or Microsoft Defender for Cloud, you can share the same Log Analytics workspace as Azure Monitor. Although creating a Log Analytics workspace is free, there may be a fee after configuring data collection.

To create an initial Log Analytics workspace in the Azure portal, follow the steps outlined at `https://bit.ly/create-log-analytics`.

Once you have created a workspace, you can configure access for specific users or groups.

If you need to configure multiple workspaces, you may want to consider using scalable methods such as Resource Manager templates. This can simplify the process of configuring workspaces and help ensure consistency across your environment. However, remember that this step may not be necessary for all environments, as many organizations may only require a minimal number of workspaces.

Ultimately, your approach will depend on your organization's specific needs and requirements. It's important to carefully consider your options and choose the approach that best meets your needs while also being mindful of scalability and ease of management.

You can deploy Azure Arc and AMA to aggregate logs into a Log Analytics workspace. Once you have your data in Log Analytics, it's time to work with the collected data.

Azure Monitor offers a range of powerful features, including **Insights**, **Visualizations**, **Analysis**, and **Response**. These features provide users with a comprehensive view of their applications and infrastructure, allowing them to gain valuable insights, create custom visualizations, analyze metrics and logs, and take proactive measures to optimize the performance and health of their systems:

- **Insights**: With Insights, you can gain visibility into the performance and health of your applications, infrastructure, and networks. Insights provides a unified view of all monitoring data and alerts, enabling users to identify and troubleshoot issues quickly.

One of the key features of Insights is its customizable dashboards, which allow users to create dashboards that display the data and insights most relevant to them. Insights also provides a range of pre-built dashboards designed to help users monitor specific Azure services, such as Azure Virtual Machines, Azure Kubernetes Service, and Azure SQL Database. With these dashboards, you can better understand your environment and identify any issues that may be impacting performance.

- **Visualizations**: You can leverage Azure Workbooks, create custom dashboards, and use Power BI and Grafana, among other tools, to create powerful visualizations.

  Workbooks provide a flexible and powerful way for users to create custom dashboards, reports, and visualizations based on their monitoring data. Workbooks are built using a drag-and-drop interface that allows users to add visualizations to their workbooks, such as charts, tables, and maps. Users can add custom text and images and embed external content such as YouTube videos or Twitter feeds.

  One of the key benefits of Azure Monitor Workbooks is their ability to aggregate data from multiple sources into a single dashboard. Users can combine data from Azure Monitor, Log Analytics, Application Insights, and other sources to comprehensively view their environment. Workbooks can also be shared with other users, allowing teams to collaborate and share insights across different departments or organizations.

  You can leverage custom dashboards in Azure Monitor, which involves identifying the data sources used to populate the dashboard, such as log data from Azure services or custom applications, and building visualizations such as charts or tables. You can customize the appearance and layout of the dashboard to suit your preferences, and once complete, you can save and share it with other users or teams. Custom dashboards provide users with a powerful tool for monitoring their systems, optimizing performance, and troubleshooting issues.

- **Analysis**: Azure Monitor enables different options for analysis, including the use of Metrics Explorer, Log Analytics, log queries, and Change Analysis.

  Metrics can be collected and analyzed in Azure Monitor to identify performance issues and monitor the health of Azure services or custom applications. You can access and explore metric data using the Metrics Explorer interface and create custom queries using the **Kusto Query Language (KQL)**.

  Another option is using Log Analytics to edit and run log queries against data stored in Azure Monitor Logs. You can write simple or advanced queries to sort, filter, and analyze log data. Log Analytics can also be used to perform statistical analysis and create visualizations to identify trends in data.

  You can leverage log queries to extract information from log data stored in Azure Monitor Logs and create custom visualizations using tools such as KQL. You can query multiple data sources, create alerts based on log query results, and use log query results to create custom dashboards in Azure Monitor.

Conversely, if you're looking to gain insights into the changes made to Azure applications, you can use Change Analysis. Change Analysis is a feature designed to provide insights into changes made to Azure applications, increasing observability and reducing the **Mean Time to Repair** (**MTTR**). While standard monitoring solutions may alert users to issues or failures, they often don't explain the cause. Change Analysis utilizes the power of Azure Resource Graph to analyze changes made to resources and identify the root cause of issues. By providing detailed information on changes made to resources, Change Analysis enables users to identify and resolve problems quickly, reducing the MTTR and improving overall system performance.

- **Response**: Azure Monitor can help you quickly respond to events and proactively solve issues. You can use alerts or enable autoscaling.

  Alerts are intended to alert users in advance when Azure Monitor data suggests a potential issue with their infrastructure or application.

  You can set up alerts to monitor any metric or log data source in the Azure Monitor data platform. By using alerts to monitor key metrics and data sources, you can quickly identify issues and take action to resolve them before they impact users.

  Autoscaling, on the other hand, can help you automatically adjust the number of resources allocated to your application based on the application's load. It can scale your application based on metrics such as CPU usage, queue length, available memory, or a pre-configured schedule. Autoscaling helps you ensure that you have the necessary resources to handle an increase in load while also reducing costs by scaling down resources when the demand decreases. To set up autoscaling, you can create rules that set your application's minimum and maximum levels of resource consumption.

As you can see, Azure Monitor is a comprehensive monitoring solution for cloud and hybrid environments that enables end-to-end observability by providing insights into the health, performance, and security of your applications and infrastructure.

Summary

This chapter reviewed the importance of having an effective strategy to monitor your resources across multiple environments, including Azure, on-premises, and other cloud providers, from a single, unified platform. This allows you to gain visibility into the entire application stack, from the user experience to the underlying infrastructure, and identify issues before they impact your business.

One of the main benefits of using Azure Monitor is that it provides a single source of truth for all your monitoring data, including logs, metrics, and traces. This allows you to correlate data across different sources and gain a holistic view of your environment. With Azure Monitor, you can quickly identify issues, troubleshoot problems, and optimize your resources for better performance and cost savings. Additionally, Azure Monitor provides advanced analytics capabilities that help you proactively detect and remediate issues before they affect your users. By enabling end-to-end observability with Azure

Monitor, you can deliver a seamless, high-quality user experience while ensuring the reliability and security of your applications and infrastructure.

In the next chapter, we will focus on the importance of containers for modern organizations and explore the various container services available in Azure. We'll take a closer look at the different types of container offerings available in Azure and how they can be used to streamline application development and deployment. If you already manage infrastructure in your organization, such as virtual machines and their components, the next chapter will help you determine how containers can complement your existing infrastructure so that you can improve your overall application performance and management.

7
Working with Containers in Azure

This chapter will bring together what you have learned in previous chapters of the book as we will discuss how to leverage core Azure services to architect your solutions.

This chapter will cover the main reasons containers are relevant for organizations and help you understand the different services available in Azure to work with containers. Then, we will dive deep into how to work with the various services in Azure related to containers and their components.

By the end of this chapter, you will better understand the multiple Azure offerings for containers and their differences. Chances are, you already manage infrastructure in your organization, such as virtual machines and their components.

We'll be covering the following main topics:

- Understanding cloud-native applications
- Understanding the difference between virtual machines and containers
- Azure Container Instances
- Working with Azure containers
- Creating Azure Container Apps

Let's get started!

Understanding cloud-native applications

The landscape of our lives and work has undergone a significant transformation. Leading industries are actively transitioning their interactions with users, shoppers, patients, and voters from physical to digital platforms. The modernization of their applications has become an imperative shared by organizations across various sectors.

In many instances, the drive to modernize our applications is born out of a competitive necessity, ensuring that businesses remain relevant and avoid being left behind amid the relentless forces of the market. Additionally, industry regulations and the need to incorporate the latest security technologies serve as compelling drivers for modernization. Furthermore, organizations are recognizing that embracing application modernization provides an excellent opportunity to shed excessive capital expenditure and reduce costs by leveraging resources in a more agile manner.

The benefits of building cloud-native applications are far-reaching and multifaceted, enabling businesses to establish a strong digital presence, enhance transparency, and meet evolving customer expectations. By adopting application modernization strategies, organizations can position themselves at the forefront of their respective industries, while also improving operational efficiency and driving cost savings.

There are various approaches available for tackling the challenge of building cloud-native applications, each with distinct advantages and drawbacks. One of the primary concerns we frequently encounter from our customers revolves around selecting the most suitable tool for hosting their applications, whether they are modernizing existing apps or developing new technology to drive their business forward.

Cloud-native refers to a modern approach to building and running applications that fully utilize the capabilities of cloud computing. It involves designing applications specifically for cloud environments, taking advantage of the scalability, elasticity, and resilience that the cloud provides.

Cloud-native applications are typically developed using containerization technologies such as Docker, which allow for the efficient packaging and deployment of software. These applications are also built using microservices architecture, where large applications are broken down into smaller, loosely coupled components that can be developed, deployed, and managed independently.

In this context, cloud-native applications leverage orchestration platforms such as Kubernetes to automate the deployment, scaling, and management of containers. This ensures that applications can dynamically adjust to changes in demand, allowing for efficient resource utilization and improved reliability.

Developers today have a range of options that span the spectrum from offering greater control to prioritizing productivity. If you opt for a microservices-based development methodology running on Kubernetes containers, it necessitates investing in additional tooling, automation, and management to fully leverage its benefits, particularly if you are transitioning from a monolithic app approach. **Infrastructure as a service (IaaS)** technologies provide you with increased control over the underlying configuration, be it virtual machines or containers, as long as you are willing to handle the management aspect yourself.

For many enterprise customers, a compelling alternative lies in managed platforms, also referred to as **platform as a service (PaaS)**. In this model, you solely focus on managing the applications and services you develop, while the cloud service provider takes care of everything else. In the upcoming discussion, we will delve deeper into the business impact, capabilities, and solution components offered by PaaS.

Cloud-native development also embraces DevOps practices, promoting collaboration and automation between development and operations teams. **Continuous integration and continuous delivery** (**CI/CD**) pipelines are commonly used to streamline the software development life cycle and enable rapid and frequent releases. Being cloud native means embracing a set of principles and practices that enable organizations to take full advantage of the cloud infrastructure. Therefore, it is important to understand the core elements of cloud infrastructure and the difference between virtual machines and containers.

Understanding the difference between virtual machines and containers

Virtual machines host a variety of workloads, from SSH servers and monitoring to identity management, and have been the way to run programs and deploy applications for the last decades instead of using physical computers. One of the top benefits of using virtual machines in previous years was providing isolation from the host operating system.

This was useful in establishing security boundaries to host applications on the same server or cluster.

While each virtual machine runs its own operating system, it requires additional computing resources, such as CPU, memory, and storage. Also, as organizations expanded their on-premises footprint based on virtual machines, the complexity of managing them at scale not only became an operational issue but also meant a very costly expense.

The advent of cloud computing impacted how applications are designed. Organizations tried to adopt microservices, a single-tiered software application in which different modules are combined into a single program instead of having a monolithic application. This approach involves each application component typically running its own service and communicating with other services.

Each microservice can be deployed independently as a standalone service or as a group of services. Usually, each separate service is hosted in a container. Lifting and shifting applications to containers can help organizations save costs and deliver value to their customers faster.

Containers represent the evolution in the virtualization of computing resources. Think of a container as a package of software that bundles the application's code, including its configuration files, libraries, and dependencies needed for the application to run. This way, applications can be moved easily from one environment to another with little or no modification.

A container virtualizes the underlying operating system and makes the app perceive that it holds the operating system and additional resources such as CPU, storage, memory, and networking components. In short, a container virtualizes the environment and shares the host operating system's kernel while providing isolation.

For practical purposes, consider a container as a virtual machine that runs on top of the host operating system and can emulate an operating system that's different from the underlying one. Containers provide the essential components for any application to run on a host operating system.

The following figure shows the difference between virtual machines and containers.

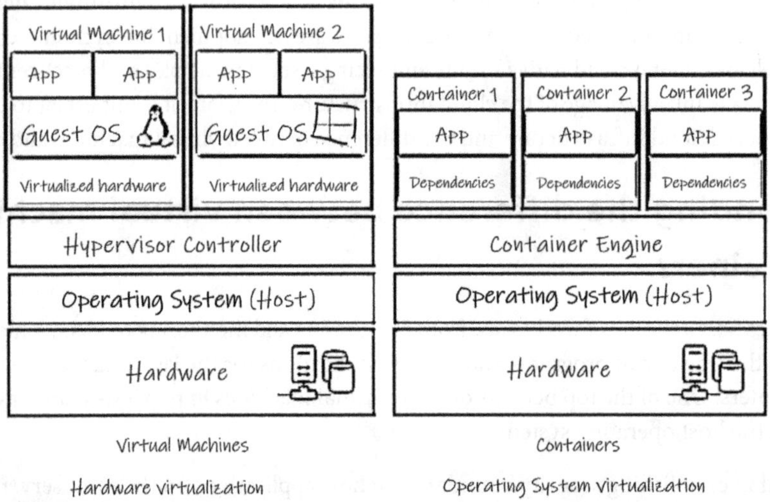
Virtual Machine 1
Virtual Machine 2
App
App
App
App
Guest OS
Guest OS
Virtualized hardware
Virtualized hardware
Hypervisor Controller
Operating System (Host)
Hardware
Virtual Machines
Hardware virtualization
Container 1
Container 2
Container 3
App
App
App
Dependencies
Dependencies
Dependencies
Container Engine
Operating System (Host)
Hardware
Containers
Operating System virtualization

Figure 7.1 – Difference between virtual machines and containers

> **Note**
>
> Containers use the host operating system and not the hypervisor as their base; therefore, containers don't need to boot an operating system or load libraries, which makes containers lightweight and much more efficient.

Applications that run in containers can start in seconds, and one crucial consideration to have in mind is that while containers are portable, they're constrained to the operating system they were built into.

Containers offer several advantages over virtual machines, such as the following:

- Increased flexibility and speed when sharing the application code
- Simplified application testing
- Accelerated application deployment
- Higher workload density with improved resource utilization

Terminology

Before moving forward, let's agree on key terms before reviewing Azure Container Instances. We will focus on the following fundamental concepts to better understand containers:

- **Container**: A container represents the execution of a single application or service. It consists of an image, an execution environment, and a set of instructions.

- **Container image**: A package that contains all the code and its dependencies to create a container.
- **Build**: This refers to the process of building a container image.
- **Pull**: This refers to downloading a container image from a container registry.
- **Push**: This is the process of pushing/uploading a container image from a container registry.
- **Dockerfile**: This represents a text file with the set of instructions to build a Docker image.

When designing a solution based on containers, we must understand the offers we can leverage in Azure. The following figure provides a decision flowchart with high-level guidance for selecting the appropriate Azure service for your scenario related to containers.

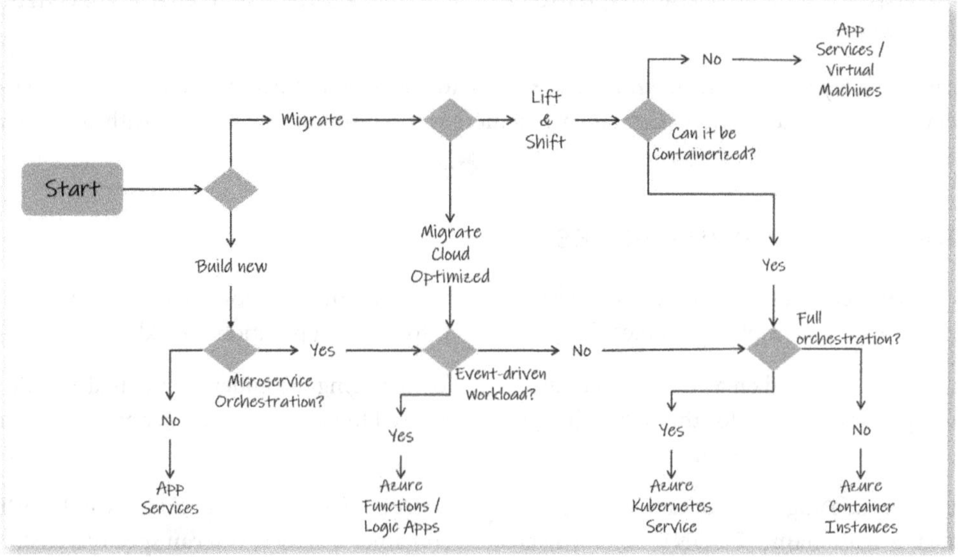
Start
Migrate
Lift & Shift
No
App Services / Virtual Machines
Can it be Containerized?
Build new
Migrate Cloud Optimized
Yes
Full orchestration?
Yes
Yes
No
Microservice Orchestration?
Event-driven Workload?
No
Yes
Yes
No
App Services
Azure Functions / Logic Apps
Azure Kubernetes Service
Azure Container Instances

Figure 7.2 – Flowchart with high-level guidance for how to select the appropriate Azure service

As shown in the preceding flowchart, there are many options for organizations to build and deploy containerized applications on Azure. The following list shows the container options on Azure that you can choose from that best suit your needs:

- **Azure Container Apps**: This is a managed service to simplify the deployment of containerized applications in a serverless environment.
- **Azure Container Instances**: This is the simplest way to deploy a container in Azure. You don't have to manage any virtual machine.
- **Azure Kubernetes Service (AKS)**: A managed Kubernetes service that can run containerized applications without the need to manage the master node and other components in the cluster.

- **Azure Red Hat OpenShift**: This is a managed OpenShift cluster where you can run your containers with Kubernetes.
- **Azure Service Fabric**: Azure Service Fabric is a powerful distributed systems platform designed to simplify the packaging, deployment, and management of scalable microservices and containers, providing flexibility in development approaches and offering efficient deployment options with processes or container images.
- **Azure Web App for Containers**: Azure Web App for Containers allows you to deploy and run containerized applications in a fully managed and scalable environment, simplifying the process of hosting and managing container-based web applications.

We will narrow the scope of this chapter to deep dive into Azure Container Apps and Azure Container Instances.

Before working with containers in Azure, it is essential to understand the components of each service. As mentioned previously, Azure Container Instances is the simplest way to work with containers in Azure.

Azure Container Instances

Azure Container Instances, or **ACI**, as we will refer to it from now on, allows organizations to run containers on Azure simply and launch them in seconds to access applications quickly.

The ACI service is billed on a per-second basis. Instead of managing the underlying virtual machine, you can specify the values for the computing power you need to run your container and define the values for CPU cores and memory.

ACI enables exposing your container groups directly to the internet with an IP address and a **fully qualified domain name** (**FQDN**). When you create a container instance, you can specify a custom DNS name label so your application is reachable, as shown in the following figure:

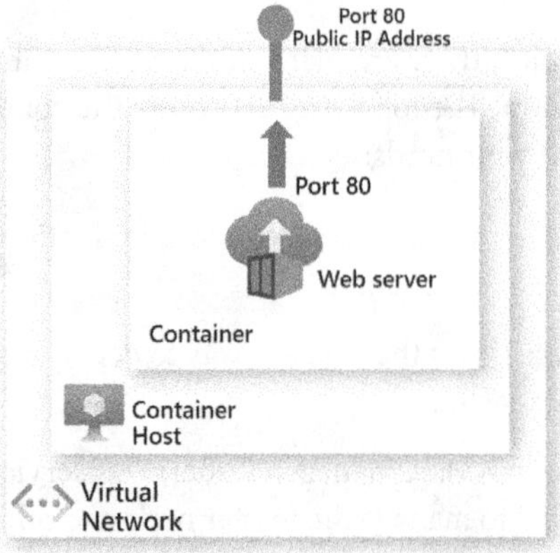
Port 80
Public IP Address
Port 80
Web server
Container
Container
Host
Virtual
Network

Figure 7.3 – Azure Container Instances

By using the ACI service, you don't need to manage virtual machines; ACI is a great solution where you can operate isolated containers and build applications, automation, and build jobs.

If we try to summarize the most relevant features of ACI, it includes the following:

- **Hypervisor-level security**: Applications will be isolated as they would be in a virtual machine
- **Public IP connectivity and DNS names**: You can expose the IP address to the internet and configure FQDNs
- **Custom size**: A contained node can be scaled dynamically on demand
- **Persistent storage**: ACI supports the direct mounting of Azure file shares
- **Co-scheduled groups**: Your container instances can be scheduled to share host machine resources
- **Virtual network integration**: Container instances can be deployed into an Azure virtual network

Consider the scenario where you work for an online clothing retailer that adopted containers for internal applications. These applications run on-premises and in Azure and can share the underlying hardware but shouldn't access resources by other applications. The company relies on you to manage these containers.

Note

ACI is not an ideal solution for long-running applications; it is better suited for batch-like tasks. ACI provides a lightweight and straightforward way to run containers without managing underlying infrastructure such as virtual machines. However, it lacks features for scalability, high availability, and cost optimization, making it less suitable for long-running apps. For example, if you have an application that requires continuous uptime, automatic scaling, or load balancing, ACI might not be the best choice. Instead, services such as AKS offer more robust features for managing long-running applications, including application life cycle management, scaling, and fault tolerance.

Note that ACI has a top-level resource called a container group. Think of a container group as a collection of containers scheduled on the same host machine.

In this context, containers in a container group will share a life cycle, resources, a network, and storage volumes. It is similar to the concept of a Pod in Kubernetes.

The following figure shows an example of a container group.

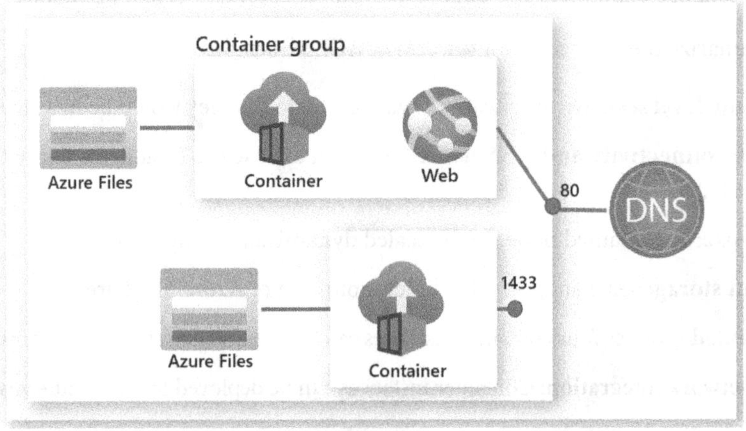
Container group
Azure Files
Container
Web
80
DNS
1433
Azure Files
Container

Figure 7.4 – Container groups

In the preceding example, a container group is scheduled on a single host machine and is assigned a DNS name label. Note that it exposes a public IP address with one exposed port. This container group consists of two containers: one listens on port `80` and the other on port `1433`.

The container group also includes two Azure file shares as volume mounts; each container mounts one share locally.

Consider using multiple container groups when you need to divide a single functional task into a small number of container images. Take the following example:

- A frontend and backend container, in which the frontend servers a web application, and the backend runs a service to retrieve data
- A web application in a container and another container instance pulling the latest content from a source control

Now, let's create, configure, and deploy a container using ACI.

Working with Azure Container Instances

In this exercise, we will perform the following tasks:

1. Create an Azure Container Registry instance.
2. Push a container image to Azure Container Registry.
3. Create an Azure Container Instance.
4. Deploy ACI for WebApp.

Let's begin!

Creating the Azure Container Registry instance

First, we will create an instance of **Azure Container Registry** (**ACR**), a managed registry service based on Docker Registry 2.0. Think of ACR as your private repository that allows you to share container images with your team.

You can leverage ACR to manage and build container images in Azure. You can pull images from ACR and target other Azure services, such as AKS, App Service, Batch, and Service Fabric.

Consider using ACR Tasks, a suite of ACR features, to automate the container image builds, updates to a container's base image, or timers. ACR Tasks support various scenarios to build and maintain your container images:

- **Quick tasks**: You can build and push a container image on demand. Think of the equivalent `docker build` and `docker push` commands for Azure.
- **Automatically triggered tasks**: You can configure triggers to build an image, for example, trigger on source code updates, on base image updates, or based on a schedule.
- **Multi-step tasks**: You can leverage the image build and push capabilities based on workflows.

ACR is available in multiple service tiers, each designed to meet specific needs: *Basic*, *Standard*, and *Premium*. The Basic tier serves as a cost-effective option for developers starting with ACR, while the Standard tier offers increased storage and image throughput for most production scenarios. The Premium tier provides the highest storage capacity and concurrent operations, catering to high-volume scenarios and offering advanced features such as geo-replication, content trust, and private links with private endpoints for enhanced security and accessibility:

1. As the first step, go to the Azure portal at `https://portal.azure.com` and look for the **Container registries** service, as shown:

container registry
All
Services (6)
Resource Groups (0)
Services
Container registries
Container Apps

Figure 7.5 – ACR – Azure portal

2. Now, in the **Basic** tab, we will provide the values for the resource group and registry names along with the location and SKU values. Note that the registry name must be unique. Then, select **Review + create**:

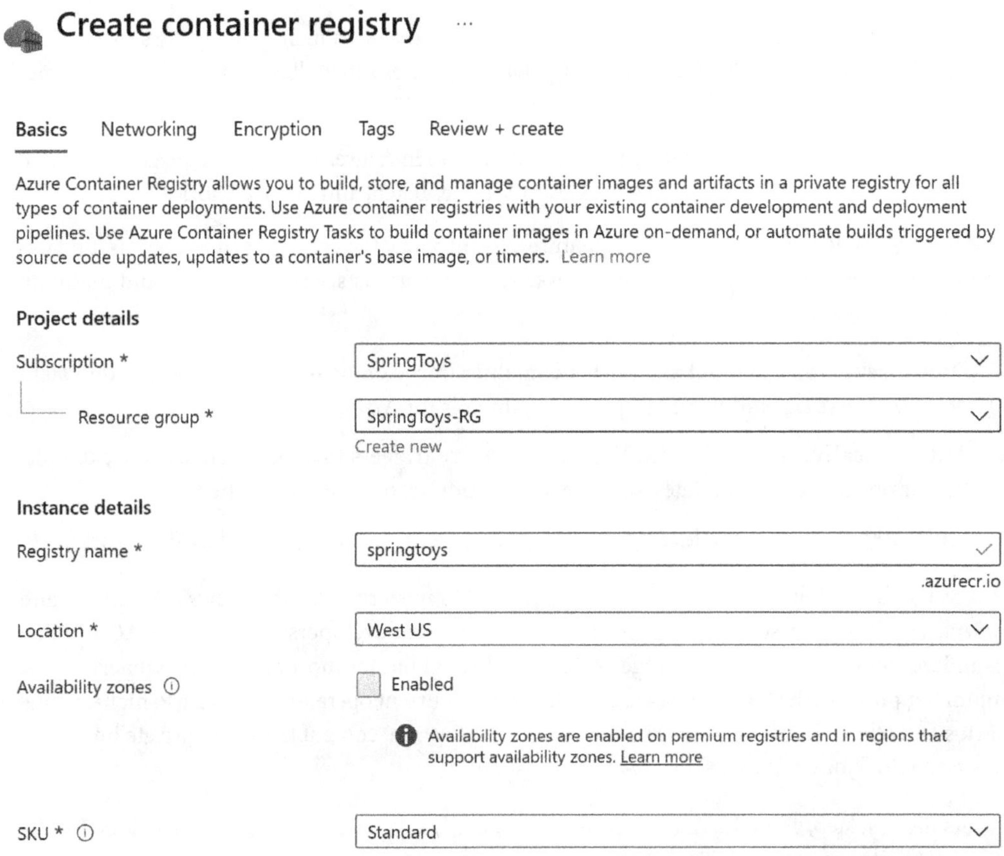
Create container registry
Basics
Networking
Encryption
Tags
Review + create
Azure Container Registry allows you to build, store, and manage container images and artifacts in a private registry for all types of container deployments. Use Azure container registries with your existing container development and deployment pipelines. Use Azure Container Registry Tasks to build container images in Azure on-demand, or automate builds triggered by source code updates, updates to a container's base image, or timers. Learn more
Project details
Subscription *
SpringToys
Resource group *
SpringToys-RG
Create new
Instance details
Registry name *
springtoys
.azurecr.io
Location *
West US
Availability zones
Enabled
Availability zones are enabled on premium registries and in regions that support availability zones. Learn more
SKU *
Standard

Figure 7.6 – ACR – Basics tab

3. Once the deployment completes, you will see the **Login server** details in the container registry, as shown:

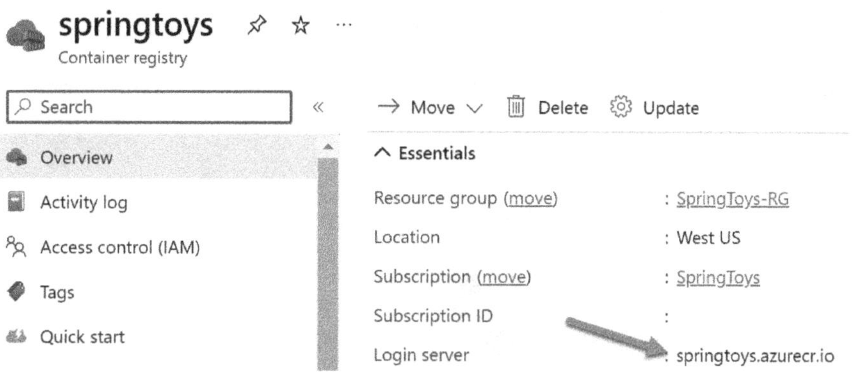
springtoys
Container registry
Search
Move
Delete
Update
Overview
Activity log
Access control (IAM)
Tags
Quick start
Essentials
Resource group (move)
: SpringToys-RG
Location
: West US
Subscription (move)
: SpringToys
Subscription ID
:
Login server
: springtoys.azurecr.io

Figure 7.7 – ACR – Overview tab

From now on, we can use ACR as our repository for container images. Before we can push and pull container images, we need to log in to the registry instance and enable the admin user.

4. Go to **Access keys** under the **Settings** section and *enable* the admin user. Once they are enabled, you will see a notification and two passwords, which can be regenerated, as shown:

Dashboard > Microsoft.ContainerRegistry | Overview > springtoys
springtoys | Access keys
Container registry
Successfully enabled admin user
Successfully enabled admin user
Search
Overview
Activity log
Access control (IAM)
Tags
Quick start
Events
Settings
Access keys
Encryption
Registry name
springtoys
Login server
springtoys.azurecr.io
Admin user
Enabled
Username
springtoys
Name
Password
Regenerate
password
password2
1
2
3

Figure 7.8 – ACR – Access keys tab

ACR offers robust support for implementing security practices within your organization by enabling **role-based access control** (**RBAC**). With RBAC, you can assign specific permissions to different users, service principals, or identities based on their roles and responsibilities.

For instance, you can grant push permissions to a service principal utilized in a build pipeline, while assigning pull permissions to a separate identity used for deployment purposes.

Now we can start working with our container registry. This completes task 1. So, let's proceed to push a container image to the container registry.

> **Note**
>
> For the upcoming task, it is recommended to install Docker. You can get it from `https://docs.docker.com/get-docker/`.

Pushing a container image to ACR

In this task, we will push a local container image to ACR:

1. First, we will log in to the container registry with the following command:

```
az acr login --name <registry-name>
```

   In this case, it is `az acr login --name springtoys'`, as shown in the following screenshot:

0.003s
❯ az acr login --name springtoys
Login Succeeded

Figure 7.9 – ACR – login

2. The next step is to download an image from Docker Hub. We will use the following command to download the container image:

```
docker pull daverendon/linux_tweet_app
```

   The following figure shows the process:

❯ docker pull daverendon/linux_tweet_app
Using default tag: latest
latest: Pulling from daverendon/linux_tweet_app
1fe172e4850f: Pull complete
35c195f487df: Pull complete
213b9b16f495: Pull complete
a8172d9e19b9: Pull complete
f5eee2cb2150: Pull complete
93e404ba8667: Pull complete
7d5c680f1eb6: Pull complete
377886776c4d: Pull complete
c2d0307c6885: Pull complete
ffc6e27a13e4: Pull complete
Digest: sha256:2c68fa3e0cc9464eb68ec217b4ef6c810d56437a55431fb823c802bb35af8477
Status: Downloaded newer image for daverendon/linux_tweet_app:latest
docker.io/daverendon/linux_tweet_app:latest

Figure 7.10 – Pull container image from Docker Hub

3. Now we will push this image to ACR. To push the image, we need to tag the container with the FQDN of the registry login server. We will use the following command to apply the tag:

```
docker tag daverendon/linux_tweet_app springtoys.azurecr.io/
springtoys-app:v1
```

Note

We are referencing the registry login server name followed by the container name and tag in the format `<login-server>/your-container-name:v1`.

4. Then, we will push the image to the container registry using the following command:

```
docker push springtoys.azurecr.io/springtoys-app:v1
```

The following figure shows the push process:

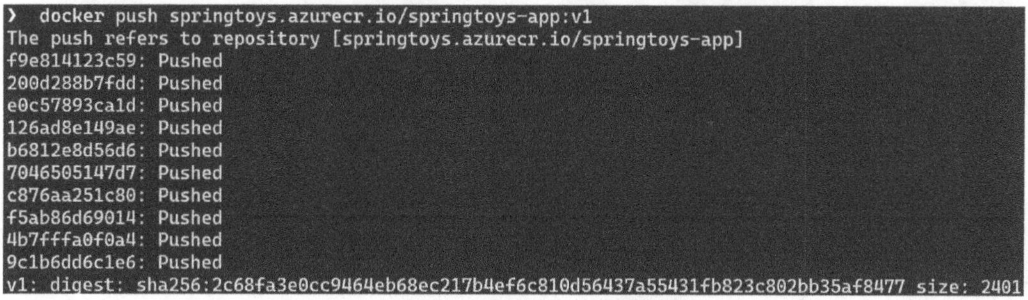
› docker push springtoys.azurecr.io/springtoys-app:v1
The push refers to repository [springtoys.azurecr.io/springtoys-app]
f9e814123c59: Pushed
200d288b7fdd: Pushed
e0c57893ca1d: Pushed
126ad8e149ae: Pushed
b6812e8d56d6: Pushed
7046505147d7: Pushed
c876aa251c80: Pushed
f5ab86d69014: Pushed
4b7fffa0f0a4: Pushed
9c1b6dd6c1e6: Pushed
v1: digest: sha256:2c68fa3e0cc9464eb68ec217b4ef6c810d56437a55431fb823c802bb35af8477 size: 2401

Figure 7.11 – Push container image to ACR

5. Next, go to the Azure portal and select the **Repositories** option to list the images in your container registry, as shown:

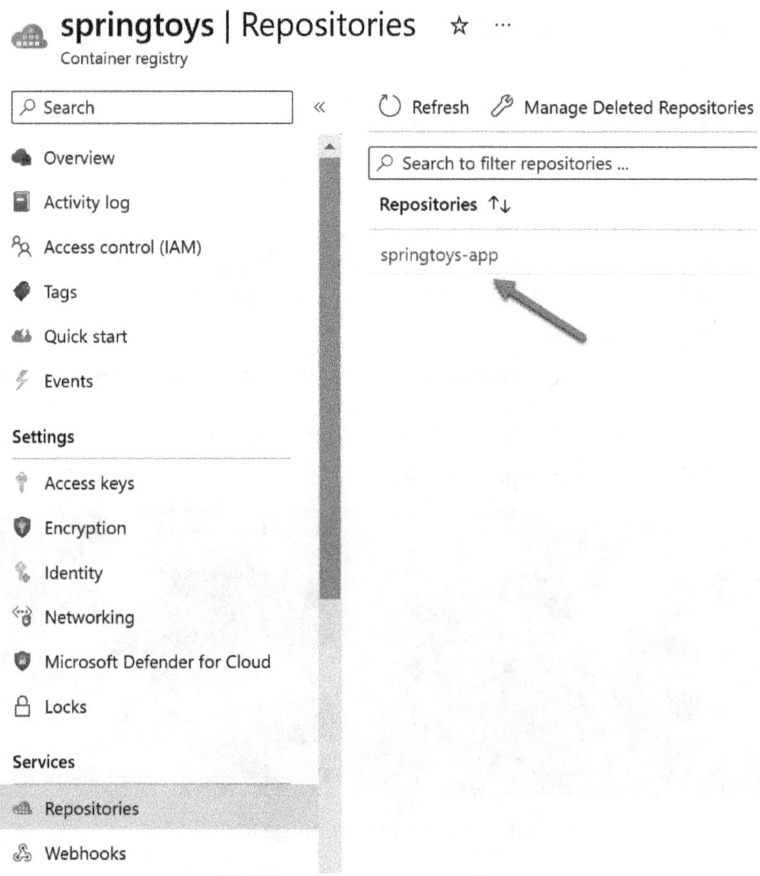
springtoys | Repositories
Container registry
Search
Overview
Activity log
Access control (IAM)
Tags
Quick start
Events
Settings
Access keys
Encryption
Identity
Networking
Microsoft Defender for Cloud
Locks
Services
Repositories
Webhooks
Refresh
Manage Deleted Repositories
Search to filter repositories ...
Repositories
springtoys-app

Figure 7.12 – Container image in ACR

This completes task 2. Now we will create an ACI instance.

Creating an Azure Container Instance

Next, we will create an Azure Container Instance:

1. From ACR, select **Repositories** and then select the container image we just pushed in the previous task. Then, choose **v1**, and in the options menu (three dots), you will be able to run this instance, as shown:

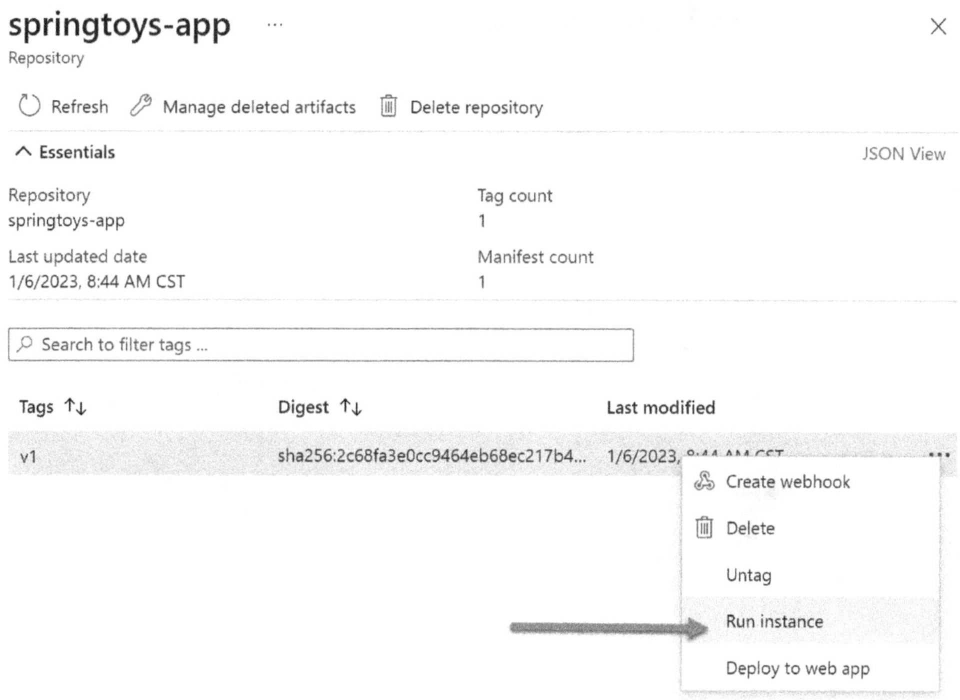
springtoys-app
Repository
Refresh
Manage deleted artifacts
Delete repository
Essentials
JSON View
Repository
springtoys-app
Tag count
1
Last updated date
1/6/2023, 8:44 AM CST
Manifest count
1
Search to filter tags ...
Tags
Digest
Last modified
v1
sha256:2c68fa3e0cc9464eb68ec217b4...
Create webhook
Delete
Untag
Run instance
Deploy to web app

Figure 7.13 – Container image in ACR – Run instance

This option will trigger the page to create an Azure Container Instance.

2. Let's provide the parameters for the creation of the container instances, as shown in the following figure:

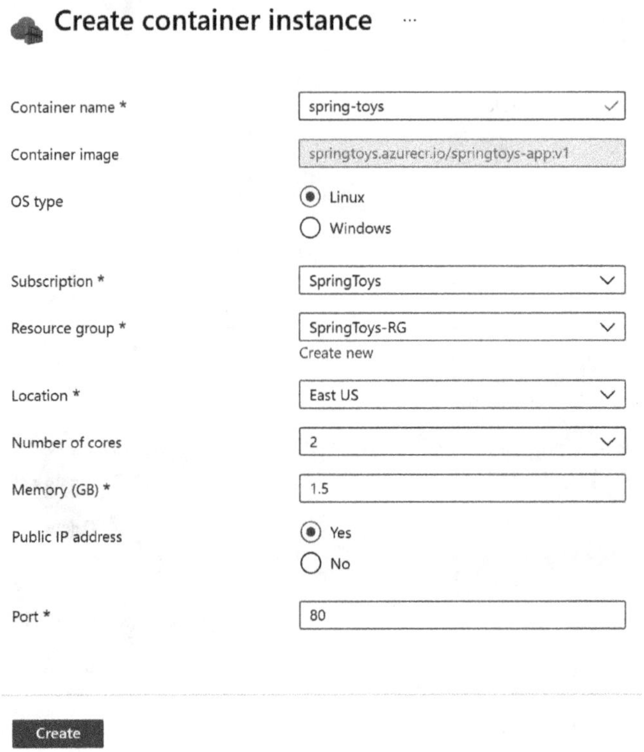
Create container instance
Container name *
spring-toys
Container image
springtoys.azurecr.io/springtoys-app:v1
OS type
Linux
Windows
Subscription *
SpringToys
Resource group *
SpringToys-RG
Create new
Location *
East US
Number of cores
2
Memory (GB) *
1.5
Public IP address
Yes
No
Port *
80
Create

Figure 7.14 – Configuration of ACI

3. Once you create the container instance, after a few moments, you should see the deployment completed. In the **Overview** tab, you will see the details of the container instance, including the public IP address, as shown in the following screenshot.

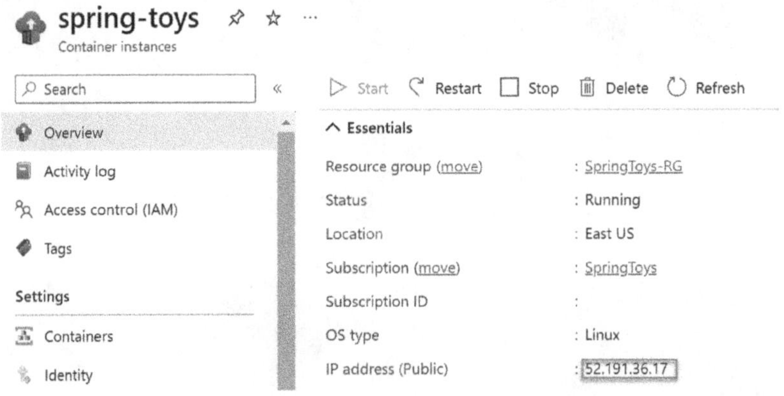
spring-toys
Container instances
Search
Start
Restart
Stop
Delete
Refresh
Overview
Activity log
Access control (IAM)
Tags
Settings
Containers
Identity
Essentials
Resource group (move)
: SpringToys-RG
Status
: Running
Location
: East US
Subscription (move)
: SpringToys
Subscription ID
:
OS type
: Linux
IP address (Public)
: 52.191.36.17

Figure 7.15 – ACI – Overview page

4. In your preferred browser, access the containerized application using the public IP address, and you should see the following screen:

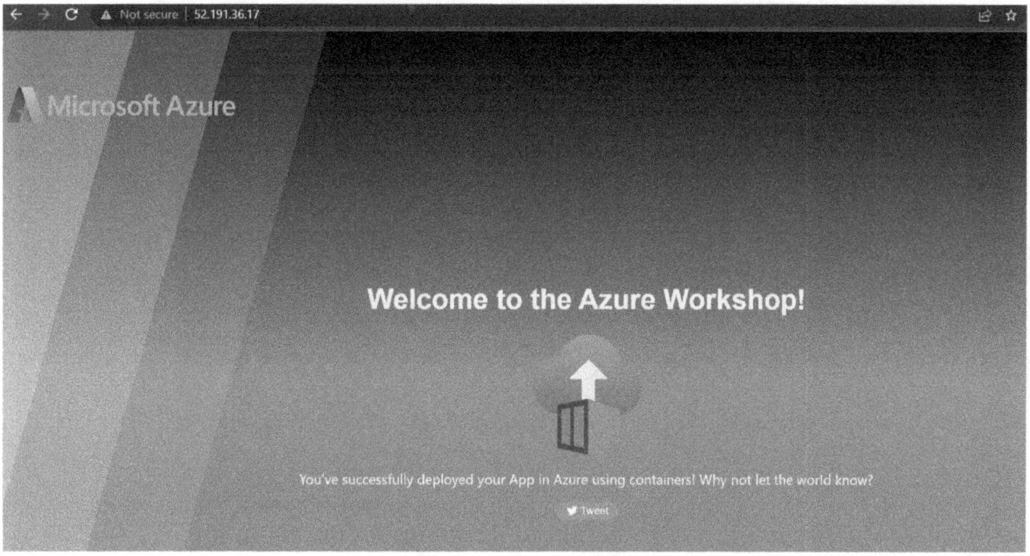
Not secure | 52.191.36.17
Microsoft Azure
Welcome to the Azure Workshop!
You've successfully deployed your App in Azure using containers! Why not let the world know?
Tweet

Figure 7.16 – ACI running

This completes task 3. In this task, we reviewed how to create an ACI. Now we will proceed to task 4 and deploy an Azure Container Instance for web app.

Deploying Azure Container Instance for web app

In this task, we will now deploy the Container Instance as a web application so that users can access it publicly:

1. Go to ACR, select **Repositories**, and then select the container image we just pushed in task 2. Then, choose **v1**, and in the options menu, choose to run this as a web app:

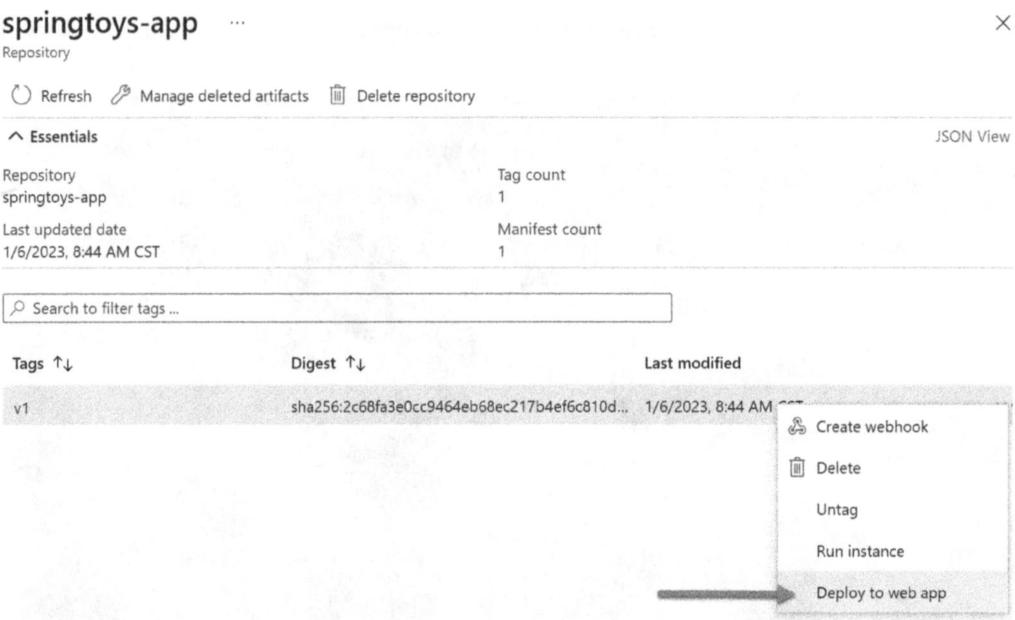
springtoys-app
Repository
Refresh
Manage deleted artifacts
Delete repository
Essentials
JSON View
Repository
springtoys-app
Tag count
1
Last updated date
1/6/2023, 8:44 AM CST
Manifest count
1
Search to filter tags ...
Tags
Digest
Last modified
v1
sha256:2c68fa3e0cc9464eb68ec217b4ef6c810d...
1/6/2023, 8:44 AM
Create webhook
Delete
Untag
Run instance
Deploy to web app

Figure 7.17 – ACR – Deploy to web app option

2. You will be taken to a page to configure the parameters to deploy this containerized application as a web application. Once done, click on **Create**:

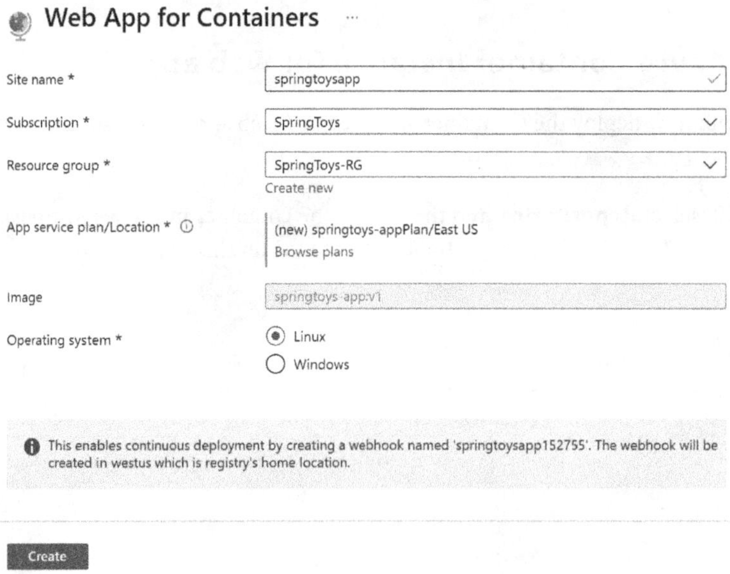
Web App for Containers
Site name *
springtoysapp
Subscription *
SpringToys
Resource group *
SpringToys-RG
Create new
App service plan/Location *
(new) springtoys-appPlan/East US
Browse plans
Image
springtoys-app:v1
Operating system *
Linux
Windows
This enables continuous deployment by creating a webhook named 'springtoysapp152755'. The webhook will be created in westus which is registry's home location.
Create

Figure 7.18 – Web App for Containers

3. Once the deployment completes, go to the resource and see its properties on the **Overview** page.

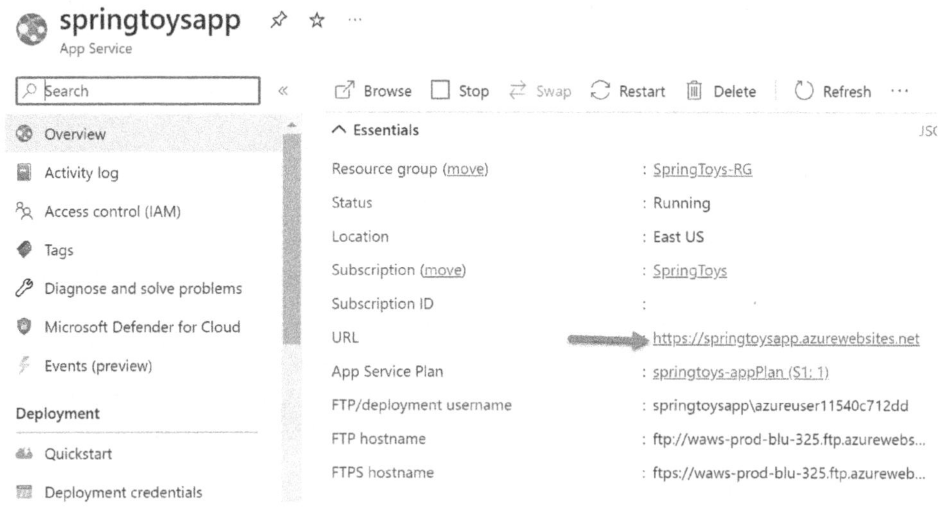
springtoysapp
App Service
Search
Browse
Stop
Swap
Restart
Delete
Refresh
Overview
Activity log
Access control (IAM)
Tags
Diagnose and solve problems
Microsoft Defender for Cloud
Events (preview)
Deployment
Quickstart
Deployment credentials
Essentials
Resource group (move) : SpringToys-RG
Status : Running
Location : East US
Subscription (move) : SpringToys
Subscription ID :
URL https://springtoysapp.azurewebsites.net
App Service Plan : springtoys-appPlan (S1: 1)
FTP/deployment username : springtoysapp\azureuser11540c712dd
FTP hostname : ftp://waws-prod-blu-325.ftp.azurewebs...
FTPS hostname : ftps://waws-prod-blu-325.ftp.azureweb...

Figure 7.19 – Web App for Containers – Overview tab

4. Now, let's verify that the application is running. In your preferred browser, open the web application URL, as shown in the following screenshot.

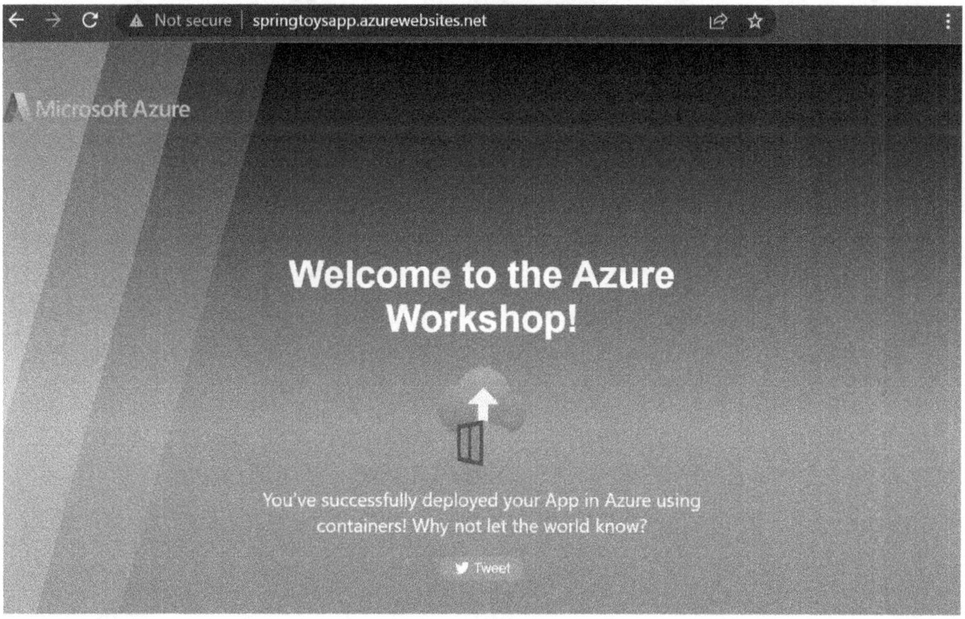
Not secure
springtoysapp.azurewebsites.net
Microsoft Azure
Welcome to the Azure Workshop!
You've successfully deployed your App in Azure using containers! Why not let the world know?
Tweet

Figure 7.20 – Web App for Containers

As you can see, Web App for Containers in Azure allows you to deploy and run your web applications inside containers, which are like packages that contain all the code and dependencies needed for your application to run, making it easy to manage and scale your application.

Now, let's explore a serverless offering to work with containers in Azure: Azure Container Apps.

Creating Azure Container Apps

Another option organizations can leverage for building and deploying containerized applications is Azure Container Apps. This fully managed serverless container service runs on top of Kubernetes infrastructure.

Azure Container Apps is a powerful service that takes care of the underlying cloud infrastructure, allowing you to focus solely on your applications. With event-driven scalability, you can even scale down to zero instances, eliminating costs when not in use. However, when high demand arises, you can effortlessly scale up to thousands of instances to handle the increased workload.

The primary purpose of Azure Container Apps is to facilitate how you build and deploy containerized applications and have the benefits of PaaS for scaling without managing Kubernetes nodes. In this approach, a developer can deploy containers as individual container applications using the Azure portal, CLI, or infrastructure as code, and there's no management from a Kubernetes standpoint.

You can build microservices, API endpoints, web applications, event-driven applications, or event background processing containers. In short, Azure Container Apps will help you deploy your containerized applications on top of Kubernetes without the need to manage the Kubernetes cluster.

Azure Container Apps include a rich set of features:

- Autoscaling
- Container revisions
- HTTP ingress
- Traffic management
- Service discovery
- Integration with multiple registries
- Integration with Dapr
- Compatibility with any runtime, programming language, and development stack

The following diagram highlights the Azure Container Apps architecture:

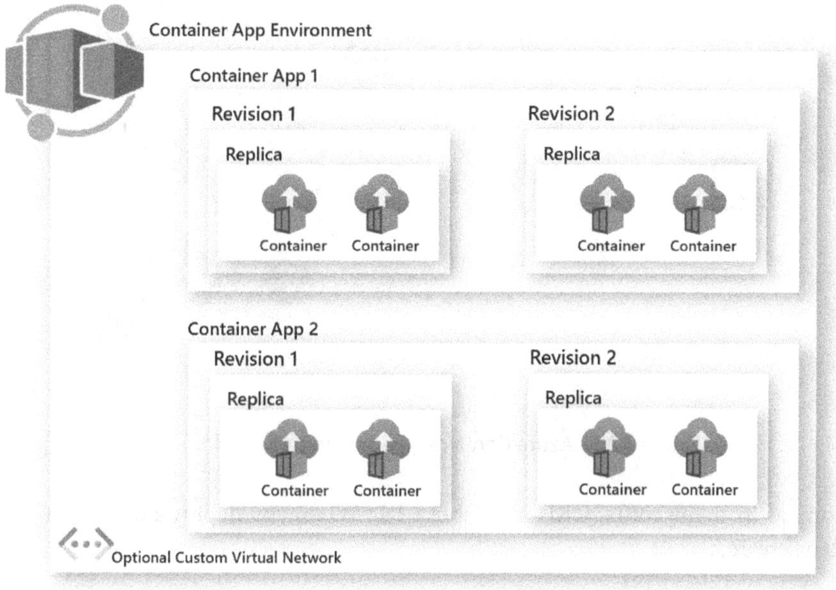
Container App Environment
Container App 1
Revision 1
Replica
Container
Container
Revision 2
Replica
Container
Container
Container App 2
Revision 1
Replica
Container
Container
Revision 2
Replica
Container
Container
Optional Custom Virtual Network

Figure 7.21 – Azure Container Apps architecture

Before deploying our application, we will need a container and an environment ready. Think of a Container Apps environment as a security boundary for groups of container apps.

Consider using the same environment for your container apps in the following cases:

- Multiple services must communicate in the same virtual network
- Services will share the same log analytics workspace for monitoring purposes
- Dapr integrations

To control versioning in Azure Container Apps, you can leverage revisions. This is a snapshot of your container app version and will help you manage releases of your application. A new revision is created each time you make a change to the application.

Let's say you have version 1.0, and then you publish version 1.1 of your application. You will have two revisions, one per release. You can also split traffic between revisions for testing purposes by using feature flags, as shown in the following figure:

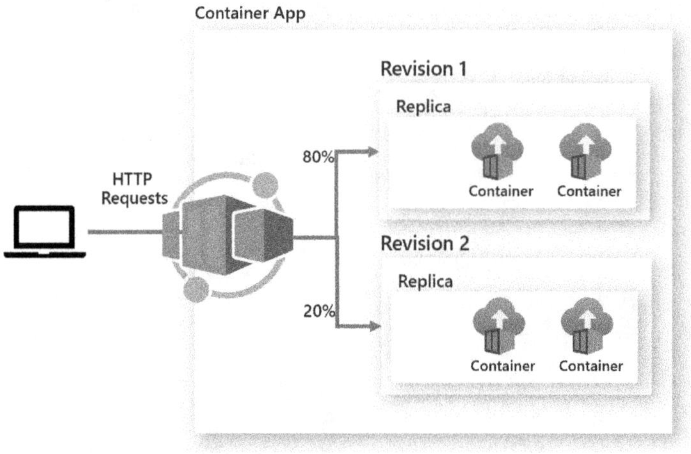
Container App
Revision 1
Replica
80%
HTTP
Requests
Container
Container
Revision 2
Replica
20%
Container
Container

Figure 7.22 – Azure Container Apps with two revisions

Each revision has its name suffix, URL, and label URL. When you deploy a container app, a first revision will be created.

The life cycle of a container app has four stages: deployment, update, deactivation, and shutdown.

Now, let's create a container app:

1. Go to the Azure portal and search for **Container Apps**, as shown in the following screenshot:

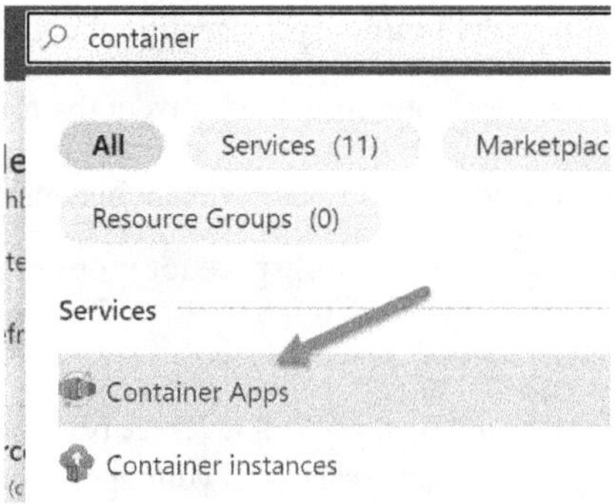
container
All
Services (11)
Marketplac
Resource Groups (0)
Services
Container Apps
Container instances

Figure 7.23 – Azure portal – search for Container Apps

2. Now we will provide the values to create our container app:

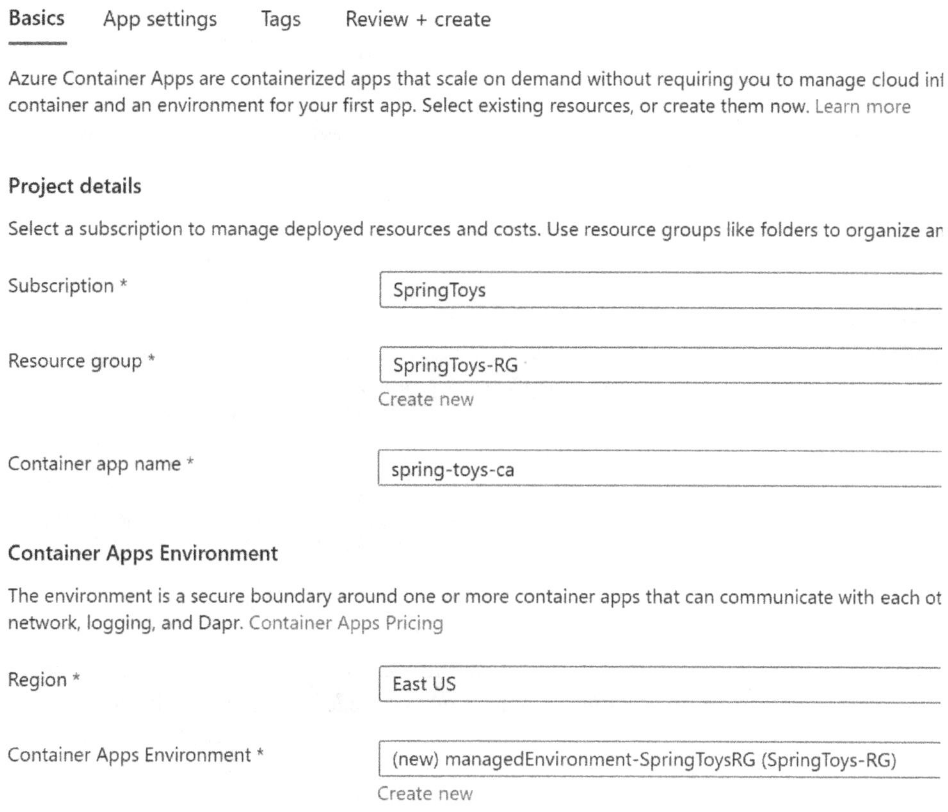
Basics
App settings
Tags
Review + create
Azure Container Apps are containerized apps that scale on demand without requiring you to manage cloud inf
container and an environment for your first app. Select existing resources, or create them now. Learn more
Project details
Select a subscription to manage deployed resources and costs. Use resource groups like folders to organize ar
Subscription *
SpringToys
Resource group *
SpringToys-RG
Create new
Container app name *
spring-toys-ca
Container Apps Environment
The environment is a secure boundary around one or more container apps that can communicate with each ot
network, logging, and Dapr. Container Apps Pricing
Region *
East US
Container Apps Environment *
(new) managedEnvironment-SpringToysRG (SpringToys-RG)
Create new

Figure 7.24 – Azure Container Apps – Basics tab

Note we are creating a new Container Apps environment. The following figure shows the configuration of the environment:

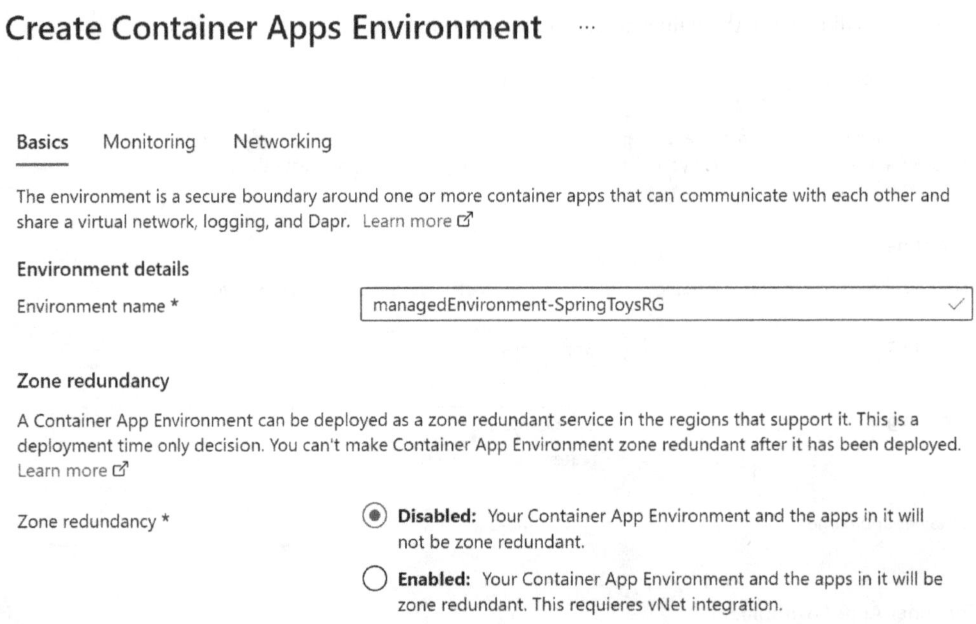
Create Container Apps Environment ...
Basics
Monitoring
Networking
The environment is a secure boundary around one or more container apps that can communicate with each other and share a virtual network, logging, and Dapr. Learn more
Environment details
Environment name *
managedEnvironment-SpringToysRG
Zone redundancy
A Container App Environment can be deployed as a zone redundant service in the regions that support it. This is a deployment time only decision. You can't make Container App Environment zone redundant after it has been deployed. Learn more
Zone redundancy *
Disabled: Your Container App Environment and the apps in it will not be zone redundant.
Enabled: Your Container App Environment and the apps in it will be zone redundant. This requieres vNet integration.

Figure 7.25 – Azure Container Apps environment – configuration

3. Next, we will configure the app settings. In this tab, we will use Docker as an image source and specify the container image and the compute resources needed to run this application, as shown:

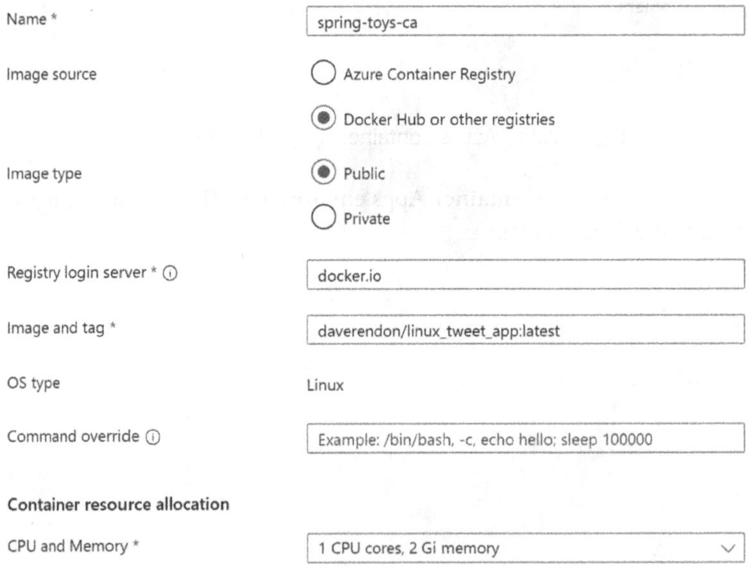
Name *
spring-toys-ca
Image source
Azure Container Registry
Docker Hub or other registries
Image type
Public
Private
Registry login server *
docker.io
Image and tag *
daverendon/linux_tweet_app:latest
OS type
Linux
Command override
Example: /bin/bash, -c, echo hello; sleep 100000
Container resource allocation
CPU and Memory *
1 CPU cores, 2 Gi memory

Figure 7.26 – Azure Container Apps – app settings

4.   In addition, you can specify environment variables and the application ingress settings. In this case, we will enable ingress and set the ingress type as HTTP and the target port as 80, as shown:

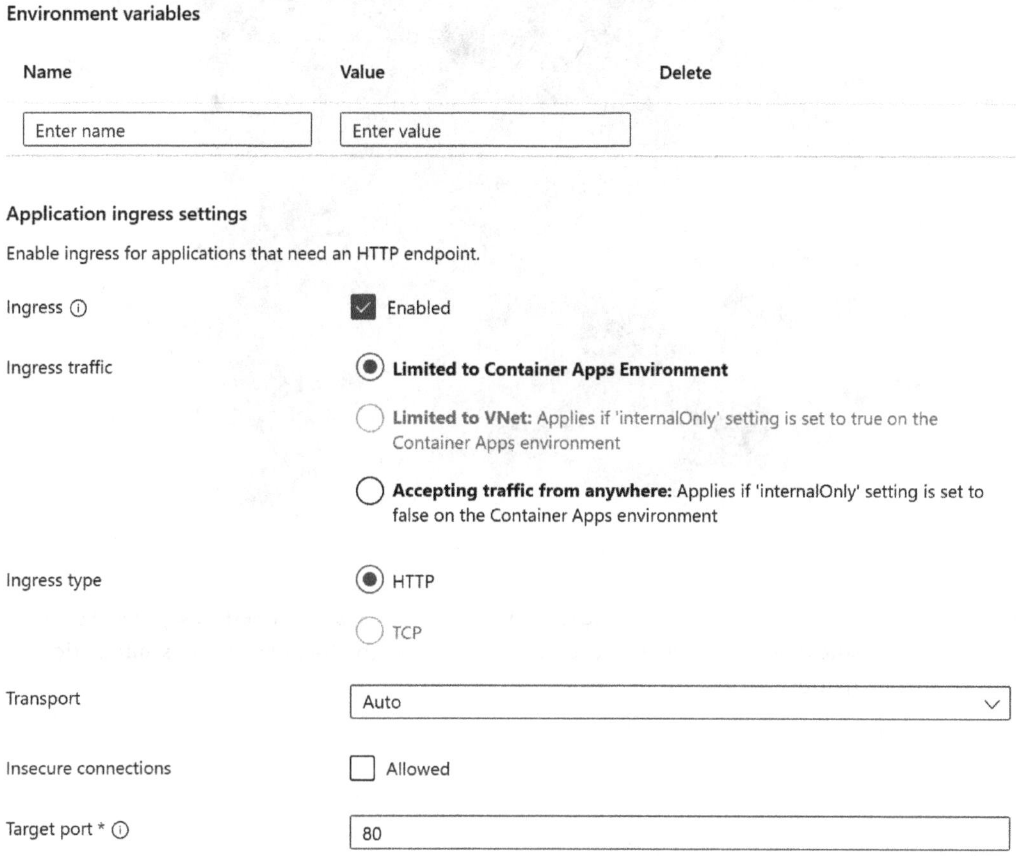

Environment variables

Name	Value	Delete
Enter name	Enter value	

Application ingress settings

Enable ingress for applications that need an HTTP endpoint.

Ingress ⓘ	✓ Enabled
Ingress traffic	● **Limited to Container Apps Environment**
	○ **Limited to VNet:** Applies if 'internalOnly' setting is set to true on the Container Apps environment
	○ **Accepting traffic from anywhere:** Applies if 'internalOnly' setting is set to false on the Container Apps environment
Ingress type	● HTTP
	○ TCP
Transport	Auto ⌄
Insecure connections	☐ Allowed
Target port * ⓘ	80

Figure 7.27 – Azure Container Apps – app settings

> **Note**
>
> You can limit the traffic to the Azure Container Apps environment or accept traffic from anywhere. If you want to test this example, set the **Ingress traffic** option to **Accepting traffic from anywhere**. Now we will proceed to create this resource.

5. After a few minutes, the container app should be running, and you can access it through the browser using the URL of the container app, as shown:

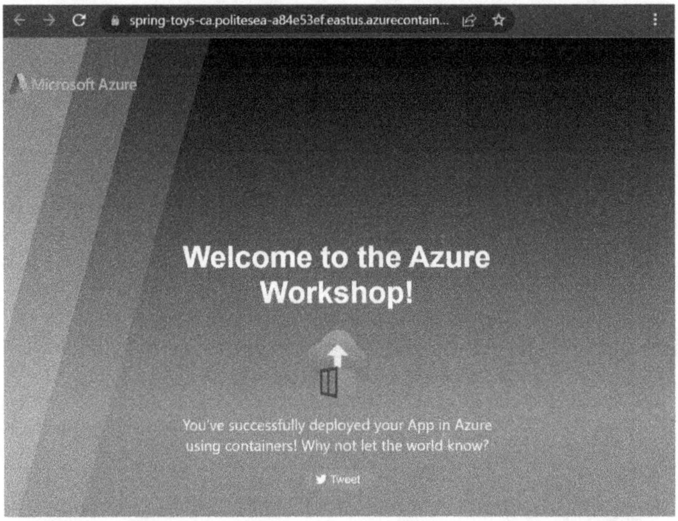
spring-toys-ca.politesea-a84e53ef.eastus.azurecontain...
Microsoft Azure
Welcome to the Azure Workshop!
You've successfully deployed your App in Azure using containers! Why not let the world know?
Tweet

Figure 7.28 – Azure Container Apps

6. Go back to the Azure portal, and in the Container Apps options under **Settings**, you will see the revisions. Note that by default, a first revision of the container deployment was automatically created by Azure.

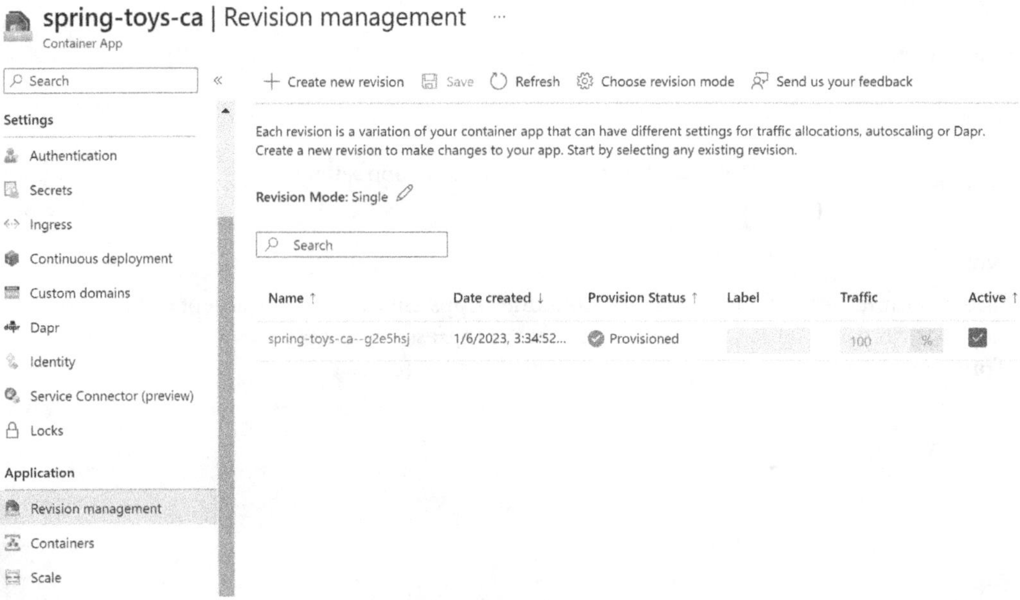
spring-toys-ca | Revision management
Container App
Search
Settings
Authentication
Secrets
Ingress
Continuous deployment
Custom domains
Dapr
Identity
Service Connector (preview)
Locks
Application
Revision management
Containers
Scale
Create new revision
Save
Refresh
Choose revision mode
Send us your feedback
Each revision is a variation of your container app that can have different settings for traffic allocations, autoscaling or Dapr.
Create a new revision to make changes to your app. Start by selecting any existing revision.
Revision Mode: Single
Search
Name
Date created
Provision Status
Label
Traffic
Active
spring-toys-ca--g2e5hsj
1/6/2023, 3:34:52...
Provisioned
100
%

Figure 7.29 – Azure Container Apps – revisions

7. Under the **Containers** option, you will see the properties of the container image. You can modify the deployment and allocate more computing resources to it if needed:

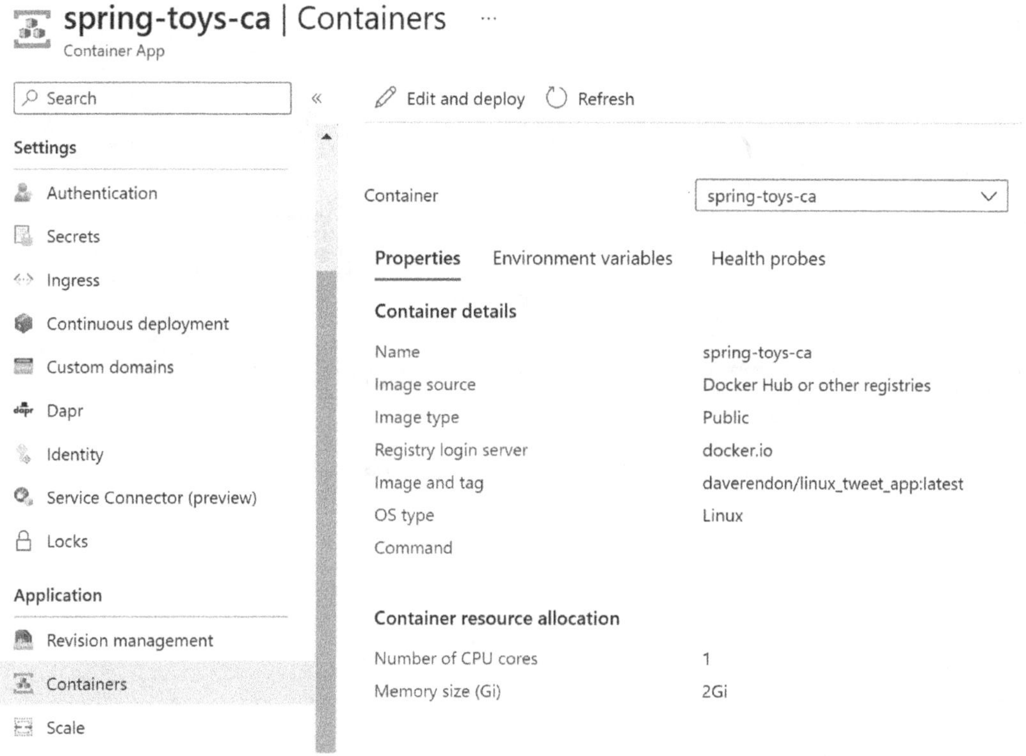
spring-toys-ca | Containers
Container App
Search
Edit and deploy
Refresh
Settings
Authentication
Secrets
Ingress
Continuous deployment
Custom domains
Dapr
Identity
Service Connector (preview)
Locks
Application
Revision management
Containers
Scale
Container
spring-toys-ca
Properties
Environment variables
Health probes
Container details
Name
spring-toys-ca
Image source
Docker Hub or other registries
Image type
Public
Registry login server
docker.io
Image and tag
daverendon/linux_tweet_app:latest
OS type
Linux
Command
Container resource allocation
Number of CPU cores
1
Memory size (Gi)
2Gi

Figure 7.30 – Azure Container Apps – Containers

Additional options for deploying Azure Container Apps include the following:

- GitHub action to build and deploy container apps (`https://github.com/marketplace/actions/azure-container-apps-build-and-deploy`)
- Azure Pipelines task to build and deploy container apps
- Build container images without a Dockerfile

As you can see, Azure Container Apps offers several benefits for deploying containerized applications.

In short, the following figure shows what you can build with Azure Container Apps and the autoscale criteria:

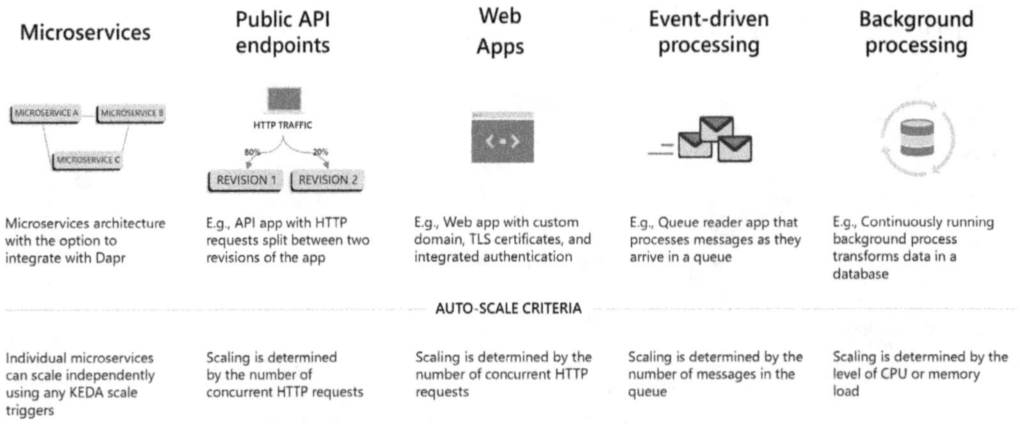
Microservices
Public API endpoints
Web Apps
Event-driven processing
Background processing
MICROSERVICE A
MICROSERVICE B
MICROSERVICE C
HTTP TRAFFIC
80%
20%
REVISION 1
REVISION 2
Microservices architecture with the option to integrate with Dapr
E.g., API app with HTTP requests split between two revisions of the app
E.g., Web app with custom domain, TLS certificates, and integrated authentication
E.g., Queue reader app that processes messages as they arrive in a queue
E.g., Continuously running background process transforms data in a database
AUTO-SCALE CRITERIA
Individual microservices can scale independently using any KEDA scale triggers
Scaling is determined by the number of concurrent HTTP requests
Scaling is determined by the number of concurrent HTTP requests
Scaling is determined by the number of messages in the queue
Scaling is determined by the level of CPU or memory load

Figure 7.31 – Azure Container Apps – use cases

Some common use cases include the following:

- Running microservices
- Deploying public API endpoints
- Hosting web apps
- Handling event-driven processing
- Hosting background processing applications

With each of those applications, you can dynamically scale based on the following:

- HTTP traffic
- Event-driven processing
- CPU or memory load
- Any KEDA-supported scaler

One major advantage of Azure Container Apps is its ease of use, providing a quick and easy way to deploy and manage containerized applications. Additionally, Azure Container Apps is a fully managed platform, meaning that users are relieved from the burden of infrastructure management, allowing them to concentrate on their applications instead. Another important benefit is the flexibility that Azure Container Apps provides as it supports multiple containerization technologies, allowing users to choose the technology that best fits their application.

Summary

Hopefully, you have a better understanding of Azure containers and can identify when to use Azure containers versus Azure virtual machines, as well as identify the features and usage cases of ACI and how to implement Azure container groups.

This chapter also reviewed the features and benefits of ACR and how you can leverage ACR Tasks to automate builds and deployments. ACI is a great offer for any scenario operating in isolated containers.

We also reviewed Azure Container Apps, its components, and how you can use this service to run microservices on a serverless platform.

In the next chapter, we will discuss how Azure networking services can help organizations optimize their network infrastructure, making it more efficient, reliable, and secure, and ultimately supporting business growth and innovation.

Further reading

- *Deploy Azure Container Apps using Bicep Language*: `https://github.com/daveRendon/azinsider/tree/main/application-workloads/azure-container-apps`
- *Working with container images and Azure Container Registry*: `https://github.com/daveRendon/azinsider/tree/main/application-workloads/deploy-container-image-to-azure-container-registry`
- *Deploy a Cognitive Services container image for Text Analytics*: `https://github.com/daveRendon/azinsider/tree/main/application-workloads/create-cognitive-services-text-analytics-container`

8
Understanding Networking in Azure

An organization is in the process of migrating infrastructure and applications to Azure. As part of the networking team, you are assigned to be part of this project. You need to allow employees to securely access resources such as databases, files, and applications on-premises and in Azure.

This chapter will review how to design and implement the core Azure networking infrastructure, including hybrid networking connections, load balancing components, routing, and network security.

Starting with the basics, we will understand the core Azure networking services such as virtual networks, DNS, and virtual network peering. We will also discuss various topics so that you can learn how to design and implement hybrid network connections such as Site-to-Site, Point-to-Site, and ExpressRoute.

Then, we will move on to how you can implement routing and load balancing components and secure your networks.

If we want to have a holistic view of all Azure networking components, they fall into one of the following categories:

- **Connectivity**: How you leverage Azure services to extend your on-premises network to Azure
- **Application protection**: How you use Azure services to protect your applications
- **Application delivery**: How you deliver applications reliably
- **Network monitoring**: How you can monitor and troubleshoot your network resources

We will start with the first pillar: connectivity. And in this category, we will explore Azure virtual networks, how to configure public IP services, DNS, and cross-virtual network connectivity.

Connectivity in Azure

The fundamental building block of your network in Azure is Azure **Virtual Networks** or **VNets**, as we will refer to from now on. VNets allow organizations to build networks as if they were on-premises. Think of VNets as the primary component for availability and isolation purposes.

On top of VNets, you can configure and manage **Virtual Private Networks** or **VPNs** and connect with other VNets across the various regions available in Azure and extend your on-premises network. Similar to an on-premises network, each VNet has its own **Classless Inter-Domain Routing** or **CIDR** block, and they can be connected with other VNets as long as their CIDR blocks don't overlap.

Similar to on-premises networks, TCP, UDP, and ICMP TCP/IP protocols are supported within VNets.

Now we will discuss how you can start designing your virtual networks in Azure.

Design considerations for VNets

Before creating a VNet, consider using address ranges enumerated in RFC 1918 to facilitate the expansion of the usable number of IP addresses. Those ranges include the following:

- `10.0.0.0` - `10.255.255.255` (`10/8` prefix)
- `172.16.0.0` - `172.31.255.255` (`172.16/12` prefix)
- `192.168.0.0` - `192.168.255.255` (`192.168/16` prefix)

> **Note**
>
> You can deploy a shared address space reserved in RFC 6598, treated as a private IP address space in Azure: `100.64.0.0` - `100.127.255.255` (`100.64/10` prefix)

Some address ranges cannot be added as Azure uses them for specific purposes; these are the following:

- `224.0.0.0/4` (multicast)
- `255.255.255.255/32` (broadcast)
- `127.0.0.0/8` (loopback)
- `169.254.0.0/16` (link-local)
- `168.63.129.16/32` (internal DNS)

Conversely, Azure subnets are a part of the Azure virtual network that allows you to segment your network resources and allocate IP addresses to different resources. Subnets enable you to group and isolate resources, control network traffic flow, and implement security policies, making it easier to manage and secure your virtual network.

Subnets

When creating VNets, you can segment them into different subnets and create as many subnets as you need. The smallest supported IPv4 subnet is `/29`, and the largest is `/2`. IPv6 subnets must be exactly `/64` in size.

Each subnet must have its unique address range in CIDR format, and you can use them to limit access to Azure resources using service endpoints. Some Azure services, such as Azure Firewall and Application Gateway, require a dedicated subnet.

Azure reserves five IP addresses within each subnet – the first four and the last IP address. For example, we can have the IP address range `X.X.X.0/24`. This address range will have the following reserved addresses:

- `X.X.X.0`: Network address
- `X.X.X.1`: Reserved by Azure for the default gateway
- `X.X.X.2`, `X.X.1.3`: Reserved by Azure to map the Azure DNS IPs to the VNet space
- `X.X.X.255`: Network broadcast address

Network security groups

Imagine a virtual network with a single subnet and multiple virtual machines. You need to allow traffic from a specific IP address range to reach a virtual machine on a certain port and protocol while blocking all other traffic to that VM.

In that scenario, you can use **Network Security Groups** or **NSGs** to secure access to your Azure resources. You can customize the rules to allow or deny traffic from and to the Azure resources.

Think of an NSG as an **Access Control List** (**ACL**), where you list rules that specify which resources are granted or denied access to a particular object or system resource.

For example, let's assume a company named SpringToys has a virtual network that contains a single subnet with several **Virtual Machines** (**VMs**) and a jumpbox virtual machine. The jumpbox VM works as a management VM for the rest of the VMs. SpringToys wants to allow only specific IP addresses to access the jumpbox and specific ports to be open on the VMs.

You could create an NSG and associate it with the subnet that contains the jumpbox virtual machine. With this configuration, you could allow traffic from a specific address range to access the jumpbox virtual machine using port `22` and block the rest of the traffic. You can associate NSGs with subnets or network interfaces.

Once you create an NSG, you need to configure the security rules to filter the type of network traffic flowing in and out of a VNet subnet and network interfaces.

To create a rule, you need to specify the following values:

- **Name**: The name of your NSG.
- **Priority**: Rules are processed based on priority order; the lower the number, the higher the priority. This value is between `100` and `4096`.
- **Port ranges**: You can add a single port, a port range, or comma-separated ports.
- **Protocol**: **Any**, **TCP**, **UDP**
- **Source**: **Any**, IP Addresses, Service tag
- **Destination**: **Any**, IP Addresses, Virtual Network
- **Action**: **Allow** or **Deny**
- **Service**: Specifies the destination protocol and port range for this rule. You can choose a predefined service, such as **HTTPS** or **SSH**.

The following figure highlights the values needed to configure a rule to allow traffic through `443`:

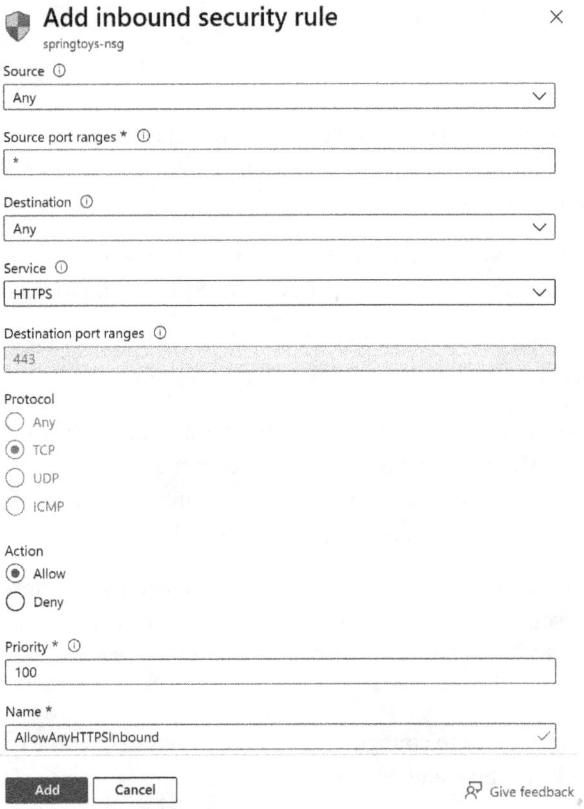
Add inbound security rule
springtoys-nsg
Source
Any
Source port ranges *
*
Destination
Any
Service
HTTPS
Destination port ranges
443
Protocol
Any
TCP
UDP
ICMP
Action
Allow
Deny
Priority *
100
Name *
AllowAnyHTTPSInbound
Add
Cancel
Give feedback

Figure 8.1 – Create inbound rule

When you create an NSG, Azure creates three inbound and three outbound default rules that cannot be removed, but you can override them by creating rules with higher priorities. As mentioned, there is the possibility to use predefined rules and custom rules as required (in the preceding example, we use a predefined rule).

The three default inbound rules deny all inbound traffic except the virtual network and Azure load balancers, as shown here:

Inbound Security Rules

65000	AllowVnetInBound	Any	Any	VirtualNetwork	VirtualNetwork	Allow
65001	AllowAzureLoadBalan...	Any	Any	AzureLoadBalancer	Any	Allow
65500	DenyAllInBound	Any	Any	Any	Any	Deny

Figure 8.2 – Default inbound rules

The three default outbound rules only allow outbound traffic to the internet and the virtual network, as shown here:

Outbound Security Rules

65000	AllowVnetOutBound	Any	Any	VirtualNetwork	VirtualNetwork	Allow
65001	AllowInternetOutBound	Any	Any	Any	Internet	Allow
65500	DenyAllOutBound	Any	Any	Any	Any	Deny

Figure 8.3 – Default outbound rules

Note that NSG rules are evaluated independently, and an *allow* rule must exist at both levels; otherwise, traffic will not be allowed. The following figure shows the inbound and outbound flow:

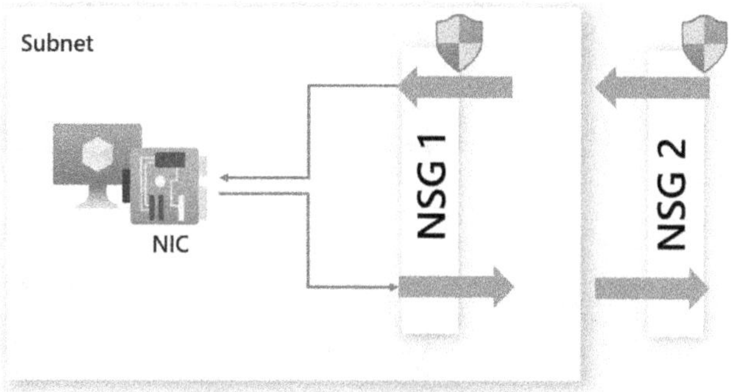
Subnet
NIC
NSG 1
NSG 2

Figure 8.4 – NSG rule precedence

In the preceding example, if there were incoming traffic on port `80`, you would need to have the NSG at the subnet level `ALLOW` port `80`. You'd also need another NSG with an `ALLOW` rule on port `80` at the NIC level.

For incoming traffic, the NSG set at the subnet level is evaluated first, then the NSG set at the NIC level is evaluated. For outgoing traffic, it is the opposite.

You can create your rules as per your needs and use application security groups to simplify a more complex configuration.

Application security groups

Instead of creating NSGs and defining policies on explicit IP Addresses, **Application Security Groups** or **ASGs** can help you group virtual machines and define network security policies based on those groups.

Instead of defining policies based on explicit IP addresses in NSGs, ASGs enable you to configure network security as a natural extension of an application's structure. You can group virtual machines and define network security policies based on those groups.

For example, you can have a group of web servers and a group of database servers and then create rules that allow only the web servers to communicate with the database servers. This makes managing security more straightforward and efficient.

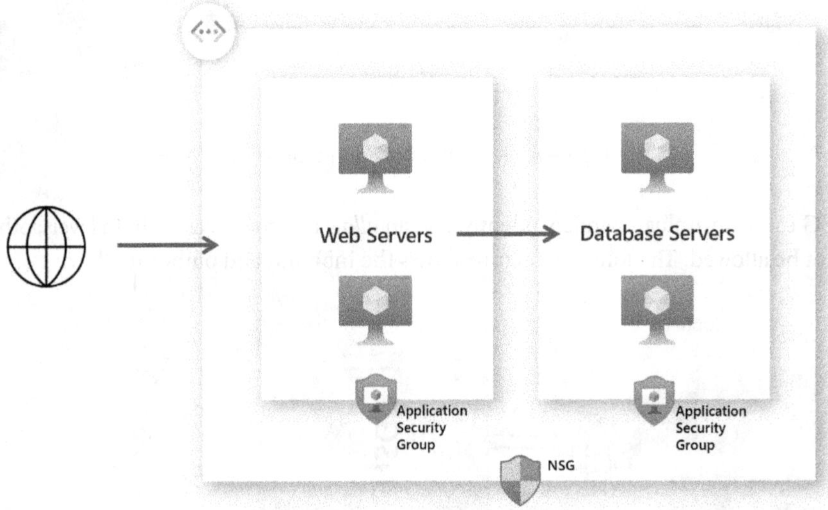
Web Servers
Database Servers
Application Security Group
Application Security Group
NSG

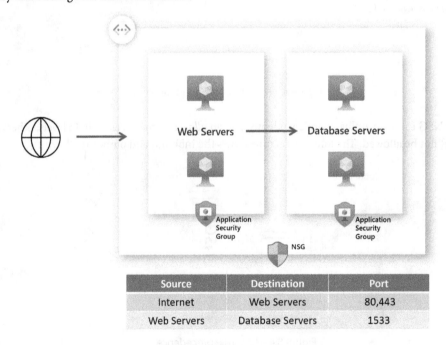

Source	Destination	Port
Internet	Web Servers	80,443
Web Servers	Database Servers	1533

Figure 8.5 – Application security groups

In the preceding figure, ASGs protect the web servers and database servers. Each ASG can have different access rules. An NSG wraps both ASGs to provide an additional level of security.

Before we move forward, we must take a formalized approach to naming the resources that will be part of our environment in Azure.

Naming conventions

In Azure, resource naming conventions are designed to ensure that names are unique and easy to identify. We strongly recommend identifying the critical information you want to reflect in the resource name.

Once you identify the relevant information for the resource type, define a standard naming convention across your organization. You can follow the best practices from the Cloud Adoption Framework for naming conventions.

Here are a few general guidelines for naming resources in Azure:

- Names can include letters, numbers, and hyphens
- Names must start with a letter
- Names must be between 3 and 63 characters long
- Names must be unique within the resource group they are created in
- Names are case-insensitive

Effective naming includes the most relevant information for the resource type; for example, if we consider a public IP address, we can have the following information:

- Resource type
- Workload/application
- Environment
- Region
- Instance

A standard naming convention is to include the resource type, location, and unique identifier in the name. For example, a virtual network could be named `vnet-westus-01`.

The following example shows the naming convention for a public IP address:

pip-hadoop-prod-westus-001
Resource type
Workload
Environment
Region
Instance

Figure 8.6 – Naming convention

You can refer to the following page and review a list of abbreviations for resource types: `https://bit.ly/az-abbreviation-examples`.

It is essential to follow these naming conventions when creating resources in Azure to ensure that your resources can be easily identified and accessed.

We recommend you check the Azure Naming Tool. This tool will help you easily handle naming conventions for your Azure resources: `https://github.com/microsoft/CloudAdoptionFramework/tree/master/ready/AzNamingTool#installation`.

Now let's understand how we can establish communication with Azure resources. Public networks such as the internet communicate by using public IP addresses. Private networks such as an Azure virtual network use private IP addresses, which are not routable on public networks.

An IP address is a numerical label assigned to each device connected to a computer network that uses the Internet Protocol for communication. An IP address serves two principal functions:

- It identifies the host or device on the network
- It provides the location of the host on the network

In Azure, an IP address is assigned to each Azure resource you create. For example, when you create a VM in Azure, it is assigned a public IP address that can be used to communicate with the VM over the internet. You can also assign a private IP address to the VM, which can be used to communicate with other resources within the same virtual network.

There are two types of IP addresses in Azure:

- **Public IP address**: This unique IP address is assigned to a resource, such as a VM, and can be accessed over the internet
- **Private IP address**: This is a non-routable IP address assigned to a resource within a virtual network and can only be accessed within the virtual network

You can use the Azure portal or the Azure CLI to manage the IP addresses of your Azure resources.

Public IP addresses

An Azure public IP address is a static or dynamic IP address assigned to a resource connected to the internet. You can assign a public IP address when you create a resource in Azure, such as a VM or an internet-facing load balancer. This public IP address is then used to connect to the resource over the internet, either from a computer or another resource on the internet.

There are two types of public IP addresses in Azure: *static* and *dynamic*. A static public IP address is a fixed address assigned to a resource and does not change. When created, a dynamic public IP address is assigned to a resource, but it can change over time.

Public IP addresses are important for a variety of reasons. They allow resources in Azure to be reached from the internet, which is necessary for many applications and services. They also allow resources in Azure to communicate with other resources on the internet, such as web servers or other Azure resources.

The appropriate SKU for a public IP address in Azure depends on your specific requirements and usage patterns.

Here are the SKUs for public IP addresses in Azure:

- **Basic**: This is the default SKU for public IP addresses in Azure. It is suitable for most workloads and offers basic features such as static or dynamic allocation and support for IPv4 and IPv6.
- **Standard**: This SKU is designed for workloads that require more advanced features, such as zone-redundant static public IPs, larger SKU sizes, and support for managed identity.

Let's test our knowledge of what we have learned so far with the help of the following example.

Exercise 1 – design and implement a virtual network in Azure

Consider the SpringToys organization, which is in the process of migrating infrastructure and applications to Azure. As a network engineer, you are tasked to plan and implement three virtual networks and subnets to comply with the requirements.

Prerequisites

Before we get started with the exercise, here are a few things you need to be ready with:

- An active Azure account
- Azure Bicep installed on your local machine
- Azure PowerShell
- A resource group in your Azure subscription – we will name it `springtoys-RG`

The goal is to plan the deployment of three virtual networks in three different Azure regions: one in the East US region, the second in West Europe, and another in Southeast Asia:

- The `ProductDevVnet` virtual network will be deployed in the East US region. This will host web services, databases, domain controllers, and DNS.
- The `SupplyServicesVnet` virtual network will be deployed in the Southeast Asia region and will host a small set of resources.
- The `ManufacturingVnet` virtual network will be deployed in the West Europe region and will host many connected devices to retrieve data from, including sensors.

We will create the virtual networks and subnets shown in the following table:

VNet Name	Region	VNet address space	Subnet	Subnet
ProductDevVnet	East US	10.20.0.0/16		
			GatewaySubnet	10.20.0.0/27
			SharedServicesSubnet	10.20.10.0/24
			DatabaseSubnet	10.20.20.0/24
			PublicWebServiceSubnet	10.20.30.0/24
ManufacturingVnet	West Europe	10.30.0.0/16		
			ManufacturingSystemSubnet	10.30.10.0/24
			SensorSubnet1	10.30.20.0/24
			SensorSubnet2	10.30.21.0/24
			SensorSubnet3	10.30.22.0/24
SupplyServicesVnet	Southeast Asia	10.40.0.0/16		
			ResearchSystemSubnet	10.40.0.0/24

The following diagram shows the planned implementation of the virtual networks and subnets.

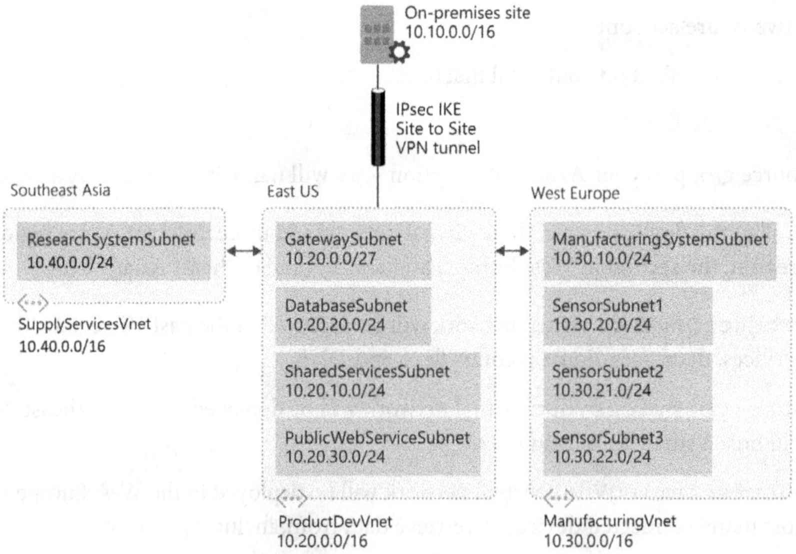
On-premises site
10.10.0.0/16
IPsec IKE
Site to Site
VPN tunnel
Southeast Asia
ResearchSystemSubnet
10.40.0.0/24
SupplyServicesVnet
10.40.0.0/16
East US
GatewaySubnet
10.20.0.0/27
DatabaseSubnet
10.20.20.0/24
SharedServicesSubnet
10.20.10.0/24
PublicWebServiceSubnet
10.20.30.0/24
ProductDevVnet
10.20.0.0/16
West Europe
ManufacturingSystemSubnet
10.30.10.0/24
SensorSubnet1
10.30.20.0/24
SensorSubnet2
10.30.21.0/24
SensorSubnet3
10.30.22.0/24
ManufacturingVnet
10.30.0.0/16

Figure 8.7 – Virtual network diagram

In the preceding diagram, you can see the three virtual networks, one virtual network per region, and the related subnets.

Instead of using the Azure portal, we will leverage Azure Bicep to deploy the resources mentioned.

Source code

You can grab the code for this exercise from the following repository: `https://github.com/PacktPublishing/Azure-Architecture-Explained/tree/main/Chapter08/exercise-1`.

In the following steps, we will deploy the infrastructure shown in *Figure 9.8* using infrastructure as code with the Bicep language:

1. Download the `main.bicep` and `azuredeploy.parameters.json` files to your machine.
2. Then, open a new instance of PowerShell or Windows Terminal in your working directory, and use the following command to deploy the environment:

```
$date = Get-Date -Format "MM-dd-yyyy"
$rand = Get-Random -Maximum 1000
$deploymentName = "AzInsiderDeployment-"+"$date"+"-"+"$rand"
New-AzResourceGroupDeployment -Name $deploymentName
-ResourceGroupName springtoys-RG -TemplateFile .\main.bicep
-TemplateParameterFile .\azuredeploy.parameters.json -c
```

The preceding code will create a new Azure resource group deployment. The script begins by setting the `$ date` variable to the current date in the format MM-dd-yyyy. It then generates a random integer between 0 and 1,000 and stores it in the `$ rand` variable.

Next, the script creates a new variable called `$deploymentName` to easily identify our deployments in Azure, a string combining the fixed string `AzInsiderDeployment-` with the current date and the random number, separated by hyphens.

Finally, the script runs the `New-AzResourceGroupDeployment` command, which creates a new Azure resource group deployment using the given name, the resource group name, the `main.bicep` template file, and the `azuredeploy.parametes.json` template parameter file.

The `-c` flag at the end of the command indicates that the deployment should be created with the `What-If` preview flag, which means that no actual resources will be created. Still, a simulation will be run to show what would happen if the deployment were to be executed.

The following figure shows a preview of the deployment:

> springtoys-bicep-deploy.ps1
Note: The result may contain false positive predictions (noise).
You can help us improve the accuracy of the result by opening an issue here: https://aka.ms/WhatIfIssues.
Resource and property changes are indicated with these symbols:
+ Create
* Ignore
The deployment will update the following scope:
Scope: /subscriptions/ /resourceGroups/SpringToys-RG
+ Microsoft.Network/virtualNetworks/ManufacturingVnet [2020-11-01]
apiVersion: "2020-11-01"
id:
"/subscriptions/ /resourceGroups/springtoys-RG/providers/Microsoft.Network/virtualNetworks/ManufacturingVnet"
location: "westeurope"
name: "ManufacturingVnet"
properties.addressSpace.addressPrefixes: [
0: "10.30.0.0/16"
]
properties.enableDdosProtection: false
properties.subnets: [
0:
name: "ManufacturingSystemSubnet"
properties.addressPrefix: "10.30.10.0/24"
properties.privateEndpointNetworkPolicies: "Enabled"
properties.privateLinkServiceNetworkPolicies: "Enabled"
1:
name: "SensorSubnet1"
properties.addressPrefix: "10.30.20.0/24"
properties.privateEndpointNetworkPolicies: "Enabled"
properties.privateLinkServiceNetworkPolicies: "Enabled"
2:
name: "SensorSubnet2"
properties.addressPrefix: "10.30.21.0/24"
properties.privateEndpointNetworkPolicies: "Enabled"
properties.privateLinkServiceNetworkPolicies: "Enabled"
3:
name: "SensorSubnet3"
properties.addressPrefix: "10.30.22.0/24"
properties.privateEndpointNetworkPolicies: "Enabled"
properties.privateLinkServiceNetworkPolicies: "Enabled"

Figure 8.8 – Deployment preview

3. Once we have confirmed the resources we want to deploy, let's confirm and perform the actual deployment operation. Then you will see the deployment output, as shown here:

```
ResourceGroupName       : springtoys-RG
ProvisioningState       : Succeeded
Timestamp               : 1/11/2023 1:54:44 PM
Mode                    : Incremental
TemplateLink            :
Parameters              :
                          Name                                        Type                       Value
                          ==========================================  =========================  ==========
                          location01                                  String                     "eastus"
                          location02                                  String                     "westeurope"
                          location03                                  String                     "southeastasia"
                          virtualNetworks_SupplyServicesVnet_name     String                     "SupplyServicesVnet"
                          virtualNetworks_ProductDevVnet_name         String                     "ProductDevVnet"
                          virtualNetworks_ManufacturingVnet_name      String                     "ManufacturingVnet"

Outputs                 :
DeploymentDebugLogLevel :
```

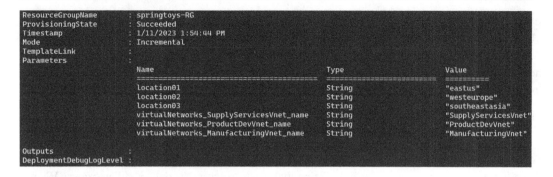

Figure 8.9 – Deployment output

4. In the Azure portal, we can review the resources deployed. Note that we have the virtual networks and subnets as required.

^ Deployment details

Resource	Type	Status
SupplyServicesVnet/ResearchSystemSubnet	Microsoft.Network/virtualNetworks/subnets	OK
ProductDevVnet/GatewaySubnet	Microsoft.Network/virtualNetworks/subnets	OK
ProductDevVnet/SharedServicesSubnet	Microsoft.Network/virtualNetworks/subnets	OK
ProductDevVnet/DatabaseSubnet	Microsoft.Network/virtualNetworks/subnets	OK
ProductDevVnet/PublicWebServiceSubnet	Microsoft.Network/virtualNetworks/subnets	OK
ManufacturingVnet/SensorSubnet2	Microsoft.Network/virtualNetworks/subnets	OK
ManufacturingVnet/ManufacturingSystemSubnet	Microsoft.Network/virtualNetworks/subnets	OK
ManufacturingVnet/SensorSubnet1	Microsoft.Network/virtualNetworks/subnets	OK
ManufacturingVnet/SensorSubnet3	Microsoft.Network/virtualNetworks/subnets	OK
ProductDevVnet	Microsoft.Network/virtualNetworks	OK
SupplyServicesVnet	Microsoft.Network/virtualNetworks	OK
ManufacturingVnet	Microsoft.Network/virtualNetworks	OK

Figure 8.10 – Deployment – Azure portal

This completes exercise 1.

Now, we will move forward and understand how organizations can create connections between different parts of their virtual network infrastructure.

Enabling cross-virtual-network connectivity

Azure VNet peering allows you to connect Azure VNets securely through the Azure backbone network. Peering enables the resources in each VNet to communicate with each other as if they were in the same network, which can be useful if you want to share resources or manage multiple VNets.

This connection allows you to establish a direct, private connection between the two VNets, enabling you to use the same protocols and bandwidth as if the VMs in the VNets were on the same local network.

For example, you might have a VNet in one Azure region that contains a database and another VNet in a different region that contains an application that needs to access the database. You can use VNet Peering to create a direct connection between the two VNets, allowing the application to access the database as if it were in the same VNet.

Overall, VNet peering can help you better organize and manage your resources in Azure, making building and maintaining distributed applications easier. With VNet peering, you can do the following:

- Connect VMs in different VNets within the same Azure region or in different regions
- Connect VNets in different Azure subscriptions as long as both subscriptions are associated with the same Azure AD tenant
- Use the same protocols and bandwidth as if the VMs were on the same local network

There are two communication categories of VNet peering:

- **Non-transitive VNet peering**: VMs in the peered VNets can communicate directly with each other. Non-transitive VNet peering does not allow communication with VMs in other peered VNets.
- **Transitive VNet peering**: VMs in the peered VNets can communicate directly, and VMs in other peered VNets can also communicate with them, indirectly. This is achieved by creating a transit VNet with peering connections to both VNets.

VNet peering can be helpful in scenarios where you need to connect VMs in different VNets, or when you need to connect an on-premises network to an Azure VNet.

It is important not to confuse the preceding two communication categories with the two types of VNet peering in Azure:

- **Regional VNet peering**: This type of peering allows you to connect two VNets in the *same* Azure *region*, allowing the VMs in the connected VNets to communicate with each other as if they were in the same network.
- **Global VNet peering**: This type of peering allows you to connect VNets in *different* Azure *regions*, allowing the VMs in the connected VNets to communicate with each other as if they were in the same network.

In both cases, the VNets remain separate and independent, but the VMs in them can communicate with each other directly over the Azure backbone network. This can be useful if you need to set up a hybrid cloud solution or if you have workloads running in multiple Azure regions that need to communicate with each other.

The following diagram highlights the two types of VNet peering:

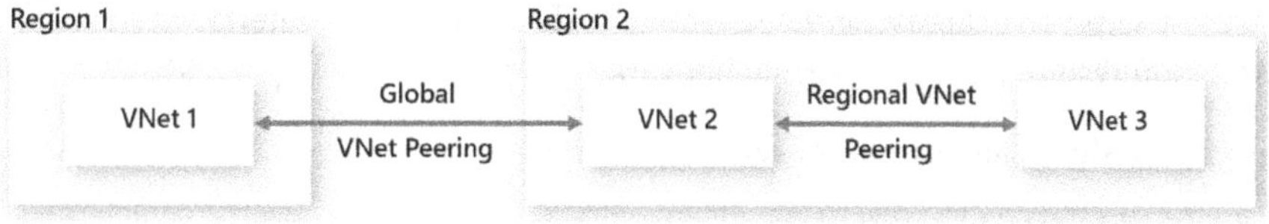
Region 1
Region 2
VNet 1
Global
VNet Peering
VNet 2
Regional VNet
Peering
VNet 3

Figure 8.11 – VNet peering types

Note that VNets can be deployed in different subscriptions and still leverage VNet peering to establish connectivity. There's no downtime in either VNets when creating the peering or after the peering is created.

When your VNets have peered, you can then configure a VPN gateway in the peered VNet and use it as a transit point. We will discuss VPN Gateway later in this chapter. The peered VNet will use the remote gateway to gain access to other resources. But note that a VNet is limited to a single gateway.

In the VNet configuration, you can set the **Allow Gateway Transit** option so the VNet can communicate to resources outside the peering. Therefore, it is essential to understand how you can leverage service chaining to direct traffic in Azure.

Using service chaining to direct traffic to a gateway

In Azure, you can use service chaining to direct traffic to a gateway, which is a service that serves as an entry point for incoming traffic. Consider service chaining when you want to route traffic from multiple sources to a single destination, such as a virtual machine or a load balancer.

To use service chaining to direct traffic to a gateway in Azure, you need to perform the following steps:

1. Create a virtual network and a gateway subnet. The gateway subnet is used to host the gateway resources.
2. Create a virtual network gateway. This gateway represents your on-premises network and routes traffic between your on-premises network and Azure.
3. Create a connection between the virtual network gateway and your on-premises network. Depending on your requirements, this connection can be either a VPN or an ExpressRoute connection.
4. Configure the routing tables for the virtual network and the on-premises network to direct traffic to the gateway.

5. Create a routing table for the gateway subnet and add a route that directs all traffic to the virtual network gateway.
6. Configure the virtual network gateway to forward traffic to the destination resources in Azure. This can be done by creating a route that specifies the destination and the next hop, the virtual network gateway.

By following these steps, you can use service chaining to direct traffic to a gateway in Azure and route it to the destination resources in your virtual network.

The following figure shows an example of service chaining:

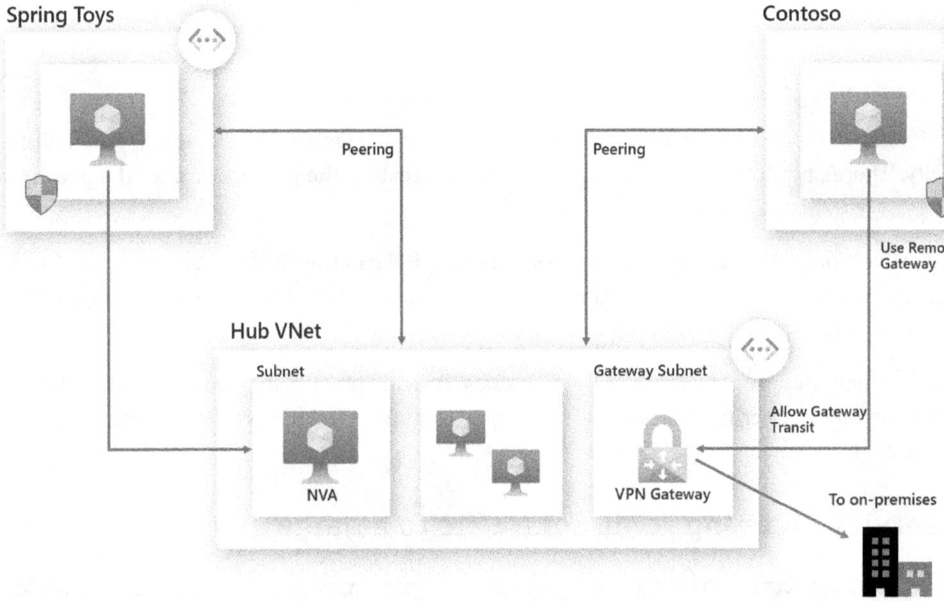
Spring Toys
Contoso
Peering
Peering
Use Remote Gateway
Hub VNet
Subnet
Gateway Subnet
Allow Gateway Transit
NVA
VPN Gateway
To on-premises

Figure 8.12 – Example of service chaining

In the preceding diagram, we have three VNets, with a hub VNet acting as a central point of connectivity to all the spoke VNets. The hub VNet hosts a network virtual appliance, virtual machines, and a VPN gateway.

The spoke VNets are peered with the hub VNet, and traffic flows through the network virtual appliance or VPN gateways in the hub VNet. The hub VNet leverages the VPN gateway to manage traffic to the on-premises network.

The hub-spoke network topology in Azure

The hub and spoke network topology is an excellent choice for enterprises seeking to strike a balance between isolating workloads and sharing crucial services such as identity and security.

This configuration centers around an Azure virtual network, serving as a central point of connectivity – the hub. Surrounding this hub are the spokes, which are virtual networks linked to the hub through peering. In this setup, shared services reside in the hub, while individual workloads are deployed in the spokes, promoting efficient resource utilization as shown in the following figure:

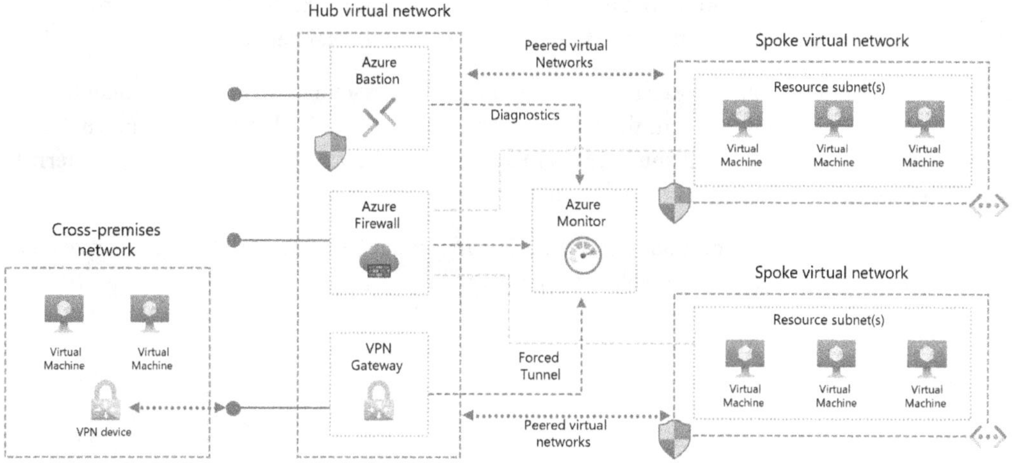

Hub virtual network
Azure Bastion
Azure Firewall
VPN Gateway
Peered virtual Networks
Diagnostics
Azure Monitor
Forced Tunnel
Peered virtual networks
Spoke virtual network
Resource subnet(s)
Virtual Machine
Virtual Machine
Virtual Machine
Spoke virtual network
Resource subnet(s)
Virtual Machine
Virtual Machine
Virtual Machine
Cross-premises network
Virtual Machine
Virtual Machine
VPN device

Figure 8.13 – Example of hub-spoke network topology in Azure

Larger enterprises can take full advantage of the hub and spoke model due to its scalability and flexibility. However, for smaller networks aiming to streamline costs and reduce complexity, a simpler design might be more appropriate. Nonetheless, the spoke virtual networks empower efficient workload isolation, with each spoke independently managed, featuring multiple tiers and subnets, seamlessly connected through Azure load balancers.

One of the greatest advantages of the hub and spoke approach is the freedom to implement these virtual networks in different resource groups and even across various subscriptions. This enables decentralized management of each workload, facilitating flexibility and adaptability. Moreover, the association of different subscriptions with the same or distinct Azure Active Directory tenants promotes efficient sharing of services hosted within the hub network, ensuring a robust yet flexible network architecture for any organization.

Now, think of the scenario where you want to allow VMs in your virtual network to connect to the internet but don't want to allow inbound connections from the internet to the VMs. Rather than configuring specific routing to and from the VMs, you can leverage the Azure NAT service.

Azure virtual NAT

Virtual **Network Address Translation** (**NAT**) is a feature of a VNet that allows outbound internet communication from VMs and other resources connected to the VNet. When you enable NAT, the VNet assigns a public IP address to the VM or resource, which can then communicate with the internet using that IP address.

Azure NAT is a feature that enables outbound-only internet connectivity for VMs in an Azure virtual network. NAT allows VMs to connect to the internet but the internet cannot initiate connections to the VMs. Azure NAT can be configured using Azure NAT Gateway, a fully managed Azure service that provides outbound-only internet connectivity for VMs in a virtual network.

The Azure NAT feature is helpful in scenarios where you want to allow outbound internet communication from your VNet. Still, you don't want the resources to be directly accessible from the internet. For example, you can use Azure NAT if you have a VM that needs to download updates from the internet but don't want to expose the VM to the internet.

The following figure shows the outbound traffic flow from **Subnet 1** through the NAT gateway, which is mapped to a public IP address or a public IP prefix.

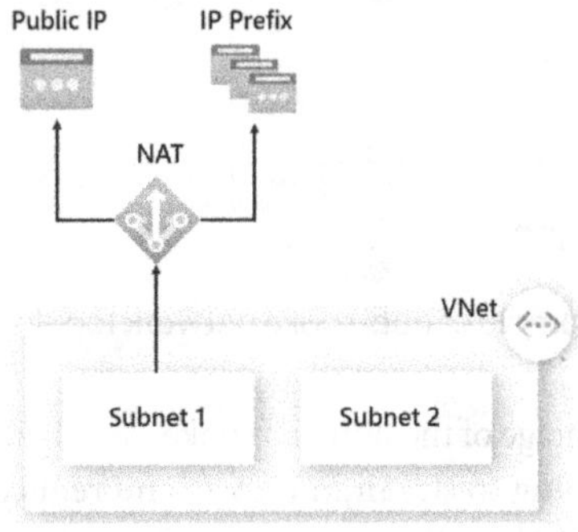
Public IP
IP Prefix
NAT
VNet
Subnet 1
Subnet 2

Figure 8.14 – Azure NAT

In summary, Azure virtual NAT allows resources connected to an Azure VNet to communicate with the internet while keeping the resources themselves private and not directly accessible from the internet.

Now that we better understand the core Azure networking services, let's review how we can enable hybrid scenarios.

Hybrid networking

Imagine SpringToys is evaluating the adoption of a global transit network architecture. This network will be utilized to connect the headquarters and multiple branch offices. Given the trend to enable remote work, SpringToys will address work-from-home scenarios. The organization needs users to access various resources, such as databases, files, and applications, in Azure and on-premises.

This section will review how you can leverage site-to-site VPN connections, point-to-site VPN connections, and how to configure Azure Virtual WAN.

When organizations connect on-premises resources to Azure, it is referred to as a hybrid network, and one way to enable this scenario is by using VPN connections. A **virtual private network** or **VPN** uses an encrypted tunnel within another network to establish connectivity.

VPNs are usually deployed to connect at least two trusted private networks, to one another and over an untrusted network such as the internet. Traffic is encrypted while traveling over the untrusted network.

How can we integrate our on-premises network with Azure? First, we need to create an encrypted connection, and we can use Azure VPN Gateway as an endpoint for incoming connections from on-premises networks.

Azure VPN Gateway

An Azure VPN gateway is a type of virtual network gateway that enables you to establish secure, cross-premises connectivity between your on-premises networks and your Azure VNets. A VPN gateway is a specific virtual network gateway that sends encrypted traffic between an Azure VNet and an on-premises location over the public internet.

Using a VPN gateway, you can create a secure connection between your on-premises network and your Azure VNet. This allows you to extend your on-premises network into the Azure cloud and to use Azure resources as if they were part of your on-premises network.

There are two types of VPN gateways in Azure:

- **Policy-based VPN**: A policy-based VPN uses policies to determine which traffic is sent over the VPN connection. It encrypts and directs packets through IPSec tunnels based on IPsec policies.
- **Route-based VPN**: A route-based VPN, previously known as dynamic routing gateways, uses routes to determine which traffic is sent over the tunnel interface. The tunnel interfaces encrypt or decrypt packets in and out of the tunnel.

In both cases, you need to set up a virtual network gateway in Azure to create the VPN connection. The gateway acts as the endpoint for the VPN connection and uses encryption to secure the traffic sent over the connection.

You can use an Azure VPN gateway to connect your on-premises network to an Azure VNet or two Azure VNets to each other. You can also use a VPN gateway to connect your Azure VNets to other VNets in other cloud platforms, such as **Amazon Web Services** (**AWS**) or **Google Cloud Platform** (**GCP**).

And before provisioning this resource type, you should determine the VPN type that best suits your scenario. There are three architectures to consider:

- Point-to-site over the internet
- Site-to-site over the internet
- Site-to-site over a dedicated network, such as Azure ExpressRoute

Major factors that impact the configuration of your VPN gateway include the following:

- Throughput – Mbps or Gbps
- Backbone – internet or private
- Availability of a public (static) IP address
- VPN device compatibility
- Multiple client connections or a site-to-site link
- VPN gateway type
- Azure VPN Gateway SKU

You can check a complete list of the Gateway SKUs by tunnel connection and throughput at the following URL:

`https://bit.ly/az-gtway-sku`

Let's review the differences between the three VPN architectures. And let's start by understanding Site-to-Site VPN connections.

Site-to-site VPN connections

A **site-to-site** (**S2S**) VPN connection is a VPN connection that allows you to connect two separate networks and establish a secure communication channel between them. An S2S VPN connection is known as a *router-to-router VPN* or *LAN-to-LAN VPN*.

S2S VPN connections are typically used to connect two remote networks and enable secure communication. For example, you might use an S2S VPN connection to connect your company's on-premises network to a virtual network in Azure or to connect two branch offices.

An Azure S2S VPN gateway is a type of VPN gateway that allows you to create a secure connection between your on-premises network and an Azure virtual network. It allows you to extend your on-premises network into the Azure cloud and enables secure communication between the two networks.

With an Azure S2S VPN gateway, you can connect your on-premises network to an Azure virtual network over an IPsec/IKE (IKEv1 or IKEv2) VPN tunnel. This tunnel is established over the internet and allows you to communicate securely between your on-premises network and the Azure virtual network.

An Azure S2S VPN gateway can connect your on-premises network to a single Azure virtual network or multiple Azure virtual networks. It is a fully managed service, which means that Azure handles the setup, maintenance, and monitoring of the VPN gateway for you.

The following figure shows a high-level overview of the S2S architecture.

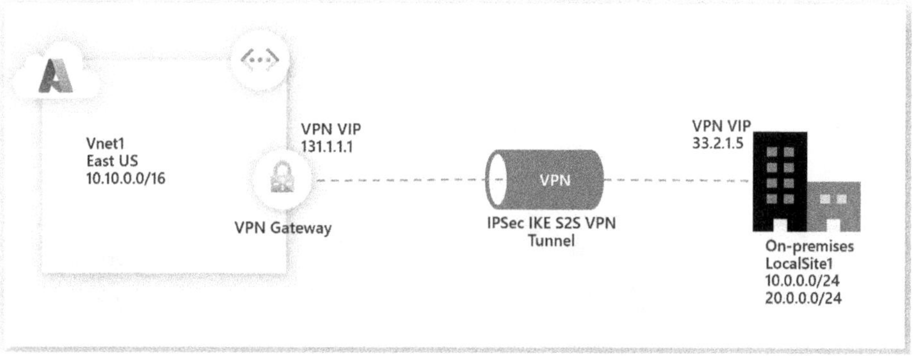
Vnet1
East US
10.10.0.0/16
VPN VIP
131.1.1.1
VPN Gateway
VPN
IPSec IKE S2S VPN
Tunnel
VPN VIP
33.2.1.5
On-premises
LocalSite1
10.0.0.0/24
20.0.0.0/24

Figure 8.15 – Azure site-to-site VPN gateway

Using an Azure S2S VPN gateway can be a cost-effective way to securely connect your on-premises network to Azure and enables you to take advantage of Azure's scale, reliability, and security while still being able to access your on-premises resources.

Now let's review the second architecture: point-to-site VPN connections.

Point-to-site VPN connections

A **point-to-site** (**P2S**) VPN connection is a VPN connection that allows you to connect a single device, such as a laptop or desktop computer, to a virtual network. A P2S connection is also known as a *client-to-site VPN* or *remote access VPN*.

With a P2S connection, you can connect to a virtual network from a remote location, such as your home or a coffee shop, and access resources on the virtual network as if you were on the local network. This is useful when you need to access resources on a virtual network but are not physically located on the organization's network.

Azure P2S VPN is a VPN connection that allows you to connect a device, such as a laptop or desktop computer, to a virtual network in Azure. With a P2S VPN connection, you can access resources on the virtual network as if you were on the organization's internal network.

To create a P2S VPN connection, you need to install a VPN client on your device and connect to the virtual network using the VPN client. The VPN client establishes a secure connection to the virtual network over the internet and allows you to access resources on the virtual network as if you were physically connected.

Azure P2S VPN is a fully managed service, which means that Azure handles the setup, maintenance, and monitoring of the VPN connection for you. Using Azure P2S VPN can be a cost-effective way to securely connect to a virtual network in Azure, enabling you to take advantage of the scale, reliability, and security of Azure while still being able to access resources on the virtual network from a remote location.

The following figure shows a high-level overview of the P2S architecture.

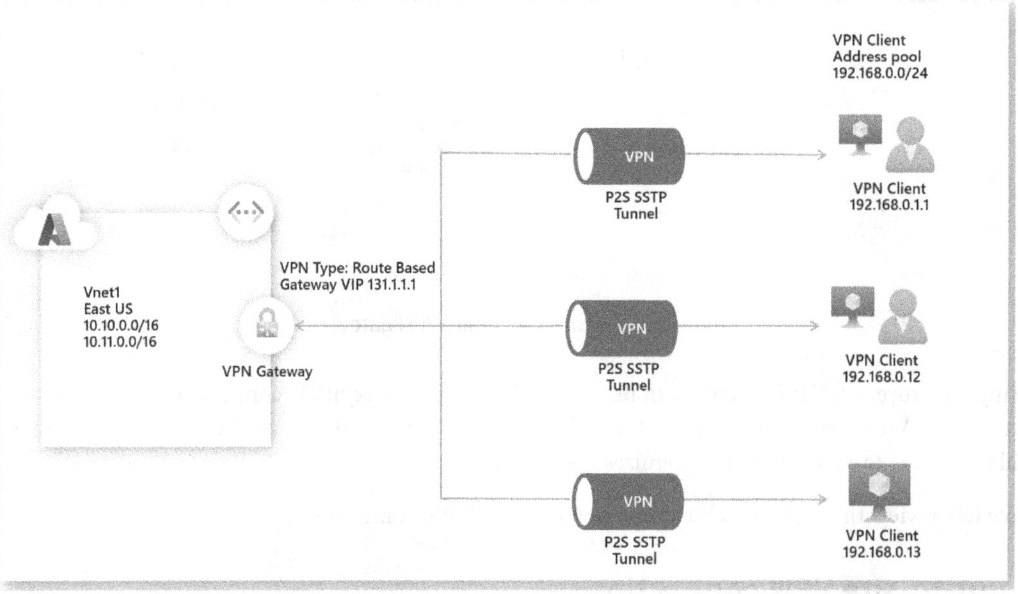
VPN Client
Address pool
192.168.0.0/24
VPN
P2S SSTP
Tunnel
VPN Client
192.168.0.1.1
VPN Type: Route Based
Gateway VIP 131.1.1.1
Vnet1
East US
10.10.0.0/16
10.11.0.0/16
VPN Gateway
VPN
P2S SSTP
Tunnel
VPN Client
192.168.0.12
VPN
P2S SSTP
Tunnel
VPN Client
192.168.0.13

Figure 8.16 – Azure point-to-site VPN connection

With Azure P2S VPN, you can use a variety of protocols to establish a secure connection to your Azure virtual network.

The following protocols are supported for Azure P2S VPN connections:

- **Secure Socket Tunneling Protocol (SSTP)**: A Microsoft proprietary protocol that uses SSL to establish a VPN connection. It is primarily used for connecting to Azure from a Windows device.
- **Internet Key Exchange version 2 (IKEv2)**: This is a secure, fast, and flexible VPN protocol that can be used on various devices and operating systems.
- **OpenVPN**: This is an open source VPN protocol that is widely used and supports a variety of authentication methods.
- **Point-to-Point Tunneling Protocol (PPTP)**: This is an older VPN protocol that is less secure and less efficient than newer protocols, but it is still supported for compatibility with legacy systems.

You can choose the appropriate protocol for your needs based on the devices and operating systems you are using and the level of security and performance you require.

Bear in mind that users must authenticate before Azure accepts a P2S VPN connection, and there are two options to authenticate a connecting user: using native Azure certificate authentication or using Azure Active Directory authentication.

> **Note**
>
> AD domain authentication allows users to connect to Azure using their organization domain credentials, but it requires a RADIUS server that integrates with the AD server.

As organizations expand their infrastructure to Azure and enable employees, partners, and customers to connect to resources from any place, it is vital to operate across regional boundaries.

Let's say SpringToys has multiple branch offices and remote users, and we need to connect to all these sites. Instead of manually configuring a combination of site-to-site VPNs to provide connectivity to the multiple branches and provisioning multiple point-to-site connections for remote users, SpringToys could leverage Azure Virtual WAN to simplify the deployment and the management of all the connectivity.

Azure Virtual WAN

Azure **Virtual WAN** (**VWAN**) is a networking service that makes it easy to connect your on-premises and Azure-based resources to the cloud. It allows you to create a **Wide Area Network** (**WAN**) across multiple Azure regions and connect it to your on-premises network using a variety of connection types, including site-to-site VPN, point-to-site VPN, and ExpressRoute.

VWAN simplifies setting up and managing a WAN by providing a single management interface for all your connections. Each Virtual WAN is implemented as a hub-and-spoke topology and can have one or more hubs, which support connectivity between different endpoints, including connectivity vendors such as AT&T, Verizon, and T-Mobile.

Then, you can use the virtual hubs to connect to your Azure resources, such as virtual machines and storage accounts, as if they were part of your on-premises network.

VWAN also provides advanced networking features, such as load balancing, routing, and traffic management, to help you optimize the performance and reliability of your WAN.

The following figure highlights the example of an organization with two Virtual WAN hubs connecting the spokes.

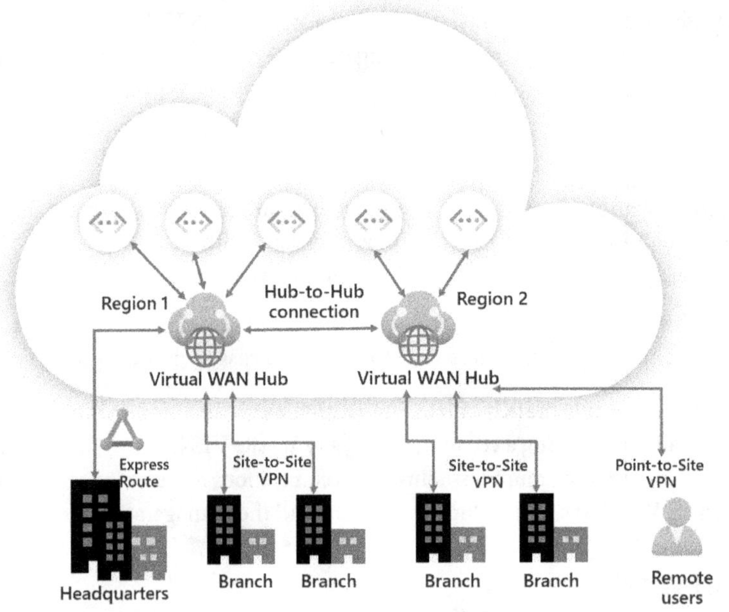

Region 1
Hub-to-Hub connection
Region 2
Virtual WAN Hub
Virtual WAN Hub
Express Route
Site-to-Site VPN
Site-to-Site VPN
Point-to-Site VPN
Headquarters
Branch
Branch
Branch
Branch
Remote users

Figure 8.17 – Azure VWAN

As you can see, Azure VWAN is a flexible and cost-effective solution for connecting your on-premises resources to the cloud and extending your network to Azure.

Another benefit of using Azure VWAN is that a hub can be associated with security and routing policies through Azure Firewall Manager. This way, you can filter traffic between Azure VNets, between VNets and branch offices, and traffic to the internet.

There are two types of Virtual WAN SKUs: *Basic* and *Standard*. The Basic SKU supports only S2S VPN connections, while the Standard SKU supports ExpressRoute, P2S, S2S, Inter-hub, VNet-to-VNet, Azure Firewall, and **Network Virtual Appliance** (**NVA**) in a VWAN.

Now let's review how we can leverage another service, called ExpressRoute, to connect on-premises infrastructure to Azure securely.

ExpressRoute

So far, we have reviewed the various networking components SpringToys can leverage to extend its network to Azure and enable access to resources both on-premises and in Azure. And now, we will dive deep into how SpringToys can extend on-premises networks into the Microsoft cloud by using a private connection using ExpressRoute.

Since ExpressRoute connections do not travel over the internet, this approach could benefit SpringToys in establishing a more reliable connection with faster speeds, consistent latencies, and higher security.

Think of ExpressRoute as your Ethernet cable to the Microsoft cloud. ExpressRoute is a dedicated network connection service that enables you to create private connections between your on-premises data centers and Azure. It uses a network service provider to establish a secure, high-bandwidth connection that bypasses the public internet, providing a more reliable and lower-latency connection than a traditional VPN.

ExpressRoute offers several benefits over other types of connections to Azure:

- **Reliability**: ExpressRoute connections use a dedicated, private network infrastructure that is more reliable than the public internet
- **Security**: ExpressRoute connections are more secure than internet connections because they use a private network rather than a public one
- **Performance**: ExpressRoute connections have lower latencies and higher speeds than internet connections, making them suitable for applications that require high bandwidth and low-latency connectivity
- **Compliance**: ExpressRoute can be used to meet compliance requirements that prohibit the use of the public internet for certain types of data

ExpressRoute is a powerful and flexible service for connecting your on-premises resources to Azure; it users Layer 3 connectivity and can be from any-to-any networks, a point-to-point Ethernet connection, or through a virtual cross-connection via an Ethernet exchange.

Note that by using ExpressRoute, you can establish connectivity to Microsoft cloud services across regions in a geopolitical region, not only to Azure resources. And it has built-in redundancy in every peering location.

Here's a brief look at how Azure ExpressRoute works:

1. You connect to an Azure ExpressRoute location by setting up a circuit. This circuit can be set up using a variety of connectivity providers, such as telecommunications companies or network service providers closer to your location.
2. Once the circuit is configured, you can create one or more virtual connections, also known as ExpressRoute circuits, between your on-premises data centers and Azure. These virtual connections are established over the physical connection you set up in the first step.
3. You can then use these virtual connections to access Azure services, such as virtual machines, storage, and databases as if they were part of your on-premises network.

Now let's review how Azure ExpressRoute enables businesses to create private connections between their on-premises infrastructure and Azure data centers, bypassing the public internet.

ExpressRoute connectivity models

There are four ways you can establish a connection between your on-premises network and the Microsoft cloud:

- **Cloud Exchange Co-Location**: In this approach, you might be already using services from a co-located facility from a third party or local provider. Co-location providers offer Layer 2 or 3 connections between your infrastructure in the co-location facility and the Microsoft cloud. You can order a connection to the Microsoft cloud through your provider.
- **Point-to-Point Ethernet Connection**: In this case, you can leverage point-to-point Ethernet links. Providers offer Layer 2 or Layer 3 manager connections between your site and the Microsoft cloud.
- **Any-to-Any Connection**: You can also integrate your WAN with the Microsoft cloud. In this case, **Internet Protocol Virtual Private Network** (**IP-VPN**) providers, usually **Multiprotocol Label Switching** (**MPLS**) VPN, can provide any-to-any connectivity between your site and data centers. You can connect your WAN to the Microsoft cloud and see it as if it were another branch office.
- **ExpressRoute Direct**: In this context, you can connect directly to Microsoft's global network at a peering location. It supports active/active connectivity at scale.

ExpressRoute offers three SKUs: *Local*, *Standard*, and *Premium*. With the local SKU, you are entitled to an unlimited data plan. With Standard and Premium, you can choose between a metered or unlimited data plan. Once you select the unlimited data plan, you will be limited to going back to a metered plan.

Creating an ExpressRoute circuit

Now let's configure an ExpressRoute gateway and circuit:

1. Sign in to the Azure portal, and search for `ExpressRoute`, as shown in the following figure:

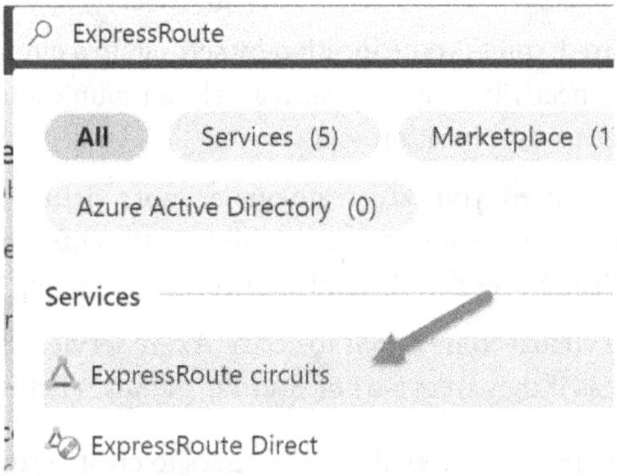
ExpressRoute
All
Services (5)
Marketplace (1
Azure Active Directory (0)
Services
ExpressRoute circuits
ExpressRoute Direct

Figure 8.18 – ExpressRoute search

2. On the **Create ExpressRoute circuit** basic page, enter the following information:
   - **Resource group**: Select an existing resource group or create a new one
   - **Region**: Select the location where the circuit will be created
   - **Name**: Enter a name for the circuit

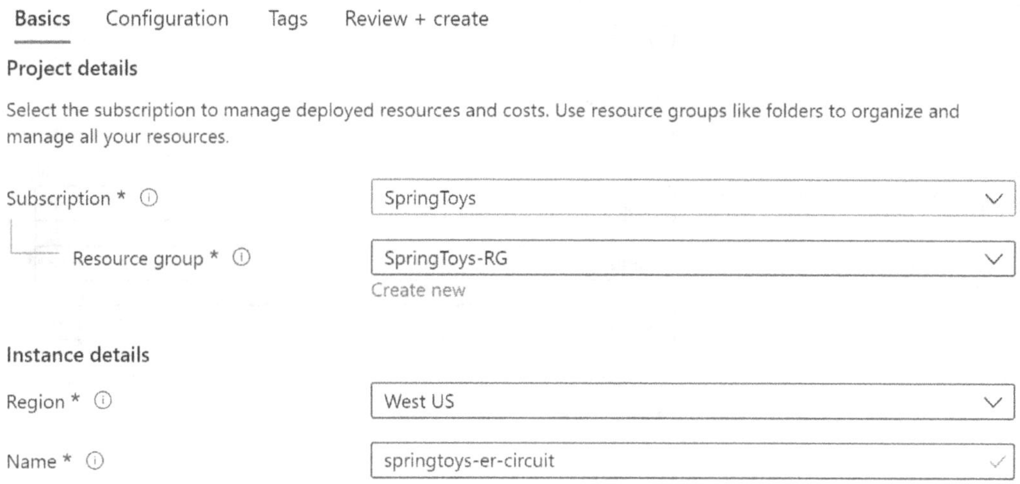
Basics
Configuration
Tags
Review + create
Project details
Select the subscription to manage deployed resources and costs. Use resource groups like folders to organize and manage all your resources.
Subscription *
SpringToys
Resource group *
SpringToys-RG
Create new
Instance details
Region *
West US
Name *
springtoys-er-circuit

Figure 8.19 – ExpressRoute circuit Basics tab

3. Now, in the **Configuration** tab, provide the following values:
   - **Service provider**: Select the service provider you want to use for the circuit
   - **Bandwidth**: Select the desired circuit bandwidth
   - **Peering location**: Select the peering location where the circuit will be connected
   - **Pricing tier**: Select the pricing tier for the circuit

Select the **Review + create** button to review the circuit configuration.

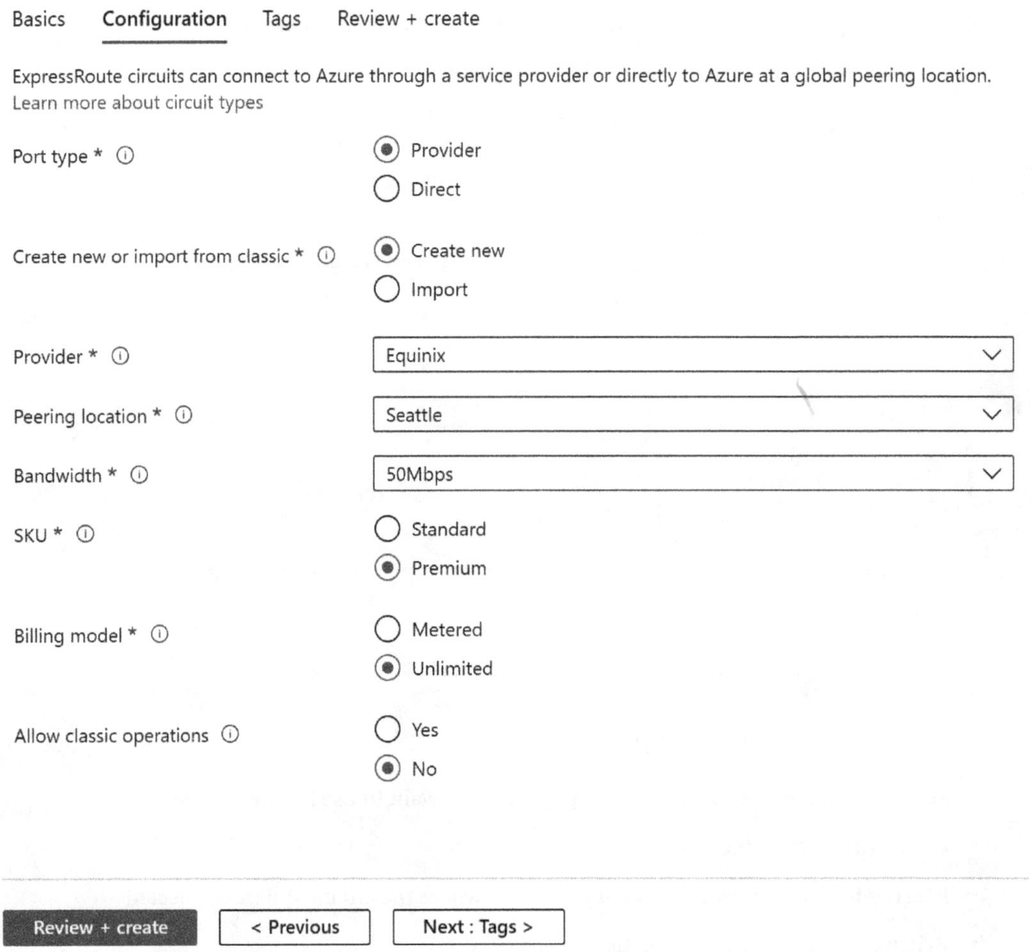

Basics   **Configuration**   Tags   Review + create

ExpressRoute circuits can connect to Azure through a service provider or directly to Azure at a global peering location. Learn more about circuit types

Port type * ⓘ	◉ Provider
	○ Direct
Create new or import from classic * ⓘ	◉ Create new
	○ Import
Provider * ⓘ	Equinix ⌄
Peering location * ⓘ	Seattle ⌄
Bandwidth * ⓘ	50Mbps ⌄
SKU * ⓘ	○ Standard
	◉ Premium
Billing model * ⓘ	○ Metered
	◉ Unlimited
Allow classic operations ⓘ	○ Yes
	◉ No

Review + create   < Previous   Next : Tags >

Figure 8.20 – ExpressRoute circuit Configuration tab

4. If the configuration is correct, select the **Create** button to create the circuit, and wait for the circuit to be created. This may take several minutes.

5. Once the circuit is created, you can view the circuit details by selecting the circuit from the list of circuits on the ExpressRoute circuits page.

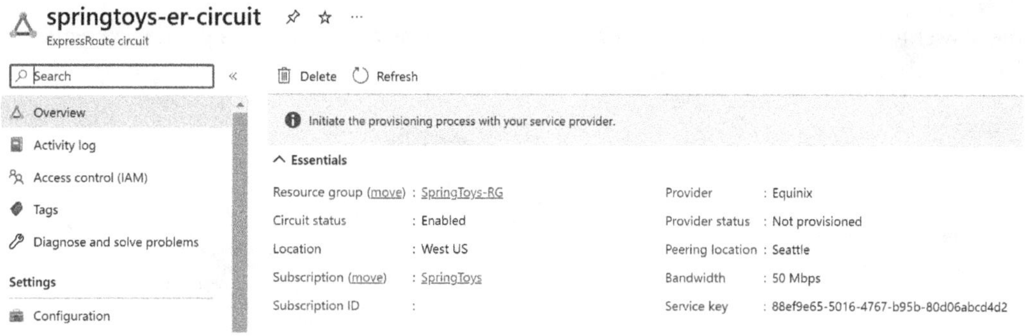
springtoys-er-circuit
ExpressRoute circuit
Search
Delete
Refresh
Overview
Activity log
Access control (IAM)
Tags
Diagnose and solve problems
Settings
Configuration
Initiate the provisioning process with your service provider.
Essentials
Resource group (move) : SpringToys-RG
Circuit status : Enabled
Location : West US
Subscription (move) : SpringToys
Subscription ID :
Provider : Equinix
Provider status : Not provisioned
Peering location : Seattle
Bandwidth : 50 Mbps
Service key : 88ef9e65-5016-4767-b95b-80d06abcd4d2

Figure 8.21 – Express Route circuit Overview tab

6. In the **Overview** tab, you will see the **Service key** value. You need to send the service key to your service provider to complete the provisioning process.

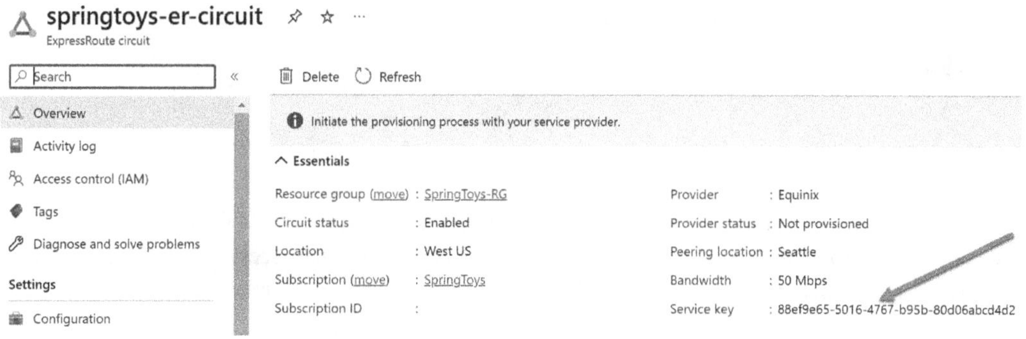
springtoys-er-circuit
ExpressRoute circuit
Search
Delete
Refresh
Overview
Activity log
Access control (IAM)
Tags
Diagnose and solve problems
Settings
Configuration
Initiate the provisioning process with your service provider.
Essentials
Resource group (move) : SpringToys-RG
Circuit status : Enabled
Location : West US
Subscription (move) : SpringToys
Subscription ID :
Provider : Equinix
Provider status : Not provisioned
Peering location : Seattle
Bandwidth : 50 Mbps
Service key : 88ef9e65-5016-4767-b95b-80d06abcd4d2

Figure 8.22 – ExpressRoute circuit Service key

At this point, you should be able to describe ExpressRoute and how it can help organizations establish a more reliable connection with faster speeds, consistent latencies, and higher security.

Decision tree on network topology

When you're considering your networking topology, start by thinking about the number of different Azure regions you need to deploy in, and whether you need native global connectivity between those regions and any on-premises servers. If you do, a few more questions can help you decide whether you need a Virtual WAN model for your network or you can stick with a traditional hub and spoke model for your network.

If you need over 100 native ISPEC tunnels, transitive routing, and over 10 Gbps for virtual network gateways, a Virtual WAN is our recommendation.

If the answer to any of those follow-up questions is no, and none of the resources you deploy require full mesh, then a hub and spoke model is the right way to go.

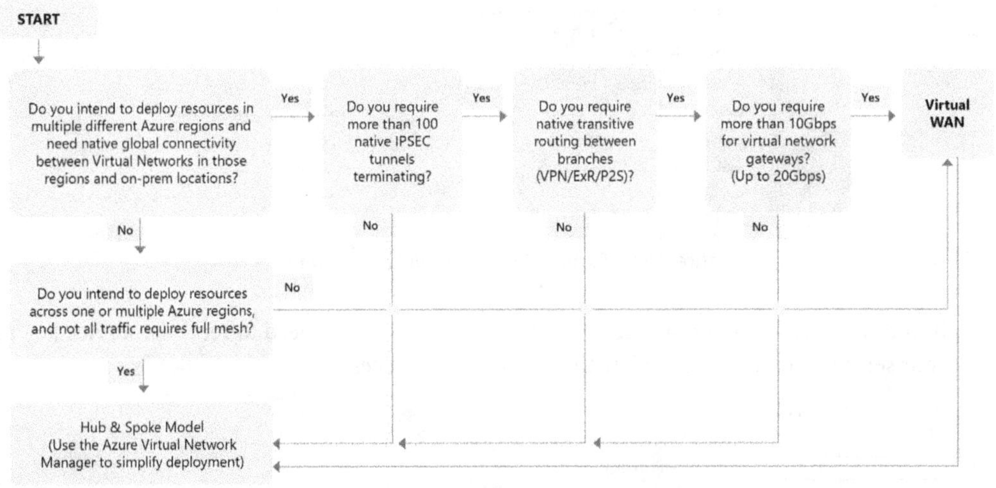
START
Do you intend to deploy resources in multiple different Azure regions and need native global connectivity between Virtual Networks in those regions and on-prem locations?
Yes
Do you require more than 100 native IPSEC tunnels terminating?
Yes
Do you require native transitive routing between branches (VPN/ExR/P2S)?
Yes
Do you require more than 10Gbps for virtual network gateways? (Up to 20Gbps)
Yes
Virtual WAN
No
No
No
No
Do you intend to deploy resources across one or multiple Azure regions, and not all traffic requires full mesh?
No
Yes
Hub & Spoke Model (Use the Azure Virtual Network Manager to simplify deployment)

Figure 8.23 – Decision tree on network topology

So far, we have worked with Azure networking services that extend connectivity from an on-premises facility to the Microsoft cloud, and SpringToys needs to ensure that users get access to network resources in Azure, such as applications, services, and data, in an efficient and timely manner.

We must ensure that applications, services, and data are always available to employees whenever and wherever possible. In the next section, we will discuss the different load balancing services that can help SpringToys distribute workloads across networks.

Load balancing

Load balancing refers to distributing incoming requests across servers or resources to optimize resource use, maximize throughput, minimize response time, and avoid overloading any one resource. In this section, we will analyze the following services in two categories:

- **Load balance non-HTTP(S) traffic**: We will discuss **Azure Load Balancer** (**ALB**) and **Azure Traffic Manager** (**ATM**)
- **Load balance HTTP(S) traffic**: We will review Azure Application Gateway and Azure Front Door.

The following diagram shows a high-level overview of the load balancing solutions available on Azure:

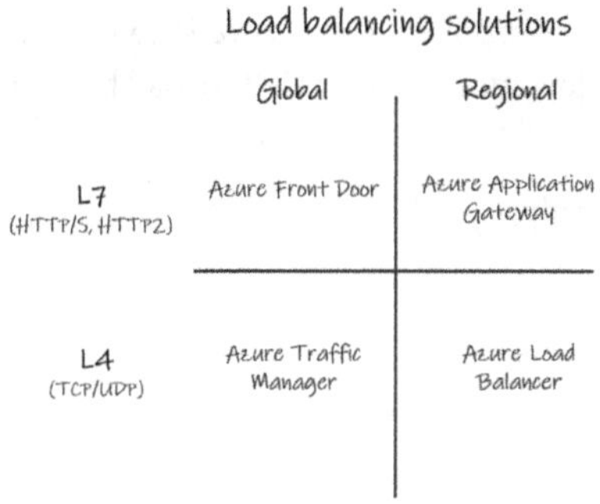
Load balancing solutions
Global
Regional
L7
(HTTP/S, HTTP2)
Azure Front Door
Azure Application Gateway
L4
(TCP/UDP)
Azure Traffic Manager
Azure Load Balancer

Figure 8.24 – Overview of the load balancing solutions available on Azure

At a very high level, we have the following solutions available on Azure:

- **Azure Load Balancer:** This is a *layer 4 (TCP, UDP)* load balancer that distributes incoming traffic among healthy instances of services in virtual machine scale sets or individual virtual machines.
- **Azure Application Gateway**: This layer 7 (HTTP, HTTPS) load balancer enables you to manage traffic to your web applications. It includes SSL termination, cookie-based affinity, and Layer 7 request routing.
- **Azure Traffic Manager**: This is a *DNS-based load balancer* that routes traffic to the most appropriate endpoint based on the routing method and the health of the endpoints.
- **Azure Front Door** is a global, scalable, secure entry point for web applications. It includes *global HTTP(S) load balancing*, DDoS protection, and CDN integration.

When choosing a load balancing option for Azure, there are some key factors to consider:

- **Type of traffic**: Consider whether you need to load balance at layer 4 (TCP, UDP) or layer 7 (HTTP, HTTPS). In short, is this for a web application?
- **Scope**: Different load balancers are optimized for different types of applications. For example, Azure Front Door is well-suited for global, highly available web applications. At the same time, Azure Traffic Manager is better for routing traffic to the most appropriate endpoint based on routing method and endpoint health. Think whether you need to load balance across regions or within a virtual network, or even both.
- **Performance and scalability**: Consider the service-level agreement and whether your application needs high performance.
- **Features:** Consider the specific features you need, such as SSL termination, cookie-based affinity, or Layer 7 request routing.

Now let's review each of them in detail.

Load balancing non-HTTP(S) traffic

The goal of enabling load balancing components in Azure is to optimize the use of resources and maximize throughput while reducing downtime. Non-HTTP(S) load balancing services are recommended for non-web workloads.

The services that can handle non-HTTP(S) traffic include the following:

- Azure Load Balancer
- Azure Traffic Manager

Azure Load Balancer

Azure Load Balancer allows you to distribute incoming network traffic across various resources, such as virtual machines, in a virtual network. This helps to improve the availability and reliability of your application.

Azure Load Balancer, or ALB as we will refer to it from now on, distributes incoming traffic using a scheduling algorithm, such as round-robin or least connections. This means it will send the first incoming request to the first virtual machine, the second request to the second virtual machine, and so on.

ALB will automatically redirect traffic to the remaining available virtual machines if a virtual machine becomes unavailable. ALB can load balance services on multiple ports, IP addresses, or both.

Load balancers can be public or internal. The public load balancer can serve outbound connections for VMs in your VNet and is used to distribute client traffic from the internet across your VMs. On the other hand, internal load balancers leverage private IP addresses as the frontend to load balance traffic from internal Azure resources to other resources inside a VNet.

The following diagram highlights the two types of load balancer.

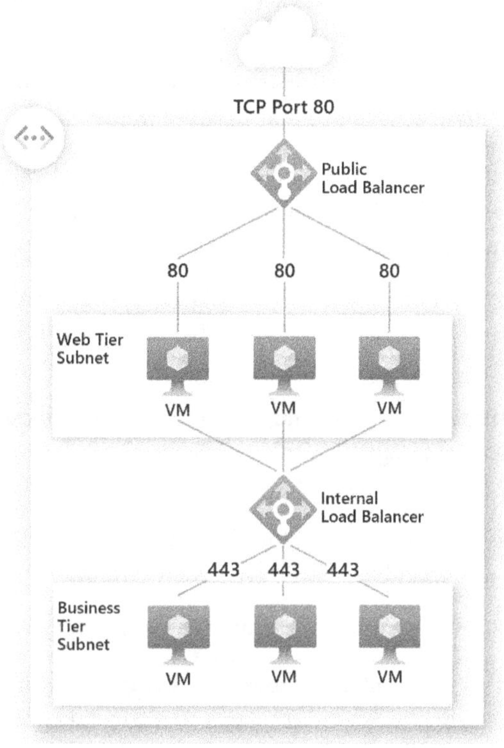
TCP Port 80
Public
Load Balancer
80
80
80
Web Tier
Subnet
VM
VM
VM
Internal
Load Balancer
443
443
443
Business
Tier
Subnet
VM
VM
VM

Figure 8.25 – Overview of the two types of load balancer in Azure

There are two SKUs for Azure Load Balancer: a *Basic* SKU and a *Standard* SKU. The Basic load balancer is suitable for simple workloads, while the standard load balancer is more feature-rich and suitable for more complex scenarios.

In terms of redundancy and scalability, Azure enables availability zones – physically separate data centers within an Azure region. Each availability zone comprises one or more data centers and has its own power, networking, and cooling infrastructure. By using availability zones, you can create highly available applications that can withstand the loss of a single data center.

Azure Load Balancer supports availability zone scenarios, and you can use the Standard ALB to increase availability and distribution across zones.

Azure services that are compatible with availability zones fall into three categories:

- **Zonal services**: A resource can be pinned to a specific zone. For example, virtual machines can be pinned to a particular zone, allowing for increased resilience by having one or more instances of resources spread across zones.

- **Zone-redundant services**: These are resources that are replicated or distributed across zones automatically. Azure replicates the data across three zones so that a zone failure does not impact its availability.
- **Non-regional services**: This refers to Azure services that are always available from Azure geographies and resilient to zone and region-wide outages.

The Standard SKU supports availability zones with zone-redundant and zonal frontends for inbound and outbound traffic and health probes, including TCP, HTTP, and HTTPS, while the Basic SKU only supports TCP and HTTP. Another benefit of using the Standard SKU is the ability to use HA ports.

Now let's move forward and understand Azure Traffic Manager's core capabilities.

Azure Traffic Manager

Imagine your application is deployed across multiple Azure regions. You can leverage **Azure Traffic Manager**, or **ATM**, to distribute traffic optimally to services across global Azure regions while providing high availability and responsiveness. In short, think of ATM as a DNS-based load balancer.

ATM uses various algorithms to determine the optimal distribution of traffic, including performance-based routing, geographic routing, and multi-value weights-based routing.

Before configuring ATM, it is important to understand a few key concepts:

- **Endpoints**: Consider an endpoint as a resource to which ATM routes traffic. Endpoints can be Azure PaaS services, such as Azure Web Apps or Azure Functions, or external endpoints, such as a website, hosted on-premises or even in another cloud.
- **Traffic routing methods**: ATM supports various traffic routing methods:
  - **Performance-based routing**: This method routes traffic based on the endpoint's performance. It periodically checks the health of each endpoint and routes traffic to the healthiest endpoints. Note that the *closest* endpoint isn't necessarily closest as measured by geographic distance. This method determines the closest endpoint by measuring network latency.
  - **Geographic routing**: This method routes traffic based on the user's geographic location. You can specify which region an endpoint belongs to, and ATM will route traffic to the nearest region.
  - **Multi-value weights-based routing**: This method allows you to specify weights for each endpoint. ATM will then distribute traffic to the endpoints based on their weights.
  - **Priority**: Consider this method if you want a primary endpoint for all the traffic. Then, you can add multiple backup endpoints if the primary endpoint is unavailable.
  - **Weighted**: This method will distribute traffic across endpoints based on their weights. And you can set the same weight on your endpoints to distribute traffic evenly.

    - **Subnet**: In this method, you can map groups of end-user IP address ranges to a specific endpoint. A request will be resolved based on the source IP address mapped.

- **Profile**: A profile is a logical container for a set of endpoints. A profile can contain endpoints in different regions and of different types.
- **Domain name**: ATM uses a domain name to route traffic to the appropriate endpoint. When you create a Traffic Manager profile, you can use a subdomain of trafficmanager.net (e.g., `myapp.trafficmanager.net`) or bring your own domain (e.g., `myapp.com`).

When a client requests a connection to a service, it must resolve the DNS of the service to an IP address. Then, the client connects to the IP address to access the service.

Let's look at the following example to give you an idea of how to use ATM to distribute traffic to multiple Azure Web Apps.

Suppose you have an Azure Web App hosted in the West US region, another in West Europe, and a third in Southeast Asia.

Then, you create an ATM profile and specify the domain name you want to use, for example, `springtoys.trafficmanager.net`.

Next, you add the web apps as endpoints to the Traffic Manager profile. And you can specify the location or region of each endpoint and assign weights to control the proportion of traffic that is routed to each endpoint.

Finally, you configure the traffic routing method for the ATM profile. For instance, you choose the **geographic** routing method to route traffic to the nearest web app or the **performance-based** routing method to route traffic to the web app with the best performance.

You can test the ATM configuration by accessing the web app using the ATM domain name `springtoys.trafficmanager.net`. ATM should route your request to the appropriate web app based on your configured routing method.

For a full description of how each routing method works, you can refer to the following URL: `https://bit.ly/atm-routing-methods`

Let's see the options to ensure SpringToys employees have consistent access to their applications, services, and data.

Load balancing HTTP(S) traffic

The services that can handle HTTP(S) traffic include the following:

- Azure Application Gateway
- Azure Front Door

Azure Application Gateway

Imagine SpringToys has a web application hosted on a group of VMs in Azure. The application is experiencing a lot of traffic, so you want to ensure that it can handle the load without downtime or performance issues.

A solution to this problem would be to use Azure Application Gateway to distribute incoming traffic across the servers in your group. This would allow you to scale out your application horizontally so that it can handle more traffic without becoming overloaded.

In addition to the load balancing capabilities included in Azure Application Gateway, you might also want to use it to perform other functions such as SSL termination or cookie-based session affinity. For example, you can use SSL termination to offload the encryption and decryption of traffic from your servers or use session affinity to ensure users are always connected to the same server during their session.

Azure Application Gateway, or **AAG** as we will refer to it from now on, will help you manage traffic to your web applications and improve their performance and security. Think of AAG as a Layer 7 web traffic load balancer that enables you to manage traffic to your web applications.

AAG can route traffic to specific servers based on additional attributes of an HTTP request, for example, URI path or host headers. It can perform other functions such as SSL termination and cookie-based session affinity.

AAG supports multiple protocols, including HTTP/S, HTTP/2, and WebSocket. It can be integrated with a web application firewall to protect against potential vulnerabilities in the web application and enables end-to-end request encryption.

There are two primary methods of routing traffic:

- **Path-based routing**: Consider this option to route traffic based on the URL path of the incoming request. For example, you can route traffic based on the incoming URL named `https://springtoys.com/images`. If `/images` is in the incoming URL, you can route traffic to a specific pool of servers configured for images. If `/video` is in the URL, then traffic is routed to a different pool optimized for videos.
- **Multiple site routing**: This option is optimal when you have multiple sites such as `springtoys.com` and `contoso.com`. And you can register multiple DNS names (CNAMEs) for the application gateway's IP address, specifying each site's name. You configure more than one site web application on the same AAG instance.

AAG has a few components that need to be configured to route requests to a poll of web servers. The following diagram shows the AAG components:

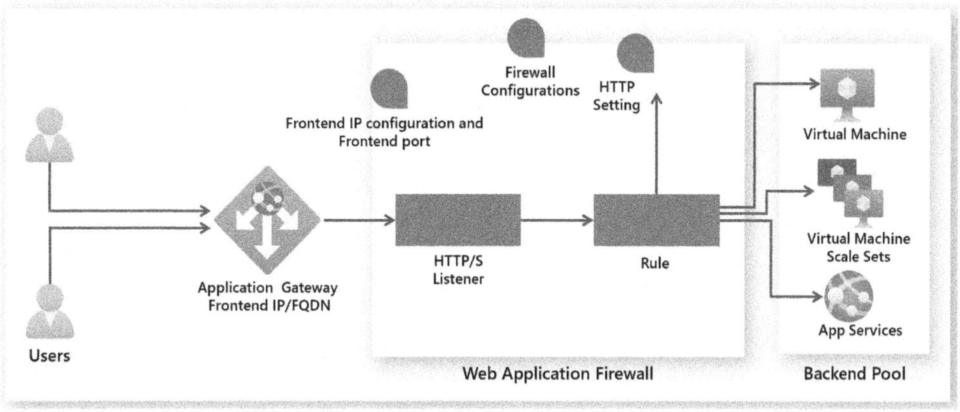
Firewall Configurations
HTTP Setting
Frontend IP configuration and Frontend port
Virtual Machine
Virtual Machine Scale Sets
HTTP/S Listener
Rule
Application Gateway Frontend IP/FQDN
App Services
Users
Web Application Firewall
Backend Pool

Figure 8.26 – Azure Application Gateway components

Here are some of the key components of Azure Application Gateway:

Frontend IP address: AAG can have a public IP address, a private IP address, or both.

- **Listeners**: This is a logical construct that verifies incoming connection requests by utilizing the port number, communication protocol, host computer, and internet protocol address.
- **Routing rules**: Rules determine how AAG routes traffic to the backend pool. You can create rules based on the URL path of the incoming request, the host header, or other criteria.
- **HTTP settings**: AAG routes traffic to the backend servers using the configuration you specify here. Then you associate it with one or more request-routing rules.
- **Backend pool**: This is a set of backend servers that the Application Gateway directs traffic to, to specific virtual machines, virtual machine scale sets, IP addresses/FQDNs, or app services. Once you create a backend pool, you must link it with one or multiple request-routing rules. Additionally, you must establish health probes for each backend pool on your application gateway. When the request-routing rule conditions are met, the application gateway redirects the traffic to the healthy servers (as determined by the health probes) within the corresponding backend pool.
- **Health probes**: These are used to check the health of the backend servers. It is highly recommended that you establish a custom probe for each backend HTTP setting in order to achieve greater control over health monitoring.

In addition, you can configure SSL certificates and Web Application Firewall:

- **SSL certificates**: SSL certificates are used to enable SSL/TLS offloading, where it handles the SSL decryption and encryption process on Application Gateway.
- **WAF (Web Application Firewall)**: This security feature allows you to protect your web applications from common attacks such as SQL injection and cross-site scripting.

The following diagram shows the relationship between AAG components:

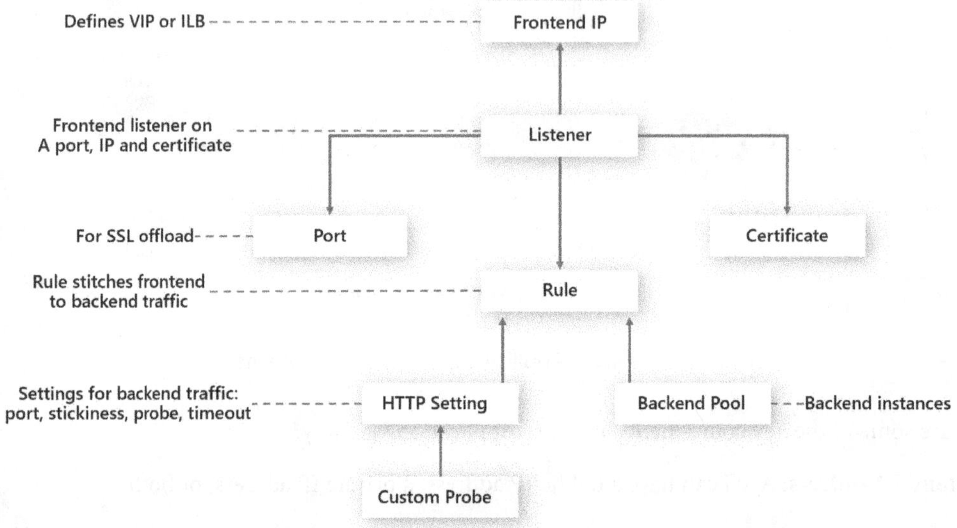
Defines VIP or ILB
Frontend IP
Frontend listener on A port, IP and certificate
Listener
For SSL offload
Port
Certificate
Rule stitches frontend to backend traffic
Rule
Settings for backend traffic: port, stickiness, probe, timeout
HTTP Setting
Backend Pool
Backend instances
Custom Probe

Figure 8.27 – Azure Application Gateway configuration

As you can see, AAG provides advanced L7 capabilities to manage traffic to your web application in Azure. All the features mentioned in this section, such as SSL termination, cookie-based session affinity, URL-based routing, health probes, and a web application firewall, authentication, and authorization, make AAG a comprehensive solution for your web application traffic management needs in Azure.

Now let's see how we can use the Azure portal to create an instance of AAG:

1. The first step to create this service is to provide the basic parameters, including name, region, tier, autoscaling, instance count, and VNet settings, as shown here:

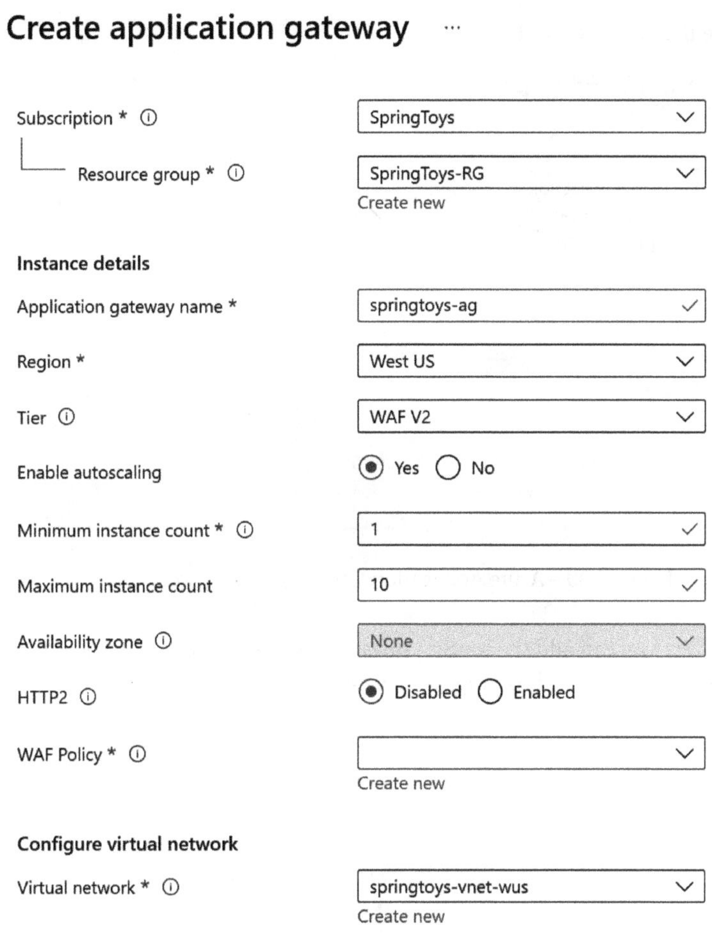

Create application gateway   ...

Subscription * ⓘ	SpringToys ⌄
└── Resource group * ⓘ	SpringToys-RG ⌄
	Create new

Instance details

Application gateway name *	springtoys-ag ✓
Region *	West US ⌄
Tier ⓘ	WAF V2 ⌄
Enable autoscaling	⦿ Yes ◯ No
Minimum instance count * ⓘ	1 ✓
Maximum instance count	10 ✓
Availability zone ⓘ	None ⌄
HTTP2 ⓘ	⦿ Disabled ◯ Enabled
WAF Policy * ⓘ	⌄
	Create new

Configure virtual network

Virtual network * ⓘ	springtoys-vnet-wus ⌄
	Create new

Figure 8.28 – Azure Application Gateway – Basics tab

We will enable WAF as we dive deep into the Azure WAF service later in this chapter.

2. Then, configure the frontend IP address. You can create a new IP address or use an existing one.

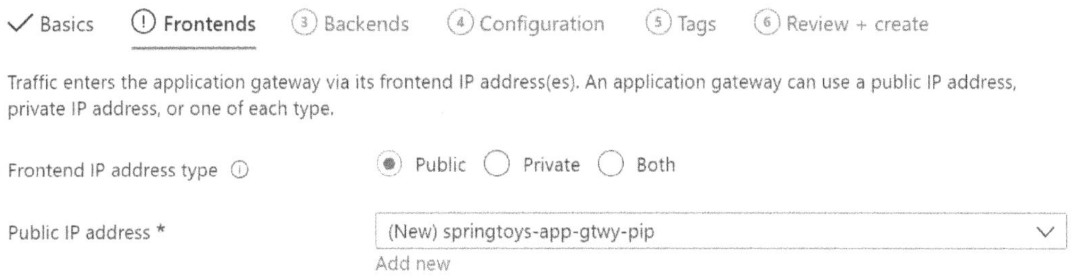

✓ Basics   ① Frontends   ③ Backends   ④ Configuration   ⑤ Tags   ⑥ Review + create

Traffic enters the application gateway via its frontend IP address(es). An application gateway can use a public IP address, private IP address, or one of each type.

Frontend IP address type ⓘ	⦿ Public ◯ Private ◯ Both
Public IP address *	(New) springtoys-app-gtwy-pip ⌄
	Add new

Figure 8.29 – Azure Application Gateway – Frontends tab

3. Next, create the backend pool.

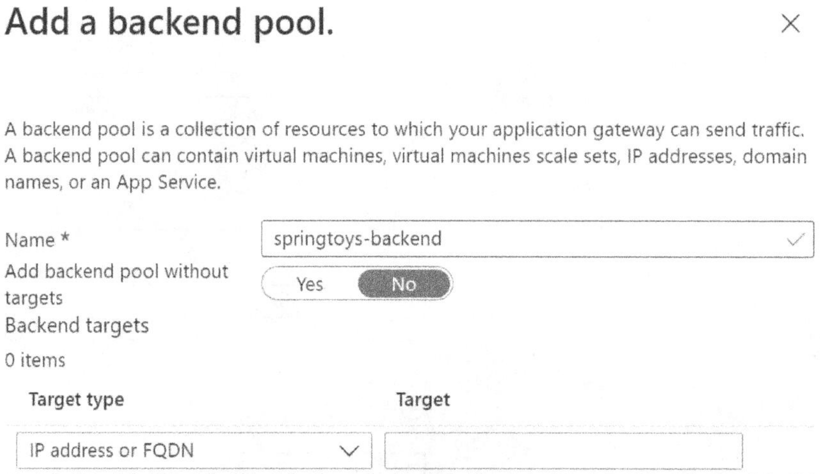
Add a backend pool.
A backend pool is a collection of resources to which your application gateway can send traffic. A backend pool can contain virtual machines, virtual machines scale sets, IP addresses, domain names, or an App Service.
Name *
springtoys-backend
Add backend pool without targets
Yes
No
Backend targets
0 items
Target type
Target
IP address or FQDN

Figure 8.30 – Azure Application Gateway – backend pool tab

4. In the next step, you can add the routing rules; the Azure portal guides the association between the frontend IP address and routing rules, as shown here:

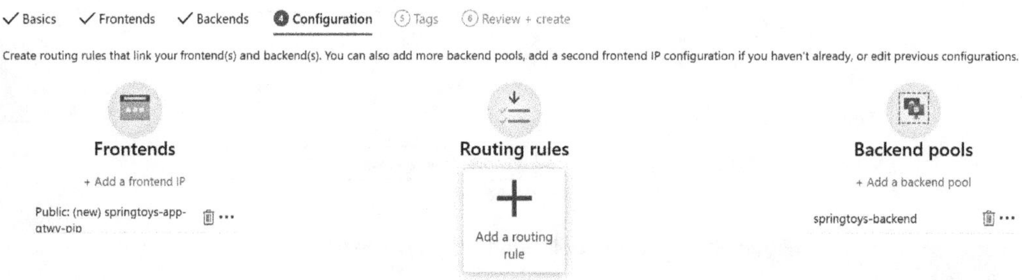
Basics
Frontends
Backends
Configuration
Tags
Review + create
Create routing rules that link your frontend(s) and backend(s). You can also add more backend pools, add a second frontend IP configuration if you haven't already, or edit previous configurations.
Frontends
+ Add a frontend IP
Public: (new) springtoys-app-gtwy-pip
Routing rules
Add a routing rule
Backend pools
+ Add a backend pool
springtoys-backend

Figure 8.31 – Azure Application Gateway – Routing rules

5. Now create a routing rule. We will keep this example simple and use port 80, as shown here:

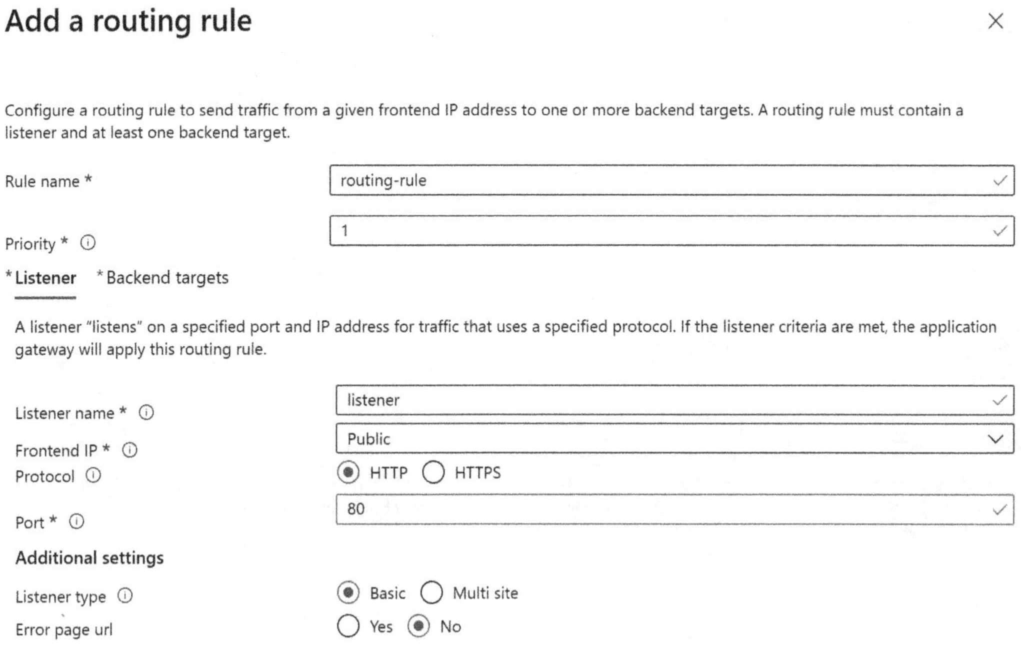

Add a routing rule   ✕

Configure a routing rule to send traffic from a given frontend IP address to one or more backend targets. A routing rule must contain a listener and at least one backend target.

Rule name *   routing-rule   ✓

Priority * ⓘ   1   ✓

* Listener   * Backend targets

A listener "listens" on a specified port and IP address for traffic that uses a specified protocol. If the listener criteria are met, the application gateway will apply this routing rule.

Listener name * ⓘ   listener   ✓

Frontend IP * ⓘ   Public   ⌄
Protocol ⓘ   ⦿ HTTP   ◯ HTTPS

Port * ⓘ   80   ✓

Additional settings

Listener type ⓘ   ⦿ Basic   ◯ Multi site
Error page url   ◯ Yes   ⦿ No

Figure 8.32 – Azure Application Gateway – routing rule configuration tab

6. Note that you need to add the HTTP settings or backend settings as shown here:

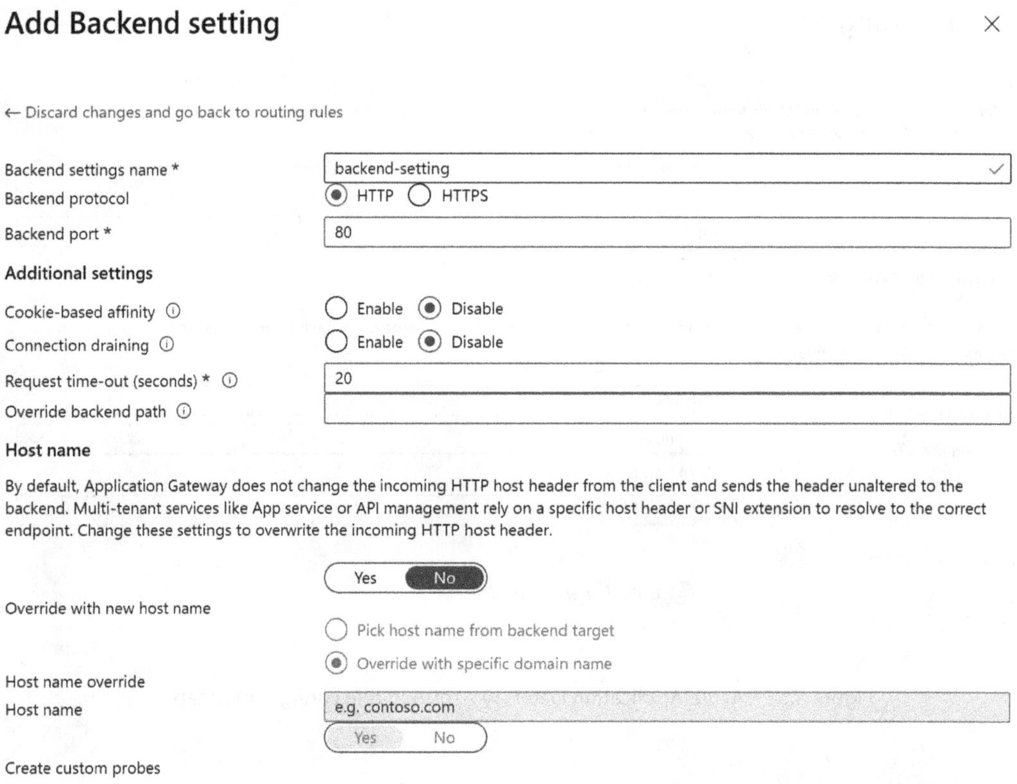

Add Backend setting   ✕

← Discard changes and go back to routing rules

Backend settings name *	backend-setting ✓
Backend protocol	⦿ HTTP ◯ HTTPS
Backend port *	80

Additional settings

Cookie-based affinity ⓘ	◯ Enable ⦿ Disable
Connection draining ⓘ	◯ Enable ⦿ Disable
Request time-out (seconds) * ⓘ	20
Override backend path ⓘ	

Host name

By default, Application Gateway does not change the incoming HTTP host header from the client and sends the header unaltered to the backend. Multi-tenant services like App service or API management rely on a specific host header or SNI extension to resolve to the correct endpoint. Change these settings to overwrite the incoming HTTP host header.

            Yes   **No**

Override with new host name

        ◯ Pick host name from backend target
        ⦿ Override with specific domain name

Host name override
Host name   e.g. contoso.com
            Yes   No

Create custom probes

Figure 8.33 – Azure Application Gateway – backend settings configuration tab

7. Then, configure the backend targets as shown here:

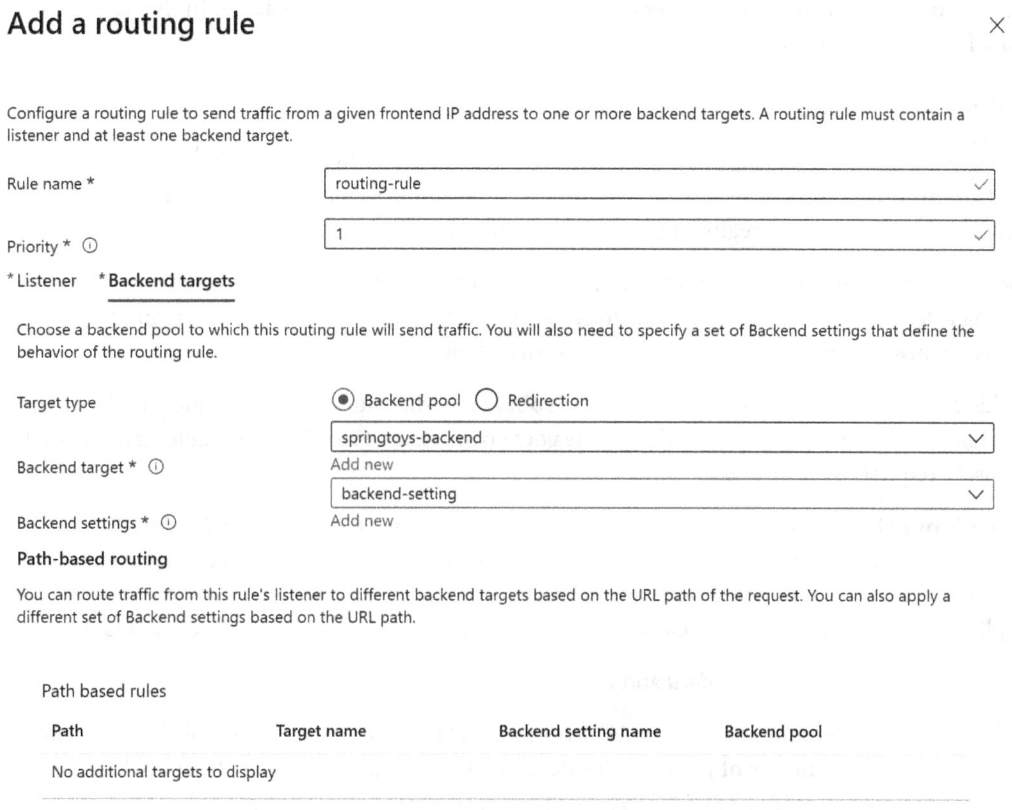

Add a routing rule   ✕

Configure a routing rule to send traffic from a given frontend IP address to one or more backend targets. A routing rule must contain a listener and at least one backend target.

Rule name * | routing-rule   ✓ |

Priority * ⓘ | 1   ✓ |

* Listener   * Backend targets

Choose a backend pool to which this routing rule will send traffic. You will also need to specify a set of Backend settings that define the behavior of the routing rule.

Target type   (●) Backend pool   (○) Redirection

Backend target * ⓘ | springtoys-backend   ∨ |
                    Add new

Backend settings * ⓘ | backend-setting   ∨ |
                    Add new

Path-based routing

You can route traffic from this rule's listener to different backend targets based on the URL path of the request. You can also apply a different set of Backend settings based on the URL path.

Path based rules

Path	Target name	Backend setting name	Backend pool
No additional targets to display			

Figure 8.34 – Azure Application Gateway – Backend targets configuration tab

8. After a few minutes, the deployment should be completed as shown here:

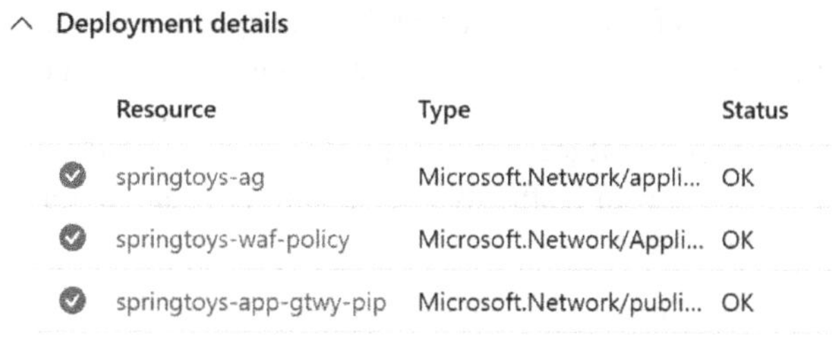

∧ Deployment details

Resource	Type	Status
✅ springtoys-ag	Microsoft.Network/appli...	OK
✅ springtoys-waf-policy	Microsoft.Network/Appli...	OK
✅ springtoys-app-gtwy-pip	Microsoft.Network/publi...	OK

Figure 8.35 – Azure Application Gateway – deployment output

With the preceding steps, we reviewed how you can create a new instance of AAG with WAF enabled.

Suppose your web application is hosted not only in a single region in Azure but across multiple Azure regions. You want to ensure consumers can access the fastest available backend. In that case, Azure Front Door would be one solution.

Azure Front Door

Let's say you have a web application hosted on two virtual machines in two Azure regions: East US and West US. You must ensure users can access the web application even if one of the virtual machines is unavailable or if one of the regions experiences an outage.

The Azure Front Door service can work as the entry point for your web application. You can configure routing rules to redirect incoming traffic to either the East US or West US virtual machines based on the user's location and the current health of the virtual machines.

In this context, if a user is in Virginia, Azure Front Door will route the user's request to the virtual machine in the East US. If that virtual machine goes down, Azure Front Door will automatically route the user's request to the virtual machine in the West US.

Azure Front Door, or **AFD** as we will refer to it from now on, works at Layer 7 using the anycast protocol with split TCP and the Microsoft global network. It includes capabilities such as SSL offloading, URL redirection, URL rewriting, and cookie-based session affinity, and it can be integrated with Web Application Firewall. AFD is a global service while Azure Application Gateway is a regional service.

AFD comes in two flavors: *Standard* and *Premium*.

Let's understand a bit more about how AFD works. AFD leverages Microsoft's global edge network with global and local **points of presence (PoPs)** globally distributed. You can check the locations at the following URL: `https://bit.ly/az-front-door-locations`.

Now, you can have a web application in the East US region, a second instance running in the West US region, and even a third instance running on-premises.

AFD adds an anycast IP address throughout the various PoPs. When a user makes a request to your application, the request is first sent to an Azure Front Door endpoint, which is one of those points of presence that are strategically placed in locations around the world. The endpoint performs several tasks:

- **Determining the best endpoint for resolution**: AFD uses a few factors to determine the best endpoint for the request, including the user's location, the health of the backend servers, and the current load on the system.
- **Routing rules**: Once the best endpoint has been determined, AFD applies routing rules that you have configured to determine how the request should be handled. A Front Door routing rule configuration comprises a "left-hand side" and a "right-hand side." AFD matches the incoming request to the left-hand side of the route. The right-hand side defines how Front Door processes the request.

- **Sending the request to the backend**: AFD sends the request to the appropriate backend server once the routing rules have been applied. The Front Door environment sends synthetic HTTP/HTTPS requests to each of the configured backends at regular intervals. These responses from the probe are then used by Front Door to determine the optimal backend resources to route client requests.
- **Caching content**: AFD can check whether the content is already cached to the user in the nearest Azure CDN edge server. If it is, it will directly return the content to the user. If not, it will forward the request to the origin server, cache the response, and return it to the user.
- **Securing the request**: AFD can be integrated with **Web Application Firewall** (**WAF**) to protect your application.
- **Collecting metrics**: AFD can collect metrics about the request and response, such as response time, status code, and other related data.

The following diagram highlights how AFD works:

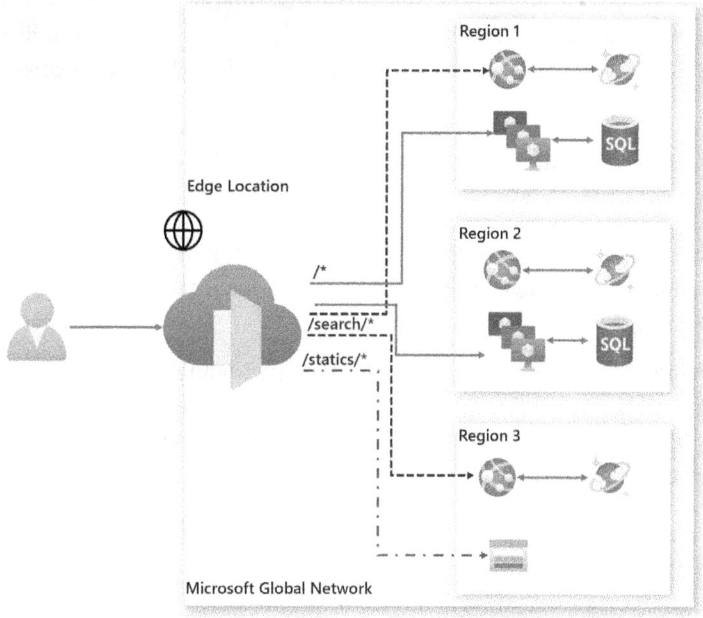
Region 1
SQL
Edge Location
Region 2
/*
/search/*
/statics/*
SQL
Region 3
Microsoft Global Network

Figure 8.36 – Azure Front Door

In the preceding diagram, you can see how Azure Front Door is deployed at a regional scope, distributing incoming traffic across multiple backend pools, such as web servers or Azure services, based on configurable routing rules and intelligent traffic management algorithms.

AFD route rule configuration structure

As mentioned previously, a routing rule configuration has two major parts: a "left-hand side" and a "right-hand side." AFD matches the incoming request to the "left-hand side" of the route. The "right-hand side" will define how AFD processes the request:

- **Incoming match**: This is the "left-hand side." The following properties determine whether the incoming request matches the routing rule:
  - HTTP protocols (HTTP/HTTPS)
  - Hosts (for example, www.`foo.com`, `*.bar.com`)
  - Paths (for example, `/`, `/users/`, `/file.gif`)

  These characteristics are internally expanded, so that every combination of protocol/host/path creates a potential match set.
- **Route data**: This is the "right-hand side." AFD speeds up the processing of requests by using caching. It uses the cached response if caching is enabled for a specific route. If there is no cached response, AFD forwards the request to the appropriate backend in the configured backend pool.

How does route matching work? AFD attempts to match to the most specific match first, looking only at the "left-hand side" of the route. It first matches based on the HTTP protocol, then the frontend host, and then the path:

- **Frontend host matching**: Azure Front Door searches for any routing with an exact match on the host. If no exact frontend host match is found, the request is rejected and a `400 Bad Request` error is sent.
- **Path matching**: AFD searches for any routing rule with an exact match on the path. If no exact path matches are found, it looks for routing rules with a wildcard path that matches. If no routing rules are identified with a matching path, the request is rejected and a `400 Bad Request` error HTTP response is sent.

The following figure shows the workflow of a routing rule.

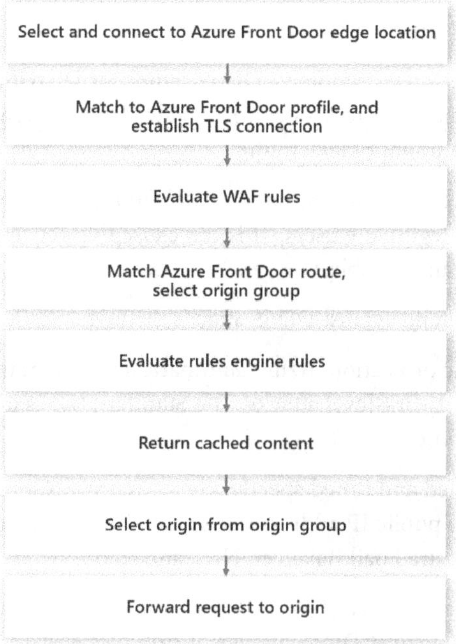
Select and connect to Azure Front Door edge location
Match to Azure Front Door profile, and establish TLS connection
Evaluate WAF rules
Match Azure Front Door route, select origin group
Evaluate rules engine rules
Return cached content
Select origin from origin group
Forward request to origin

Figure 8.37 – Azure Front Door – route rule configuration structure

So far, we have reviewed the Azure load balancing core for distributing incoming requests across a group of servers or resources to optimize resource utilization, increase throughput, reduce response time, and prevent overloading any resource.

In the next section, we will review a comprehensive set of tools and services to help you protect your Azure-based resources and applications from network-based attacks.

Network security

Imagine SpringToys is spread across multiple Azure regions, and their infrastructure includes multiple virtual networks and connections to an on-premises network. The SpringToys IT team is looking for options to protect its assets against malicious actors trying to infiltrate the network and web applications.

In this section, we will discuss the core services available in Azure that help secure network connections and communications. We will highlight the following:

- Azure DDoS Protection
- Azure Firewall
- Azure WAF

Let's start by reviewing how SpringToys can leverage Azure DDoS Protection to improve its security posture.

Azure DDoS protection

First, let's set the stage and agree on terminology. **DDoS** stands for **Distributed Denial of Service** – a type of cyber-attack in which many devices, often infected with malware, are used to flood a targeted website or server with a huge amount of fake traffic.

The goal of a DDoS attack is to overwhelm the targeted web application or server with so much traffic that it becomes inaccessible to legitimate users. This can cause the web application or server to slow down or even crash, making it unavailable.

Azure DDoS protection can defend against DDoS attacks through two tiers: *Basic* and *Standard*.

Azure, by default, enables Basic protection. Azure automatically enables traffic monitoring and real-time mitigation of common network-level attacks. The Standard tier is ideal if you want additional mitigation and analysis capabilities.

The Standard tier can be enabled and tuned specifically to Azure VNet resources. Once you enable this tier, policies are applied to public IP addresses associated with resources in your VNet, such as Azure Application Gateway, Azure Load Balancer, NIC, and so on.

You can leverage real-time telemetry using Azure Monitor and detect and analyze traffic patterns to better understand and protect against DDoS attacks. It provides a real-time, historical view of the network traffic across Azure Virtual Network.

Before you can enable DDoS protection on a new or existing virtual network, you need to create a DDoS plan, as shown here:

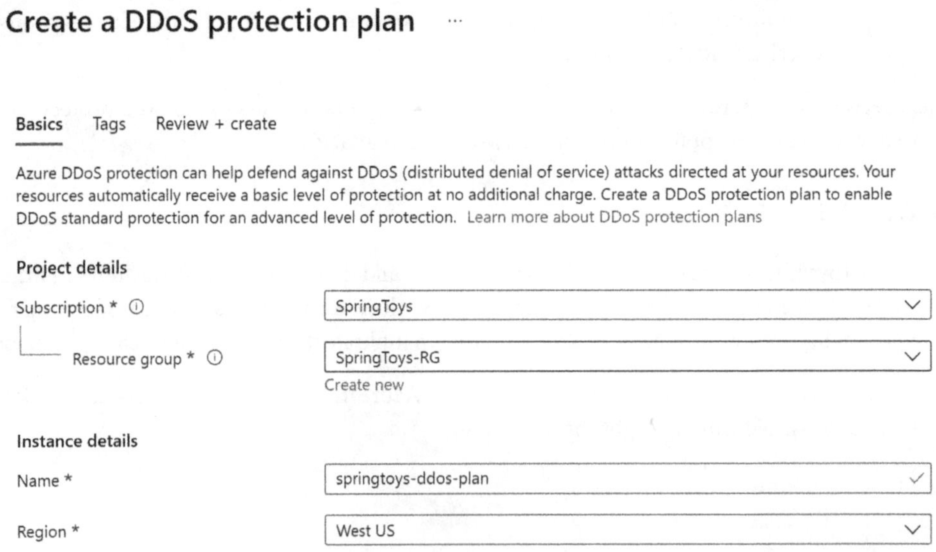
Create a DDoS protection plan ...
Basics
Tags
Review + create
Azure DDoS protection can help defend against DDoS (distributed denial of service) attacks directed at your resources. Your resources automatically receive a basic level of protection at no additional charge. Create a DDoS protection plan to enable DDoS standard protection for an advanced level of protection. Learn more about DDoS protection plans
Project details
Subscription *
SpringToys
Resource group *
SpringToys-RG
Create new
Instance details
Name *
springtoys-ddos-plan
Region *
West US

Figure 8.38 – Create a DDoS protection plan

Once you create a DDoS protection plan, you enable DDoS protection on a new or existing VNet.

We will create a new VNet, and under the security settings, you will see the option to enable **DDoS Protection Standard**, and then you can select the plan just created, as shown here:

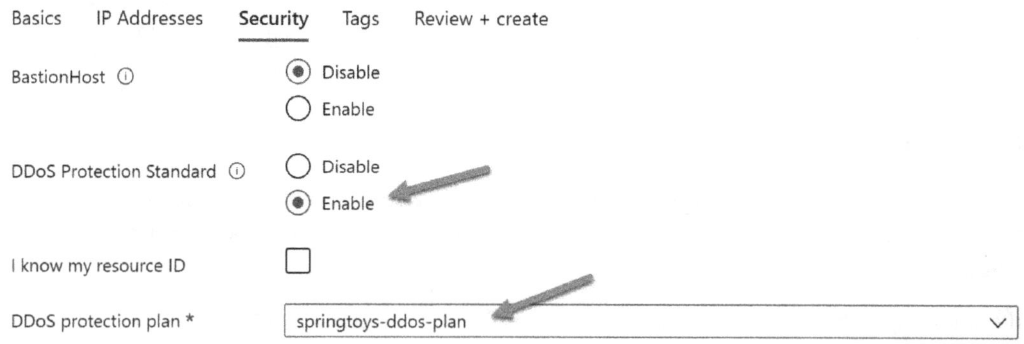
Basics
IP Addresses
Security
Tags
Review + create
BastionHost
Disable
Enable
DDoS Protection Standard
Disable
Enable
I know my resource ID
DDoS protection plan *
springtoys-ddos-plan

Figure 8.39 – Associate DDoS protection plan to new VNet

Once you associate a DDoS protection plan, you can configure DDoS telemetry, diagnostic logs, and alerts using Azure Monitor.

The following figure shows the VNet as one protected resource in the DDoS protection plan:

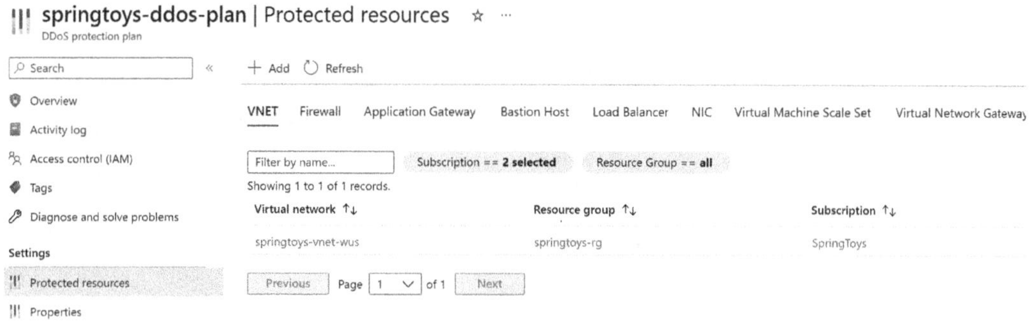
springtoys-ddos-plan | Protected resources
DDoS protection plan
Search
Add
Refresh
Overview
Activity log
Access control (IAM)
Tags
Diagnose and solve problems
Settings
Protected resources
Properties
VNET
Firewall
Application Gateway
Bastion Host
Load Balancer
NIC
Virtual Machine Scale Set
Virtual Network Gateway
Filter by name...
Subscription == 2 selected
Resource Group == all
Showing 1 to 1 of 1 records.
Virtual network
Resource group
Subscription
springtoys-vnet-wus
springtoys-rg
SpringToys
Previous
Page
1
of 1
Next

Figure 8.40 – DDoS protection plan – protected resources

Azure DDoS Protection features include the following:

- Turnkey protection
- Always-on traffic monitoring
- Adaptive tuning
- Attack metrics and alerts
- Multi-layered protection

DDoS protection in the Standard tier continuously monitors traffic utilization and compares it against thresholds defined in the DDoS policy. If the threshold is exceeded, then the DDoS mitigation is automatically triggered. If traffic returns to the normal threshold, the mitigation is stopped.

Consider the scenario where SpringToys has multiple Azure virtual networks and wants to secure communication between them and access to resources. Azure DDoS Protection can help them, and they can use Azure Firewall to protect resources from external threats.

Azure Firewall

Azure Firewall is a firewall-as-a-service that protects a network from incoming and outgoing threats. It uses firewall rules to control access to resources and services in a virtual network and supports threat intelligence-based filtering and logging features.

You can leverage a hub-and-spoke topology and implement Azure Firewall in the hub VNet, our central network. Then, SpringToys can create security rules to allow traffic to flow between the VNets, but only on specific ports and protocols and from specific IP addresses or ranges.

SpringToys can also use threat protection and flow logging to protect against potential threats. The following figure provides a high-level overview of Azure Firewall capabilities.

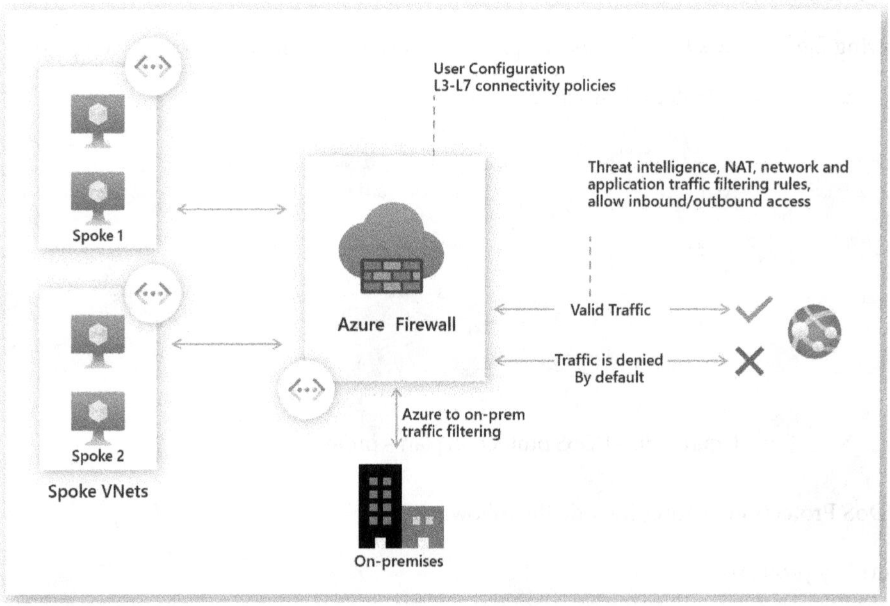
User Configuration
L3-L7 connectivity policies
Threat intelligence, NAT, network and application traffic filtering rules, allow inbound/outbound access
Spoke 1
Spoke 2
Spoke VNets
Azure Firewall
Valid Traffic
Traffic is denied By default
Azure to on-prem traffic filtering
On-premises

Figure 8.41 – Azure Firewall overview

Azure Firewall features include the following:

- **Built-in high availability**: Azure Firewall is a PaaS service, and no additional resources are required to achieve high availability.
- **Scalability**: Azure Firewall can scale up or down as required to adapt to fluctuating network traffic patterns.
- **Application FQDN filtering rules**: You can restrict outbound HTTP/S traffic or Azure SQL traffic to a particular list of **fully qualified domain names** (**FQDNs**) that includes wildcards. This feature does not necessitate TLS termination.
- **Network traffic filtering rules**: You can centrally create rules for network filtering by source and destination IP address, port, and protocol – whether to allow or deny them. Azure Firewall is fully stateful and able to identify legitimate packets for different connection types. The rules are enforced and logged across various subscriptions and virtual networks.
- **FQDN tags**: Tags allow you to easily permit well-known Azure service network traffic through your firewall. For instance, to enable Windows Update network traffic through your firewall, you create an application rule and include the Windows *Update* tag. This enables network traffic from Windows Update to pass through the Azure Firewall.
- **Service tags**: A service tag represents a group of IP address prefixes to minimize the complexity of creating security rules. Microsoft manages the address prefixes included in the service tag and automatically updates the service tag as addresses change.
- **Threat intelligence**: This capability allows your firewall to alert and block traffic from/to known malicious IP addresses and domains, sourced from the Microsoft Threat Intelligence feed.
- **Outbound SNAT support**: When using Azure Firewall, all outbound virtual network traffic IP addresses are translated to the public IP of the firewall. This process is known as **Source Network Address Translation** (**SNAT**). This means that all outbound traffic will appear to be coming from the Azure Firewall's public IP address, providing an added layer of security and making it easier to identify and manage the traffic. You can identify and allow traffic from your virtual network to remote internet destinations.
- **Inbound DNAT support**: Inbound internet network traffic directed to your firewall's public IP address is translated (Destination Network Address Translation) and filtered to the private IP addresses on your VNets.
- **Multiple public IP addresses**: You can associate multiple public IP addresses (up to 250) with Azure Firewall, to enable specific DNAT and SNAT scenarios.
- **Azure Monitor integration**: All events are integrated with Azure Monitor. There are various options available for managing logs with Azure Firewall, such as archiving them to a storage account, streaming events to an event hub, or sending them to Azure Monitor logs. This gives you flexibility in how you choose to store and analyze your logs, depending on your organization's specific needs and requirements.

- **Forced tunneling**: With Azure Firewall, you have the ability to direct all internet-bound traffic to a specific next hop, rather than allowing it to go directly to the internet. This can be useful if you have an on-premises edge firewall or another **network virtual appliance** (**NVA**) that you want to use to process the traffic before it reaches the internet. This feature allows you to add an extra layer of security to your network and control the traffic flow to the internet.
- **Web categories**: Administrators can use web categories to control user access to various types of websites, such as gambling or social media sites. Both Azure Firewall Standard and Azure Firewall Premium Preview offer web category capabilities, but the Premium Preview version offers a more advanced level of control. In the Standard SKU, web categories are matched based on the FQDN, while the Premium SKU uses the entire URL, including both HTTP and HTTPS traffic, for more precise categorization.
- **Certifications**: Azure Firewall is a well-certified solution, holding various certifications, such as **Payment Card Industry** (**PCI**), **Service Organization Controls** (**SOC**), **International Organization for Standardization** (**ISO**), and **International Computer Security Association** (**ICSA**) Labs compliance. These certifications demonstrate that Azure Firewall has met strict security standards and is suitable for use in various industries.

As mentioned previously, you can configure different types of rules in Azure Firewall: NAT rules, network rules, and application rules. Azure Firewall denies all traffic by default. You configure rules to allow traffic.

Azure Firewall rules are configured with classic rules or using Firewall Policy. If you use classic rules, rule collections are processed in priority order according to the rule type, from lower numbers to higher numbers, from 100 to 65,000. A rule collection name can have only letters, numbers, underscores, periods, or hyphens. If you use Firewall Policy, rules are organized inside rule collections, which are contained in rule collection groups: DNAT, network, and application.

Now let's review the following exercise on how to deploy and configure Azure Firewall using infrastructure-as-code.

Exercise 2 – Azure Firewall – implement secure network access using the Bicep language

In the following exercise, we will leverage infrastructure-as-code to deploy our network security infrastructure:

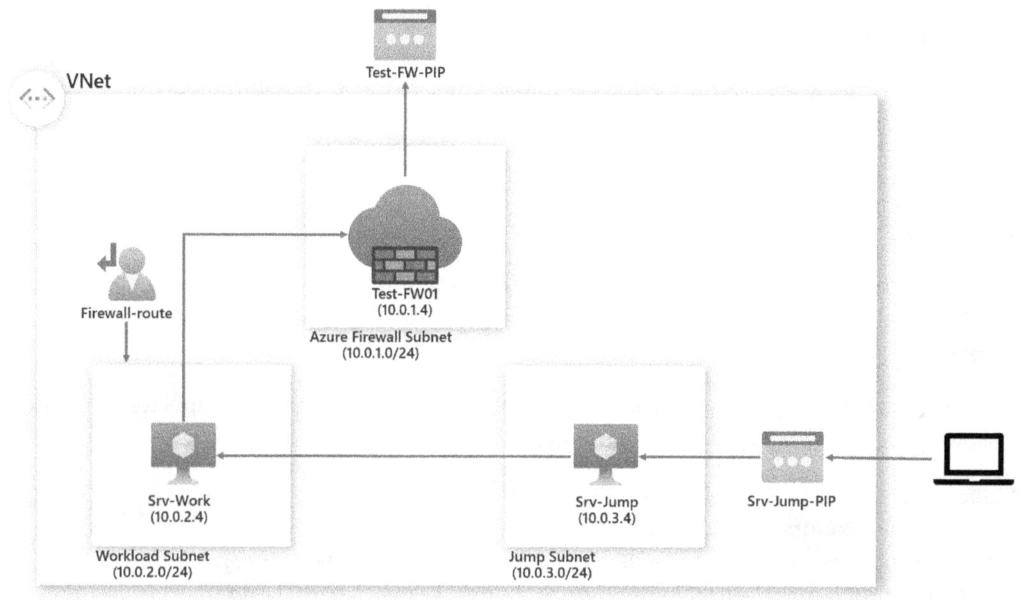
Test-FW-PIP
VNet
Firewall-route
Test-FW01
(10.0.1.4)
Azure Firewall Subnet
(10.0.1.0/24)
Srv-Work
(10.0.2.4)
Srv-Jump
(10.0.3.4)
Srv-Jump-PIP
Workload Subnet
(10.0.2.0/24)
Jump Subnet
(10.0.3.0/24)

Figure 8.42 – Azure Firewall deployment

The preceding diagram highlights the resources we will deploy as part of the environment using the Bicep Language:

- A virtual network with a workload subnet and a jump host subnet
- A virtual machine in each subnet
- A custom route that ensures all outbound workload traffic from the workload subnet uses the firewall
- Firewall application rules that only allow outbound traffic to `www.bing.com`
- Firewall network rules that allow external DNS server lookups

Prerequisites

Before we get started with the exercise, here are a few things you need to be ready with:

- An active Azure account
- Azure Bicep installed on your local machine
- Azure PowerShell
- A resource group in your Azure subscription

The solution will include the following files:

- The `main.bicep` file: This is the Bicep template
- The `azuredeploy.parameters.json` file: This parameter file contains the values for deploying your Bicep template.

Source code

You can download the files from the following GitHub repository: `https://github.com/PacktPublishing/Azure-Architecture-Explained/tree/main/Chapter08/exercise-2`.

Now, open a new instance of PowerShell or Windows Terminal in your working directory, and use the following command to deploy the environment:

```
$date = Get-Date -Format "MM-dd-yyyy"
$rand = Get-Random -Maximum 1000
$deploymentName = "AzInsiderDeployment-"+"$date"+"-"+"$rand"
New-AzResourceGroupDeployment -Name $deploymentName -ResourceGroupName
springtoys-RG -TemplateFile .\main.bicep -TemplateParameterFile .\
azuredeploy.parameters.json -c
```

The following image shows the output of the deployment.

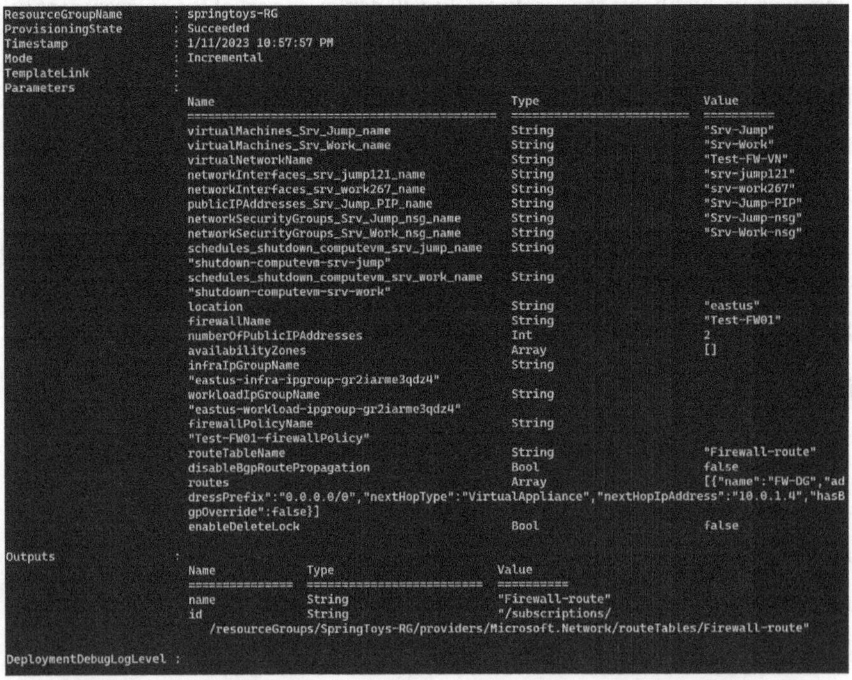
ResourceGroupName : springtoys-RG
ProvisioningState : Succeeded
Timestamp : 1/11/2023 10:57:57 PM
Mode : Incremental
TemplateLink :
Parameters :
Name Type Value
virtualMachines_Srv_Jump_name String "Srv-Jump"
virtualMachines_Srv_Work_name String "Srv-Work"
virtualNetworkName String "Test-FW-VN"
networkInterfaces_srv_jump121_name String "srv-jump121"
networkInterfaces_srv_work267_name String "srv-work267"
publicIPAddresses_Srv_Jump_PIP_name String "Srv-Jump-PIP"
networkSecurityGroups_Srv_Jump_nsg_name String "Srv-Jump-nsg"
networkSecurityGroups_Srv_Work_nsg_name String "Srv-Work-nsg"
schedules_shutdown_computevm_srv_jump_name String
"shutdown-computevm-srv-jump"
schedules_shutdown_computevm_srv_work_name String
"shutdown-computevm-srv-work"
location String "eastus"
firewallName String "Test-FW01"
numberOfPublicIPAddresses Int 2
availabilityZones Array []
infraIpGroupName String
"eastus-infra-ipgroup-gr2iarme3qdz4"
workloadIpGroupName String
"eastus-workload-ipgroup-gr2iarme3qdz4"
firewallPolicyName String
"Test-FW01-firewallPolicy"
routeTableName String "Firewall-route"
disableBgpRoutePropagation Bool false
routes Array [{"name":"FW-DG","ad
dressPrefix":"0.0.0.0/0","nextHopType":"VirtualAppliance","nextHopIpAddress":"10.0.1.4","hasB
gpOverride":false}]
enableDeleteLock Bool false
Outputs :
Name Type Value
name String "Firewall-route"
id String "/subscriptions/
/resourceGroups/SpringToys-RG/providers/Microsoft.Network/routeTables/Firewall-route"
DeploymentDebugLogLevel :

Figure 8.43 – Azure Firewall deployment output

In the Azure portal, you can go to Firewall Policy and review the **Rule collections** section, as shown here:

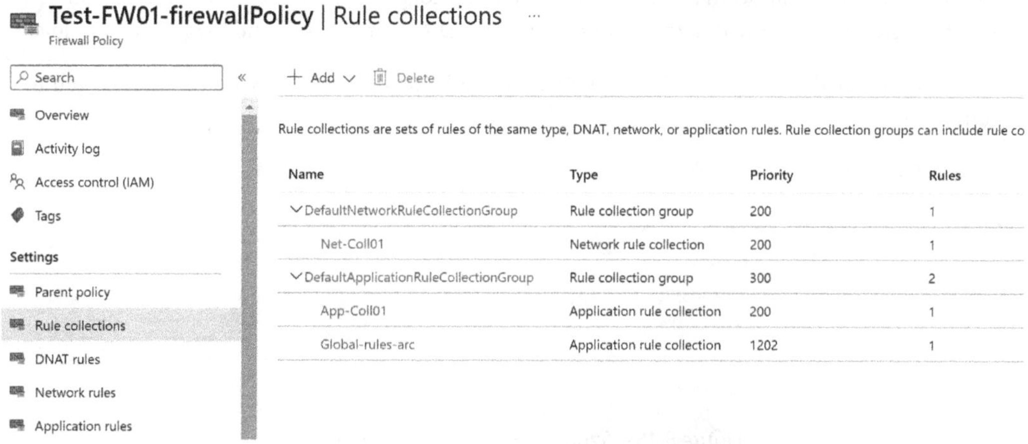
Test-FW01-firewallPolicy | Rule collections
Firewall Policy
Search
Overview
Activity log
Access control (IAM)
Tags
Settings
Parent policy
Rule collections
DNAT rules
Network rules
Application rules
Add
Delete
Rule collections are sets of rules of the same type, DNAT, network, or application rules. Rule collection groups can include rule co
Name
Type
Priority
Rules
DefaultNetworkRuleCollectionGroup
Rule collection group
200
1
Net-Coll01
Network rule collection
200
1
DefaultApplicationRuleCollectionGroup
Rule collection group
300
2
App-Coll01
Application rule collection
200
1
Global-rules-arc
Application rule collection
1202
1

Figure 8.44 – Firewall Policy – Rule collections

On this page, you should see two rule collections, one with a network rule and the second with two application rules.

To test this environment, open a Remote Desktop session to `Srv-Work`, start Internet Explorer, and browse to `https://www.bing.com`. The website should display without errors. Then try a different site, and you should see that Azure Firewall blocks access to the website.

Now let's understand how we can use network security services to protect web applications using Azure Web Application Firewall.

Consider the SpringToys e-commerce platform that runs on app services and microservices. The company wants to protect its store from common web attacks such as SQL injection, **cross-site scripting** (**XSS**), and session hijacking. In the next section, we will understand how SpringToys can leverage Azure WAF to enhance the protection of their web-based applications.

Azure WAF

Azure WAF helps protect web applications from malicious attacks by monitoring and blocking suspicious traffic. It is a service provided by Microsoft Azure that helps protect web applications from a variety of attacks, such as SQL injection and cross-site scripting.

Suppose SpringToys is using Azure Application Gateway or Azure Front Door to improve the performance and availability of their e-commerce platform. They could improve their application security by enabling Azure WAF as it can be integrated with Azure Application Gateway or Azure Front Door, which act as a reverse proxy, handling the traffic going to your web application.

Azure WAF can be enabled in two modes: *Prevention* and *Detection*. In Detection mode, WAF does not block any requests; requests matching the WAF rules are logged in WAF logs. In Prevention mode, requests that match rules defined in the **Default Rule Set** (**DRS**) are blocked and logged in WAF logs.

The following screenshot shows SpringToys WAF using Detection mode. Note you can switch to Prevention mode:

springtoys-waf-policy
Application Gateway WAF policy
Search
Disable
Switch to prevention mode
Delete
Refresh
Overview
Activity log
Access control (IAM)
Tags
Essentials
Resource group (move) : SpringToys-RG
Location : West US
Subscription (move) : SpringToys
Subscription ID :
Policy state : Enabled
Policy mode : Detection
Associated application g... : 1

Figure 8.45 – Azure WAF – Detection mode

When there's a client request to SpringToys' e-commerce application, Azure WAF will inspect the incoming traffic and block any request that appears to be malicious. It can also monitor and log incoming traffic for security and compliance purposes.

Azure WAF can be configured with predefined rules (Azure-managed rules) or custom rules to inspect and block incoming traffic that contains known attack patterns.

The Azure-managed rulesets are based on the OWASP ModSecurity **Core Rule Set** (**CRS**) and are updated automatically by Azure to protect against new and emerging threats. The managed rules protect against the following threat categories:

- Cross-site scripting
- Java attacks
- Local file inclusion
- PHP injection attacks
- Remote command execution
- Remote file inclusion
- Session fixation
- SQL injection protection
- Protocol attackers

The following screenshot shows an example of managed rules with OWASP 3.2:

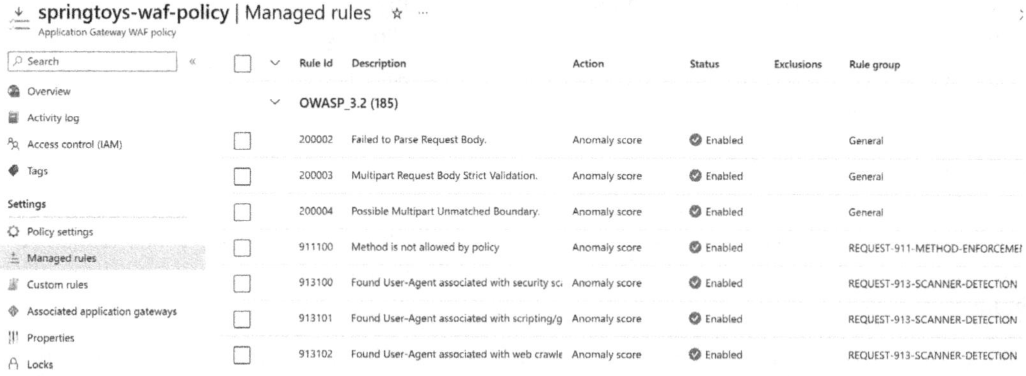
springtoys-waf-policy | Managed rules
Application Gateway WAF policy
Search
Overview
Activity log
Access control (IAM)
Tags
Settings
Policy settings
Managed rules
Custom rules
Associated application gateways
Properties
Locks
Rule Id
Description
Action
Status
Exclusions
Rule group
OWASP_3.2 (185)
200002 Failed to Parse Request Body. Anomaly score Enabled General
200003 Multipart Request Body Strict Validation. Anomaly score Enabled General
200004 Possible Multipart Unmatched Boundary. Anomaly score Enabled General
911100 Method is not allowed by policy Anomaly score Enabled REQUEST-911-METHOD-ENFORCEME
913100 Found User-Agent associated with security sc Anomaly score Enabled REQUEST-913-SCANNER-DETECTION
913101 Found User-Agent associated with scripting/g Anomaly score Enabled REQUEST-913-SCANNER-DETECTION
913102 Found User-Agent associated with web crawle Anomaly score Enabled REQUEST-913-SCANNER-DETECTION

Figure 8.46 – Azure WAF – Managed rules

On the other hand, you can create custom rules, and they will be evaluated for each request that goes through Azure WAF. A custom rule consists of a priority number, rule type, match conditions, and an action. And you can create *match rules* and *rate limit rules*:

- **Match rules**: These rules control access based on a set of conditions
- **Rate limit rules**: These rules control access based on matching conditions plus the rates of incoming requests

The following screenshot shows you can add a custom rule in the Azure portal.

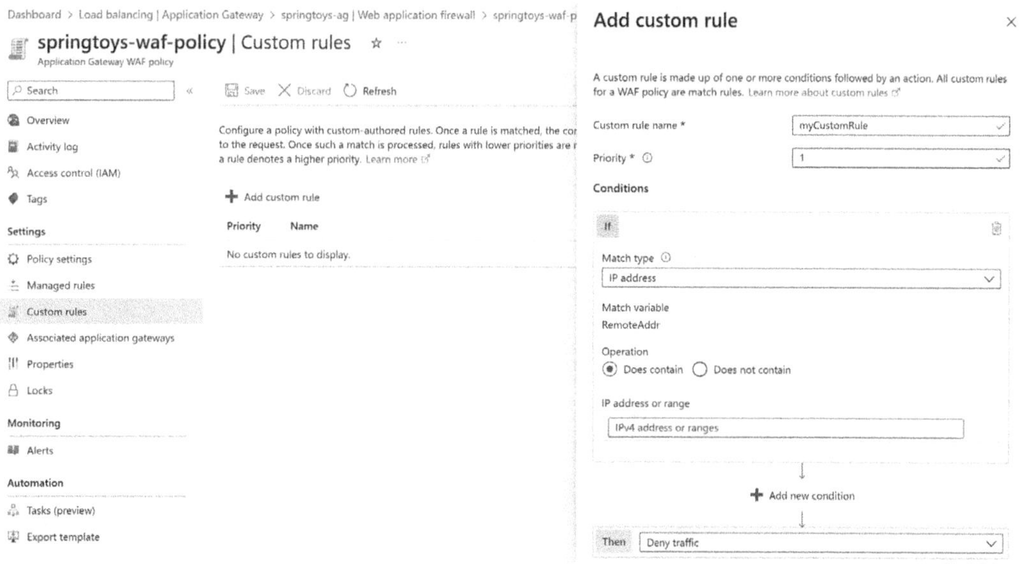
Dashboard > Load balancing | Application Gateway > springtoys-ag | Web application firewall > springtoys-waf-p
springtoys-waf-policy | Custom rules
Application Gateway WAF policy
Search
Save
Discard
Refresh
Overview
Activity log
Access control (IAM)
Tags
Settings
Policy settings
Managed rules
Custom rules
Associated application gateways
Properties
Locks
Monitoring
Alerts
Automation
Tasks (preview)
Export template
Add custom rule
Priority
Name
No custom rules to display.
Add custom rule
A custom rule is made up of one or more conditions followed by an action. All custom rules for a WAF policy are match rules. Learn more about custom rules
Custom rule name *
myCustomRule
Priority *
1
Conditions
If
Match type
IP address
Match variable
RemoteAddr
Operation
Does contain
Does not contain
IP address or range
IPv4 address or ranges
Add new condition
Then
Deny traffic

Figure 8.47 – Azure WAF – Custom rules

SpringToys can also leverage Azure WAF to monitor the traffic to their e-commerce application and generate reports on the types of attacks and where they are coming from, which helps them to identify and mitigate vulnerabilities in their code. You can refer to the following decision tree to decide which service might suit your load balancing needs: `https://bit.ly/az-load-balancing-decision-tree`.

Summary

As your organization transitions to Azure, it's important to design a secure network that prevents unauthorized access or attacks by implementing controls on network traffic and only allowing legitimate requests.

In this chapter, we discussed various network security options that can be implemented to meet the specific security needs of your organization. In the next chapter, we will analyze how SpringToys can manage and control its Azure environment, ensuring that resources are used efficiently and securely and that compliance requirements are met.

9

Securing Access to Your Applications

Ensuring the confidentiality, integrity, and availability of applications and data hosted on cloud platforms such as Microsoft Azure is crucial in today's digital age.

In this chapter, we will explore various Azure security tools and capabilities that can be employed to design a secure environment for your applications. By leveraging these features, you can effectively protect your applications from unauthorized access and potential security threats.

Specifically, we will look at the following areas:

- Designing for security
- Securing traffic
- Securing keys and secrets
- Using managed identities

Technical requirements

This chapter will use the Azure portal (`https://portal.azure.com`) throughout for the examples. You will also need Visual Studio for Windows installed. The code examples for this chapter can be found in this book's GitHub repository at `https://github.com/PacktPublishing/Azure-Architecture-Explained`.

Designing for security

Azure provides a range of tools to help us ensure our systems are secure. Some of these tools, such as Azure Key Vault, are independent services to be consumed as part of our solution. Other tools are component-specific, such as an Azure SQL Server's firewall or threat protection capabilities.

In many cases, some options may seem to be duplicated or overlap in services – this isn't by accident. When designing cloud applications, we often want to deploy and combine multiple tools that seem to serve the same purpose, or at the very least provide additional layers.

This multi-layered approach is called **defense in depth** and is an important subject in cloud platforms such as Azure. This concept essentially states that we should expect one or more of our security measures to fail – and then design additional measures to compensate when they do.

It may seem odd to assume something will fail – after all, years of system design is often about providing redundancy around such events.

However, modern applications employ different patterns when we build them and leverage cloud services that are meant to be flexible enough to be used for multiple use cases. For example, an application built in an Agile manner will be continually developed, updated, and deployed into Azure – we can check and test as much as possible, but there's always a chance an alteration has opened a hole in our security.

In addition, data services such as Azure SQL and Azure storage accounts can be configured to be publicly available, privately available, or a mixture of both. It's possible to accidentally deploy a storage account that is open to the public when we meant it to be private. This may seem like a silly mistake that can easily be caught, but you'd be surprised how often it can happen – and this is just a very simple example.

More complex solutions will have multiple components, each configured differently and sometimes managed by different teams. And even the short history of the cloud is littered with examples of mistakes that inadvertently exposed sensitive information.

Another area to consider is the concept of **least privilege access**, which states that users should have the bare minimum access rights they need to do their job and nothing more. In development scenarios, one common issue that arises is that our developers have all the access keys to the databases that contain our data.

In development, this is fine, but at some point, a system goes live and may hold sensitive data. A developer does not need to know the logins and passwords required to access production data, but then if they don't know this information, how can the application be configured to access the database?

Finally, we need to continually monitor and be alerted to security threats. Your application may be secure today, but, in the words of Mark Twain:

> *"It ain't what you don't know that gets you into trouble. It's what you know for sure that just ain't so."*

In other words, we can try our best to secure our application against threats we know about, but it's the ones we didn't think about or didn't know about that will expose us. Threat monitoring is about being on the lookout for suspicious activity and responding to it.

Luckily, Azure provides us with a whole range of tools and configuration options – however, that's all they are, *options*, and we need to ensure we design our systems to use them.

Of these options, **Identity Access Management** (**IAM**), which involves being able to securely authenticate our users, and **role-based access control** (**RBAC**), which involves being able to authorize our users, are key components in any solution design. We covered IAM, RBAC, and using Sentinel to monitor users in the first three chapters of this book, so in this chapter, we'll cover securing our applications at the network level, as well as learning how to secure authentication between different components.

Securing traffic

Securing traffic at the network level is often a given in an on-premises network. When you create systems in a corporate network, they will usually be secure by default. In other words, anything you deploy would be inside your network and to expose it to the internet, you would need to specifically allow traffic out through a firewall.

In contrast, within Azure, many services are often public by default, and if you wanted to secure them so that they're internal, only would you need to configure this aspect.

> **Important note**
>
> Many applications, even those built for internal use only, are often exposed and consumed over the internet (for example, a SaaS product). Zero trust is a common pattern that means we control access to applications via identity and conditional access policies that are applied to end devices to control access rather than firewalls.
>
> Therefore, you may think that any form of network-level security is no longer relevant – however, firewalls are still used in hybrid networks when we only want to allow access from a corporate network, or even set IP ranges. Also, although a frontend app may be exposed to the internet, its backend databases and storage accounts will still need protection, so network-level controls are still important.

There are several different network-level controls we can implement to secure our networks, such as SQL-level firewall rules, web application firewall rules, VNet firewall rules, and even separate firewall devices. Some of these may seem to overlap, so it is important to consider the differences

These different controls are often used at different stages as data moves from a backend database to the user interface on a web app and ultimately a user's device. Having multiple roadblocks along the path data takes is a key design consideration for the security-in-depth ethos.

Essentially, our data, which is in a database or storage account, should be completely locked down to prevent any unauthorized access. Only upstream services, such as a web application, should be able to access it.

Next, the web application itself might need some protection – a firewall in front of a web app can protect us by limiting who can access the application at the IP level – that is, only allowing certain IP addresses to access it. The web app itself would be configured to only allow traffic from the firewall.

Finally, a **web application firewall** (**WAF**) can provide further protection by scanning the requests coming into the app for known attacks, such as SQL injection. Again, the Azure firewall would be configured to only send traffic to the web app that originated in the WAF.

The resultant architecture could look something like what is shown in *Figure 9.1*:

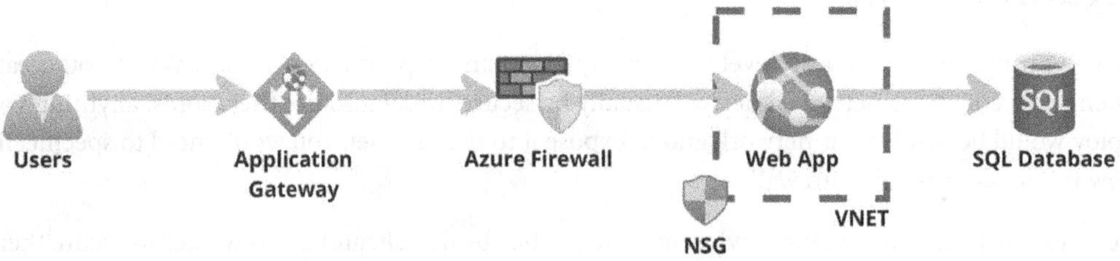
Users
Application Gateway
Azure Firewall
Web App
SQL
SQL Database
NSG
VNET

Figure 9.1 – Multi-layered security

This is a robust solution, with each layer performing a specific task or tasks and configured to force traffic from set sources. Any attempt to access the database or web app directly, for example, would be denied.

It's a bit overkill for some use cases, but for other scenarios, another type of WAF called Azure Front Door would be better instead of Azure Application Gateway.

To understand which options are best for which scenarios, let's have a look at each component.

SQL database firewalls

Azure SQL servers have network firewalls built in. They allow you to restrict access via IP addresses or specific VNets – in this way, we can configure them to only allow access from specific web apps.

Web application VNet integration

By default, Azure web apps are not connected to a VNet; instead, they are publicly available. However, the underlying IP address is hidden and they must be accessed via a URL. To provide tighter network-level integration with other services, such as the Azure SQL firewall, you must turn on VNet integration. This attaches the web app to a VNet in your subscription, which can then be used to tighten access.

> **Note**
>
> Technically, the IP address for a web app isn't hidden – they use a wide range of IP addresses. Because web apps are managed services, the actual VM, and therefore its associated IP, come from a range that is configured internally. You can view that range, but if you want to lock down access properly, VNet integration is the best way.

We shall see an example of how to do this shortly.

Azure Firewall

Azure Firewall is a stateful, managed firewall service. It enables you to filter traffic based on IP addresses, protocols, and ports, ensuring that only authorized traffic can access your applications. With features such as threat intelligence-based filtering and integration with Azure Monitor, Azure Firewall can help you identify and respond to potential security threats.

Application Gateway

Azure Application Gateway is a web traffic load balancer and WAF that helps secure your applications from common web-based attacks. It offers features such as SSL termination, cookie-based session affinity, and URL path-based routing. The integrated WAF provides centralized protection against various attack vectors, including SQL injection, **cross-site scripting** (**XSS**), and other OWASP Top 10 vulnerabilities.

Application Gateway also provides a feature called **path-based routing** – this isn't a security setting but it may influence your choice of technology. Path-based routing allows you to configure multiple backend services under the same URL. For example, you can have an application under the `https://mydomain.com` domain but two separate services at the backend – for example, one that manages orders and another that manages accounts. Path-based routing allows you to send all traffic to `https://mydomain/oders` to one service, and `https://mydomain.com/accounts` to a different service.

Azure Front Door

Azure Front Door is both a WAF and a **content delivery network** (**CDN**). In essence, it is a global, scalable entry point that optimizes the delivery of your web applications by caching data at the user's closest location (the CDN part). It also provides features such as global load balancing, SSL offloading, DDoS protection, and, just like Application Gateway, a WAF.

What to use and when?

With so many different options, it can be tricky to know which solution is best. Of course, it always comes down to your specific requirements.

If you want to protect your backend database (highly recommended), then at the very least, it is advisable to use web app VNet integration and Azure Firewall to lock down the database.

Using a WAF with web apps is also recommended, and therefore the first choice is between Azure Front Door or Application Gateway. Application Gateway can only connect to and load balance between services in the same region. If your application is multi-region, then you can either use a traffic manager in front of multiple application gateways (one in each region) or use a single Azure Front Door service that combines the WAF with a multi-region load balancer, but with the added benefit of being a CDN as well to speed up access to your applications by providing cached data close to the user.

Finally, we need to decide if we need a firewall. Application Gateway and Azure Front Door only support the HTTP/HTTPS protocols, so if you need to use anything else, you'll need an Azure firewall. Application Gateway and Front Door are also for inbound traffic, so if you want to control the outbound flow of traffic, such as if your web application needs to call out to other services and you want to manage what it can and can't access, again, you'll need an Azure firewall.

It's also worth noting that all three types of firewalls are relatively expensive, so you will often wish to use them across multiple solutions. This essentially means the choice of which combination to use needs to be considered as an overall cloud security strategy. If this is a consideration, you also need to plan how to use them because if you use both Azure Firewall and Application Gateway or Front Door, you also need to choose whether to set them up in parallel or sequentially.

Figure 9.1 showed an example of a sequential configuration, whereby traffic goes through both firewalls, whereas *Figure 9.2* shows a similar setup in parallel:

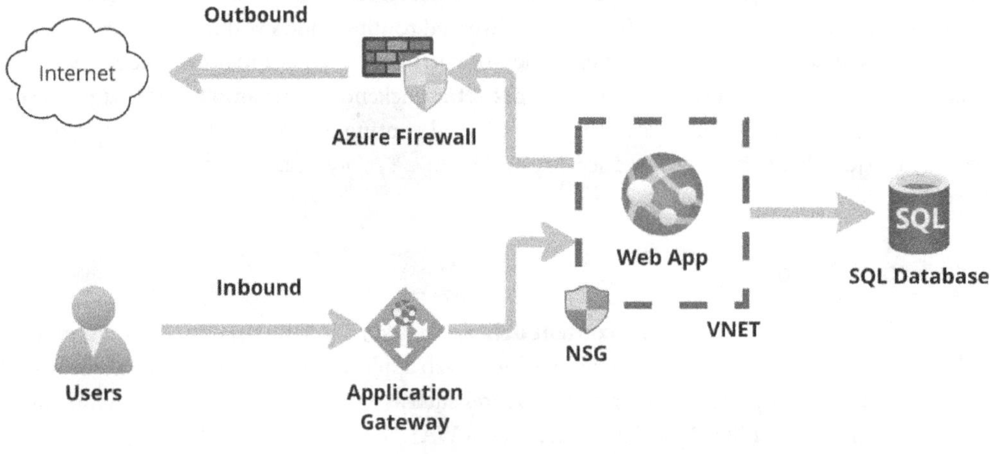
Outbound
Internet
Azure Firewall
Web App
SQL
SQL Database
Inbound
NSG
VNET
Users
Application Gateway

Figure 9.2 – Parallel firewall configuration

As we can see in the preceding diagram, we use a parallel configuration when we want inbound traffic going through the application gateway but outbound traffic going out through the Azure firewall.

Configuring network-level security

In this section, we will walk through how to set up a simple web app with a backend SQL database, secure it using VNet integration, and then add an Azure Front Door.

The first step, therefore, is to create a SQL Server and database, which we will work through now.

Creating the SQL database

First, we need to create the SQL database. Follow these steps:

1. Go to the Azure portal and click **+ Create Resource**, then click **SQL Database** under **Popular resources**.
2. Add the following values:
   - **Subscription**: Pick a subscription
   - **Resource group**: Create a new one and call it `SecurityInDepth`
   - **Database name**: `PacktSecureDatabase`
   - **Server**: Create a new one and add the following values:
     - **Server name**: `packtsecuresqlserver`
     - **Authentication method: Use both SQL and Azure AD Authentication**
     - **Set Azure AD Admin:** Click **Set admin** and choose your account
     - **Server admin login**: `PacktAdmin`
     - **Password**: `P@ssword123`
     - **Location**: **East US**
   - **Want to use SQL elastic pool**: No, we are going to use a single database
   - **Compute + storage**: Keep the default setting here
3. Click **Review + create**, and then **Create**.

The database is now being created. Once we've done this, we need to get the connection string, which we will use later.

Find the deployed SQL database in the Azure portal and, from the **Overview** page, click **Show database connection strings**. Different options will appear. Copy the `ADO.NET` (SQL authentication) string.

You will need to replace `{your_password)` with `P@ssword123!` or whatever password you set when you created the SQL database in the previous steps.

See *Figure 9.3* for an example:

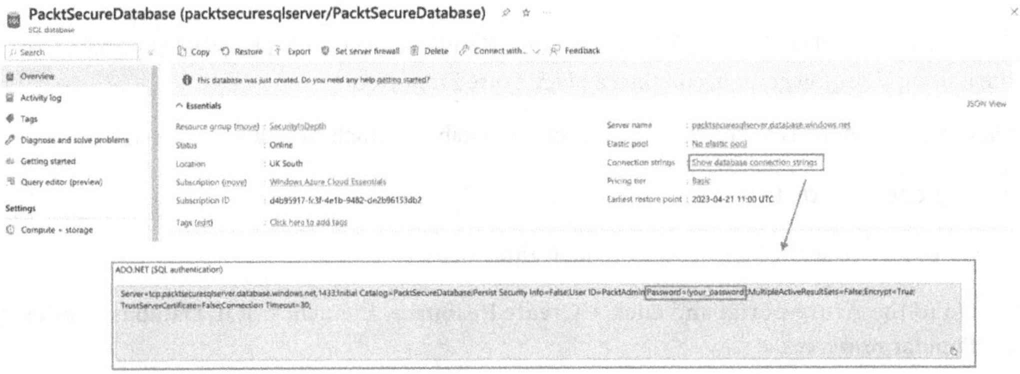

Figure 9.3 – Getting the database connection string

With our SQL database set up, next, we need to create a web app.

Creating a web application

Let's learn how to create a web app:

1. Go back to the Azure portal home page, click **+ Create Resource**, then click **Web App** under **Popular resources**.
2. Add the following values:
   - **Subscription**: Pick a subscription
   - **Resource group**: select `SecurityInDepth`
   - **Name**: `PacktSecureWebApp` (or any other available name)
   - **Publish**: Select **Code**
   - **Runtime stack**: Select **.NET 7 (STS)**
   - **Operating System**: **Linux**
   - **Region**: **East US**

3. Click **Review + create**, then **Create**.

With that, we have deployed our components, but they're unsecure. Before we continue, we need to deploy some code. We will do this using Visual Studio and use a web app I have already created to demonstrate the tasks.

Updating and deploying the web app

First, we need to get the code. If you haven't already, download the source code for this book from `https://github.com/PacktPublishing/Azure-Architecture-Explained`.

Launch **Visual Studio** and open the project in the `SecurityInDepth\src` folder.

The first thing you need to do is set the connection string to your SQL Server.

In **Solution Explorer**, look for the `appsettings.json` file; it should look like this:

```
{
 "Logging": {
  "LogLevel": {
   "Default": "Information",
   "Microsoft.AspNetCore": "Warning"
  }
 },
 "AllowedHosts": "*",
 "ConnectionStrings": {
  "ItemsContext": "Server=(localdb)\\
mssqllocaldb;Database=ItemsDatabase;Trusted_
Connection=True;MultipleActiveResultSets=true"
 }
}
```

In the `ConnectionsStrings` section, you'll see `ItemsContext` – this is where the code picks up the database connection details, and currently, it is set to run on a local instance of SQL. Replace this with the connection string you copied in the previous section. Save the file and close it.

Now, we need to publish or deploy our code to our web app in Azure.

In **Solution Explorer**, right-click on the `SecurityInDepth` project and choose **Publish...** from the tear-off menu. You will see a screen similar to the one shown in *Figure 9.4*:

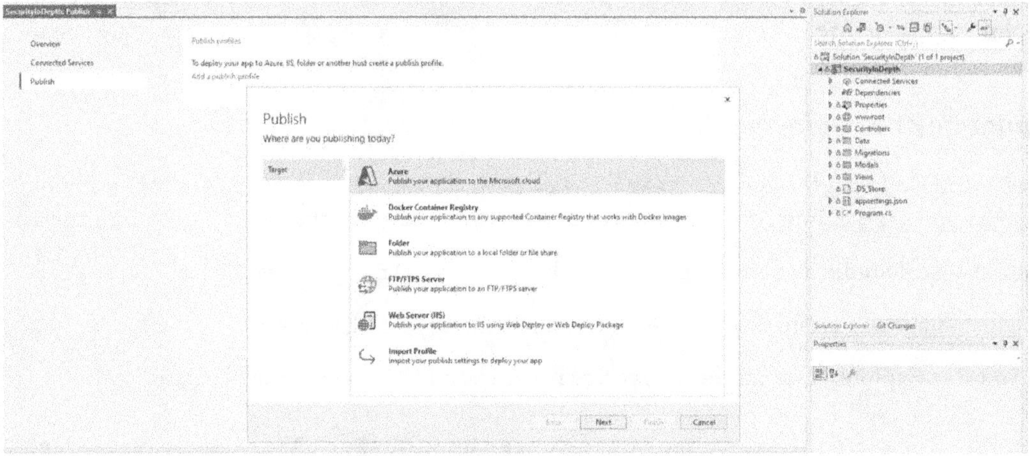
Overview
Connected Services
Publish
Publish
Where are you publishing today?
Target
Azure
Publish your application to the Microsoft cloud
Docker Container Registry
Folder
FTP/FTPS Server
Web Server (IIS)
Import Profile
Next
Cancel
Solution Explorer
Properties

Figure 9.4 – Creating a publishing profile

In the publish wizard, do the following:

1. Select **Azure** for the target and click **Next**.
2. Select **Azure App Service (Linux)** and click **Next**.
3. You will see a folder of resource groups in the Azure tenant; `SecurityInDepth` should be visible here. Expand it, select `PackSecureWeb`, and click **Next**.
4. Choose **Publish (generates pubxml file)** and click **Finish**.
5. A publish profile will be created and opened and will look similar to what's shown in *Figure 9.5*. At the top right, click **Publish** to deploy your web app to Azure.

This will start the publishing process and deploy your app to Azure:

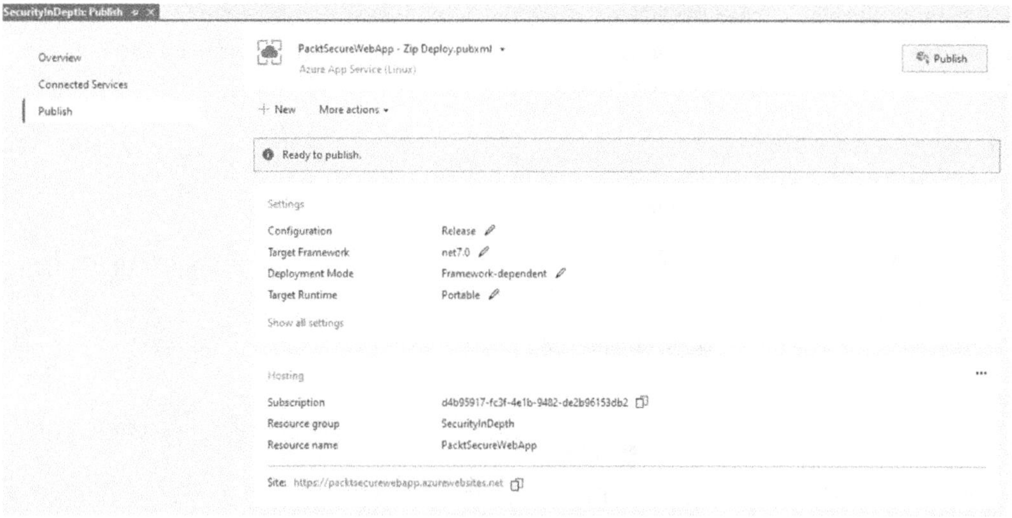
SecurityInDepth: Publish
Overview
Connected Services
Publish
PacktSecureWebApp - Zip Deploy.pubxml
Azure App Service (Linux)
Publish
New
More actions
Ready to publish.
Settings
Configuration
Release
Target Framework
net7.0
Deployment Mode
Framework-dependent
Target Runtime
Portable
Show all settings
Hosting
Subscription
d4b95917-fc3f-4e1b-9482-de2b96153db2
Resource group
SecurityInDepth
Resource name
PacktSecureWebApp
Site: https://packtsecurewebapp.azurewebsites.net

Figure 9.5 – Publishing the web app

Once deployed, your browser will automatically open the website.

Testing and securing the app

There are two options on the page, and both require network connectivity to SQL Server. The **Create the database** button attempts to connect to the SQL database and create some tables and data. However, if run that at the moment, we'll see an error, as shown in *Figure 9.6*:

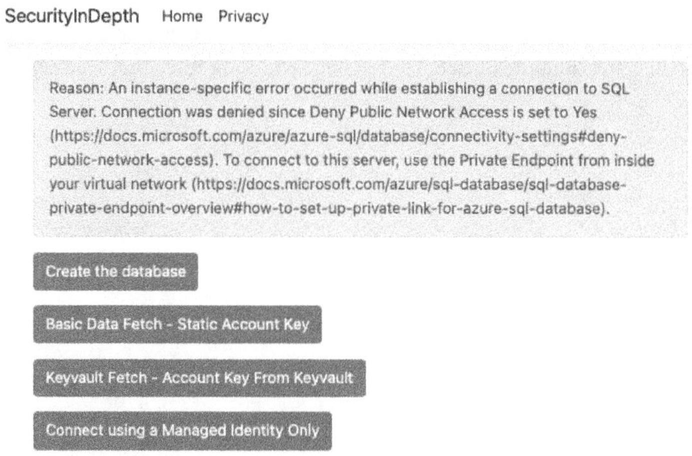
SecurityInDepth
Home
Privacy
Reason: An instance-specific error occurred while establishing a connection to SQL Server. Connection was denied since Deny Public Network Access is set to Yes (https://docs.microsoft.com/azure/azure-sql/database/connectivity-settings#deny-public-network-access). To connect to this server, use the Private Endpoint from inside your virtual network (https://docs.microsoft.com/azure/sql-database/sql-database-private-endpoint-overview#how-to-set-up-private-link-for-azure-sql-database).
Create the database
Basic Data Fetch - Static Account Key
Keyvault Fetch - Account Key From Keyvault
Connect using a Managed Identity Only

Figure 9.6 – Database connection error

This is demonstrating that, at present, we cannot connect from the web app to the SQL database. This is because when we created the SQL database, the default network rule was to block all network connections. Therefore, we must configure SQL Server to allow connections from the web app. We are going to do this using VNet integration.

Securing the database

The first step is to create a VNet that our web app can connect to:

1. Go back to the Azure portal home page and click + **Create Resource**. In the search bar, type `virtual network` and select **Virtual Network**.
2. From the list of services, click **Virtual Network**, then **Create**.
3. Add the following values:
   - **Subscription**: Select your subscription
   - **Resource Group**: Select `SecurityInDepth`
   - **Virtual network name**: Enter `SecureVNET`
   - **Region**: **East US**
4. Click **Review** + **create**, then **Create**.

Next, we must attach our web app to the VNet using VNet integration:

1. In the Azure portal, navigate to the web app you created earlier.
2. From the left-hand menu, click **Networking** under **Settings**. We want to control the outbound traffic to other Azure services, so in the **Outbound Traffic** box, click **VNET integration**.
3. This will take you to a new screen. Click **Add VNet**, then select `SecureVNET`, which we created earlier, and the default subnet. Your screen will look like what's shown in *Figure 9.7*:

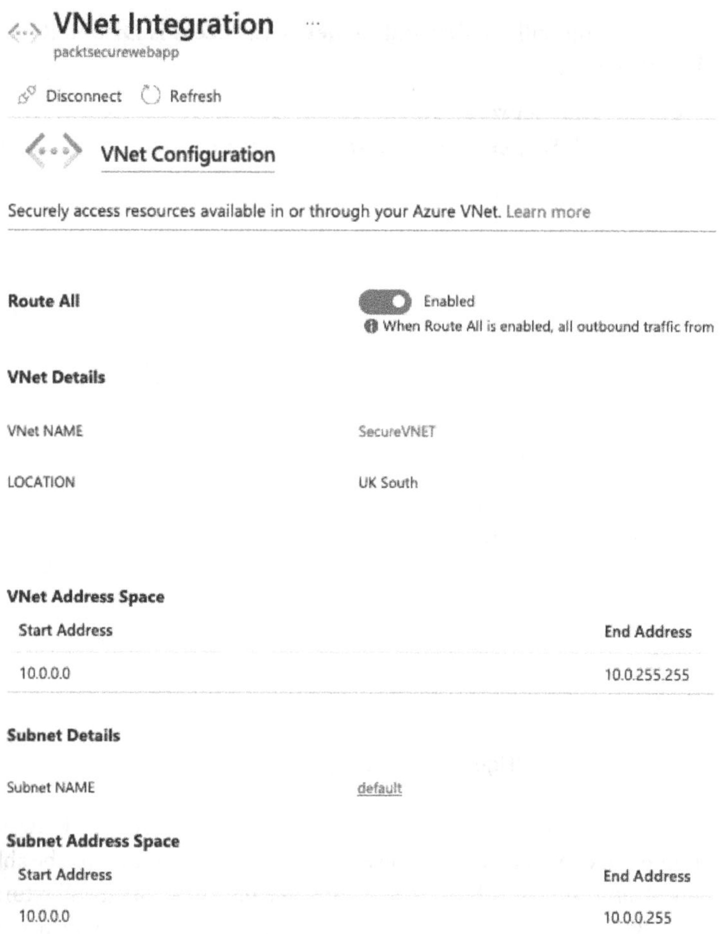
VNet Integration
packtsecurewebapp
Disconnect
Refresh
VNet Configuration
Securely access resources available in or through your Azure VNet. Learn more
Route All
Enabled
When Route All is enabled, all outbound traffic from
VNet Details
VNet NAME
SecureVNET
LOCATION
UK South
VNet Address Space
Start Address
End Address
10.0.0.0
10.0.255.255
Subnet Details
Subnet NAME
default
Subnet Address Space
Start Address
End Address
10.0.0.0
10.0.0.255

Figure 9.7 – The VNet Integration page

4. With that set, all outbound traffic will go via the `SecureVNET` VNet. Now, find your SQL database in the Azure portal and, from the overview screen, click **Set server firewall**, as per *Figure 9.8*:

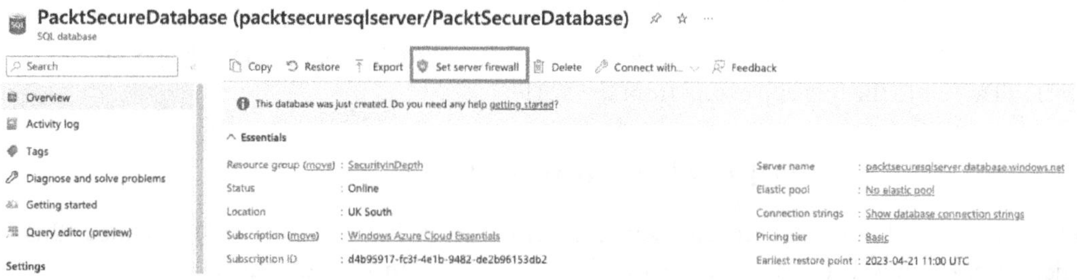
PacktSecureDatabase (packtsecuresqlserver/PacktSecureDatabase)
SQL database
Search
Copy
Restore
Export
Set server firewall
Delete
Connect with...
Feedback
Overview
Activity log
Tags
Diagnose and solve problems
Getting started
Query editor (preview)
Settings
This database was just created. Do you need any help getting started?
Essentials
Resource group (move) : SecurityInDepth
Status : Online
Location : UK South
Subscription (move) : Windows Azure Cloud Essentials
Subscription ID : d4b95917-fc3f-4e1b-9482-de2b96153db2
Server name : packtsecuresqlserver.database.windows.net
Elastic pool : No elastic pool
Connection strings : Show database connection strings
Pricing tier : Basic
Earliest restore point : 2023-04-21 11:00 UTC

Figure 9.8 – Setting the database firewall

5. On the next screen, you will see that **Public network access** is **Disabled** – click **Selected networks**, then click **Save**.
6. Now, click + **Add a virtual network rule** and, again, select the `SecureVNET` VNet and subnet you created earlier. Click **Enable**, wait a few seconds, then click **OK**. Your screen should look like *Figure 9.9*:

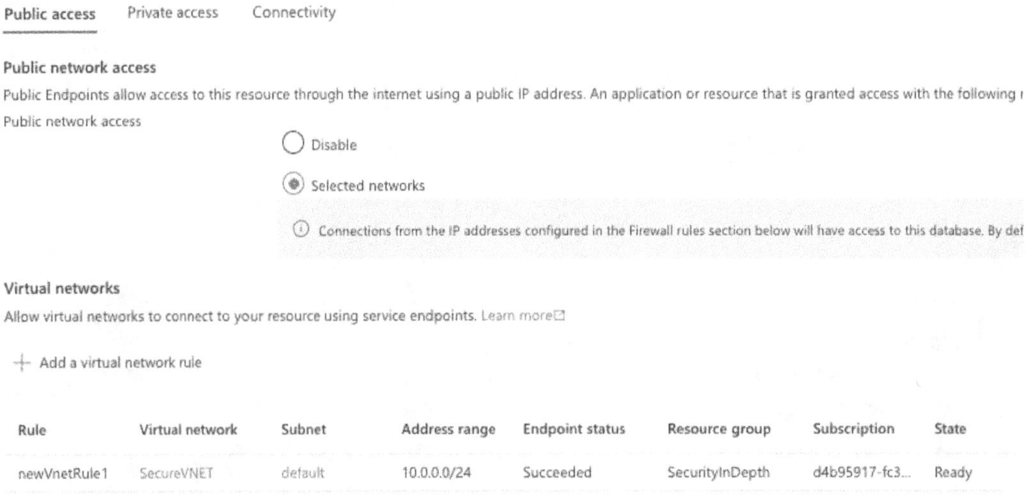

Public access
Private access
Connectivity
Public network access
Public Endpoints allow access to this resource through the internet using a public IP address. An application or resource that is granted access with the following r
Public network access
Disable
Selected networks
Connections from the IP addresses configured in the Firewall rules section below will have access to this database. By def
Virtual networks
Allow virtual networks to connect to your resource using service endpoints. Learn more
Add a virtual network rule
Rule
Virtual network
Subnet
Address range
Endpoint status
Resource group
Subscription
State
newVnetRule1
SecureVNET
default
10.0.0.0/24
Succeeded
SecurityInDepth
d4b95917-fc3...
Ready

Figure 9.9 – Adding the VNet

Although it says everything has been completed, in the background, the firewall changes can take 10-20 minutes to come into effect. However, when they do, you should be able to return to the deployed web app and click the **Create a database** button – which will return a **Database create successfully** response – then click **Basic Data Fetch**, which will return a row of records.

What this exercise has shown is that with public access disabled, we cannot access the database, but by enabling VNet integration on the web app, and then configuring the SQL firewall to allow access from that VNet, we can securely allow access.

You could test it by trying to connect to the database directly from your computer using a tool such as SQL Studio Manager, but the connection will fail.

The next step in securing our application will be to place a WAF in front of the web app, and then disable all access to the web app except from the WAF.

Creating an Azure application gateway

In this walkthrough, we will build and configure an Azure application gateway. We'll choose this option because our web app is only hosted, and will be accessed in a single region.

When setting up an app gateway, you need to understand some of the components we'll need to configure:

- **Frontend IP**: This is the entry point for requests. This can be an internal IP or a public IP or both.
- **Listeners**: These are configured to accept requests to specific protocols, ports, hosts, or IPs. Each listener is configured to route traffic to a specified backend pool based on a routing rule.
- **Routing rules**: These bind listeners to backend pools. You specify rules to interpret the hostname and path elements of a request and then direct the request to the appropriate backend pool.
- **Backend pools**: These define the backend pools that are your services. They can be virtual machines, scale sets, Azure apps, or even on-premises servers.

Now that we understand what we need to configure, let's build one and integrate it with our web app. First, we will need a dedicated subnet to host our application gateway. We've already created a VNet, so we'll reuse that and add a new subnet:

1. In the Azure portal, find the `SecureVNET` VNet you created earlier.
2. From the left-hand menu, click **Subnets**.
3. Click + **Subnet**.
4. Next to Name, enter `ApplicationGateway`.
5. Leave everything else as-is and click **OK**.

You should now see two subnets – the default, which contains your web app, and the new `ApplicationGateway` subnet, as shown in *Figure 9.10*:

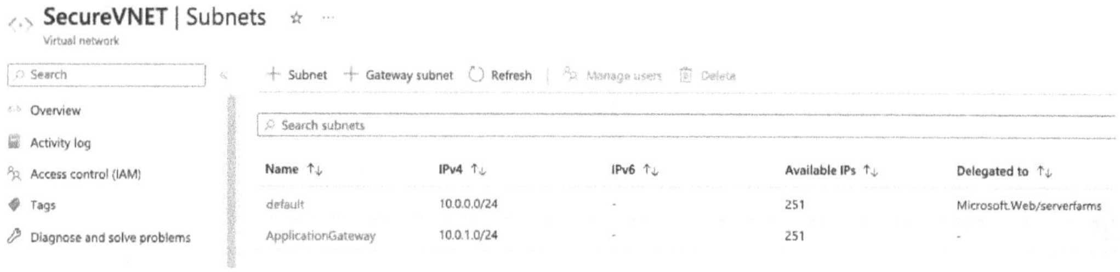
SecureVNET | Subnets
Virtual network
Search
Overview
Activity log
Access control (IAM)
Tags
Diagnose and solve problems
+ Subnet
+ Gateway subnet
Refresh
Manage users
Delete
Search subnets
Name
IPv4
IPv6
Available IPs
Delegated to
default
10.0.0.0/24
251
Microsoft.Web/serverfarms
ApplicationGateway
10.0.1.0/24
251

Figure 9.10 – Creating the ApplicationGateway subnet

Next, we can create the application gateway by performing the following steps:

1. Go to the Azure portal home screen and click + **Create Resource**.
2. Search for and select **Application Gateway**.
3. Click **Create**.

4. Enter the following details:
    - **Subscription**: Your subscription
    - **Resource Group**: Select `SecurityInDepth`
    - **Application Gateway name**: `PacktAppGateway`
    - **Region: East US**
    - **Tier: WAF V2**
    - **Enable autoscaling: No**
    - **Instance Count**: `1`
    - **Availability Zone: None**
    - **HTTP2: Disabled**
    - **WAF Policy**: Click **Create new**:
        - Name it `WAFPolicy`
        - Click **OK**
    - **Virtual Network**: `SecureVNET`
    - **Subnet**: `ApplicationGateway`
5. Click **Next: Frontends**.
6. Choose **Public** for **Frontend IP address**.
7. Next to **Public IP address**, click **Add new** and name it `PacktAGW-pip`.
8. Click **Next: Backends**.
9. Click **Add a backend pool**.
10. Name it `WebApp`.
11. Under **Backend targets**, set **Target type** to **App Services** and then set **Target** to `PacktSecureWebApp`.
12. Click **Add**.
13. Click **Next: Configuration**.
14. The final step is to create a rule to route traffic from the frontend to the backend pool. To do this, click **Add a routing rule**.
15. Name the rule `WebAppRule`.
16. Set **Priority** to `1`.

17. Under **Listener**, enter the following details:
    - **Listener name**: `WebAppListener`
    - **Frontend IP**: **Public**
    - **Protocol**: **HTTP**
    - **Port**: `80`
18. Click **Backend targets** and fill in the following details:
    - **Target type**: **Backend pool**
    - **Backend target**: `WebApp`
19. Next to **Http settings**, click **Add new** and fill in the following:
    - Set **Http settings name** to `HttpSettings`
    - **Backend protocol**: **HTTPS**
    - **Use well known CA certificate**: **Yes**
    - **Override with new host name**: **Yes**
    - **Host name override**: Pick the hostname from the backend target
20. Click **Add**, then click **Add** again.
21. This will add the rule; click **Next: Tags**.
22. Click **Next: Review + create**.
23. Click **Create**.

There are a couple of points we need to highlight here.

The first is the listener setup. In a production environment, we should have set this to HTTPS on port `443`, not HTTP on port `80`. We only chose this for simplicity; otherwise, we would have had to set up a custom domain and purchase a certificate to use HTTPS.

Similarly, we set the override with a new hostname rule – again, this is for simplicity. What we should do is set the override to *no*, but configure our web app to use a custom domain, and then set the IP of that custom domain to our gateway address.

The gateway can take some time to set up, but once it has been created, find the resource in the Azure portal and select it.

From the left-hand menu, click **Frontend IP configurations** under **Settings**. You will see the public IP address that has been created for us. Copy that IP address and paste it into a new browser window.

This should take us to our web app, but we could bypass the gateway address and go to the web app directly. So, the final step is to disable access to the web app to only allow access from the application gateway.

In the Azure portal, navigate to the web app you created earlier and, from the left-hand menu, click **Networking** under **Settings**. This time, we want to control inbound traffic, so under the **Inbound traffic** box, click **Access restriction**.

By default, all access is allowed because we have a default firewall rule that allows all traffic, and we're allowing anything that *doesn't* match any rules.

Change **Unmatched rule action** to **Deny**, then click **+Add** just above the rules (see *Figure 9.11*) and enter the following details:

- **Name**: `AppGWAllow`
- **Action**: **Allow**
- **Priority**: `100`
- **Type**: **VirtualNetwork**
- **VirtualNetwork**: `SecureVNET`
- **Subnet**: `ApplicationGateway`

Click **Add Rule**.

Your screen should look like *Figure 9.11*:

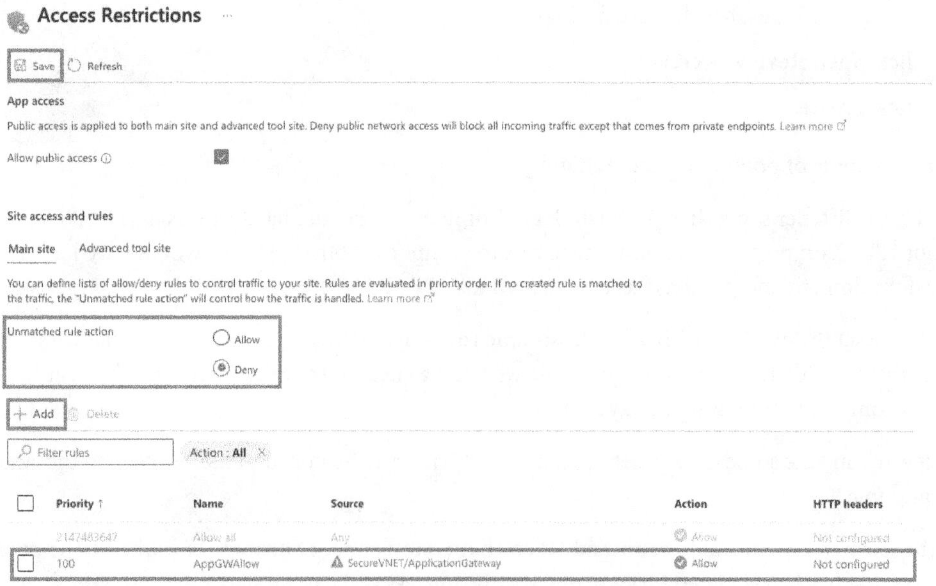
Access Restrictions
Save
Refresh
App access
Public access is applied to both main site and advanced tool site. Deny public network access will block all incoming traffic except that comes from private endpoints. Learn more
Allow public access
Site access and rules
Main site
Advanced tool site
You can define lists of allow/deny rules to control traffic to your site. Rules are evaluated in priority order. If no created rule is matched to the traffic, the "Unmatched rule action" will control how the traffic is handled. Learn more
Unmatched rule action
Allow
Deny
Add
Delete
Filter rules
Action : All
Priority
Name
Source
Action
HTTP headers
2147483647
Allow all
Any
Allow
Not configured
100
AppGWAllow
SecureVNET/ApplicationGateway
Allow
Not configured

Figure 9.11 – Securing a web app with inbound traffic

If all looks OK, click **Save** (highlighted in *Figure 9.11*).

The result should be that if you now browse the web app on its internal Azure URL, such as `https://packtsecurewebapp.azurewebsites.net/`, you will see a forbidden message. However, if you browse the application gateway's public IP, you will get to the site. This proves that you can force traffic to use the application gateway, thus enforcing the protection it provides.

Before we leave this topic, it's worth looking at the ruleset the application gateway's WAF policy provides. When we created the gateway, we simply asked it to create a policy, but let's have a look at what it is.

In the Azure policy, find and select your application gateway. Then, from the left-hand menu, under **Settings**, click **Web application firewall**.

You will see just one policy: `WAFPolicy`. Click on it to be taken to the **Policy settings** pages. You will that the `OWASP_3.2` ruleset has been applied – if you click the arrow by the ruleset, you'll see all the rules that are being applied. Your screen should look like *Figure 9.12*:

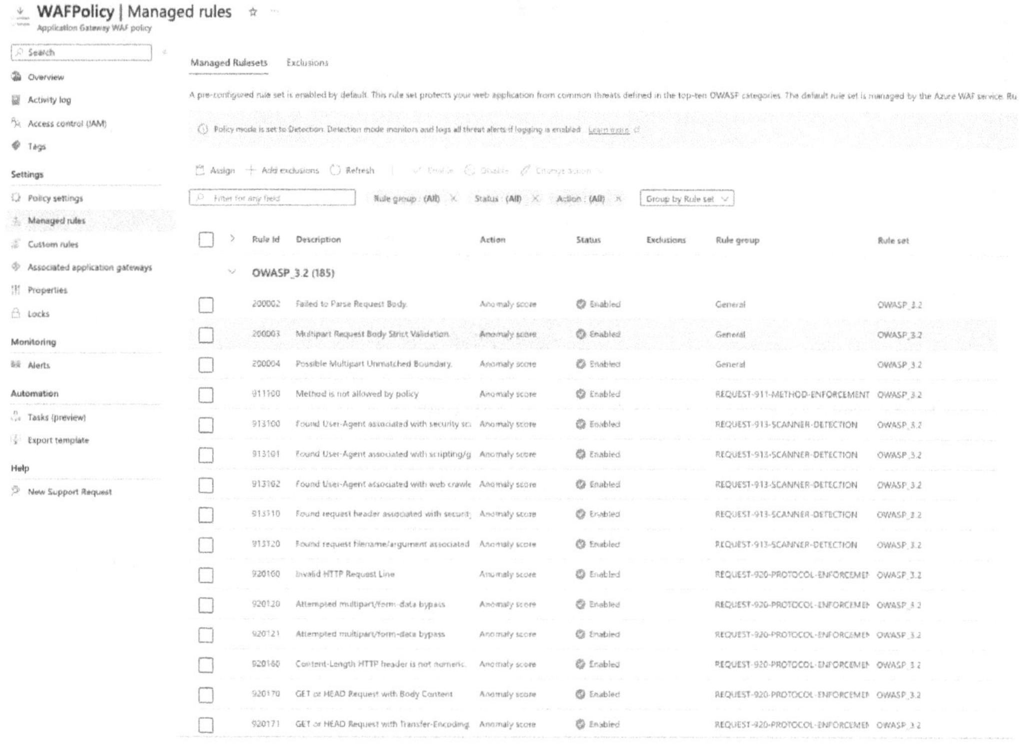
WAFPolicy | Managed rules
Application Gateway WAF policy
Search
Overview
Activity log
Access control (IAM)
Tags
Settings
Policy settings
Managed rules
Custom rules
Associated application gateways
Properties
Locks
Monitoring
Alerts
Automation
Tasks (preview)
Export template
Help
New Support Request
Managed Rulesets
Exclusions
Assign
Add exclusions
Refresh
Rule group : (All)
Status : (All)
Action : (All)
Group by Rule set
Rule Id
Description
Action
Status
Exclusions
Rule group
Rule set
OWASP_3.2 (185)

Figure 9.12 – OWASP rules

The OWASP ruleset is an industry-standard set of known attacks, and these are all built into the ruleset provided.

By default, all are included, but you can disable individual rules if needed.

Also, it's important to know that, again by default, those rules are set to detection only – this means that if somebody tries to exploit the attacks, the WAF will *not* block them, it will only report them. This is to give you the chance to test your application for a certain period first to make sure that the rules OWASP applies don't have a negative effect.

Once you are happy that everything is fine, you can switch detection to prevent mode by going to the **Overview** area of the WAF policy and clicking **Switch to prevention mode**. See *Figure 9.13* for an example:

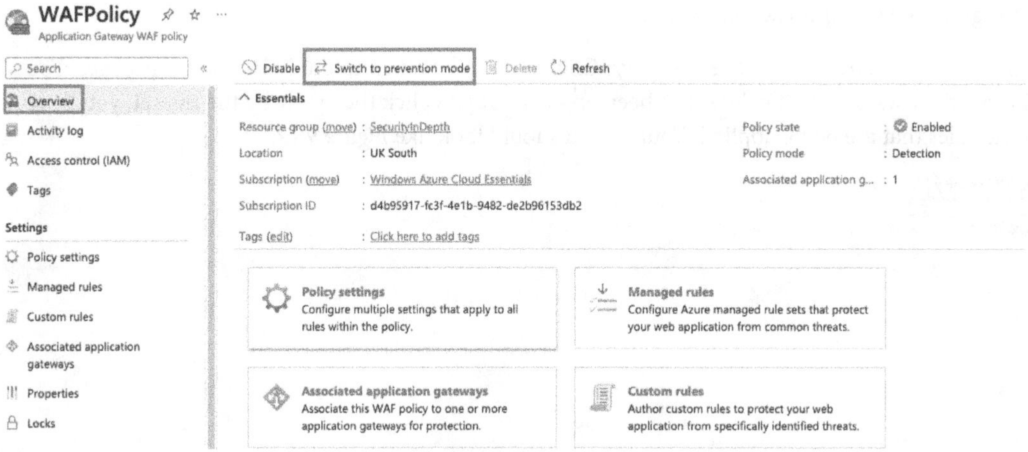
WAFPolicy
Application Gateway WAF policy
Search
Overview
Activity log
Access control (IAM)
Tags
Settings
Policy settings
Managed rules
Custom rules
Associated application gateways
Properties
Locks
Disable
Switch to prevention mode
Delete
Refresh
Essentials
Resource group (move) : SecurityInDepth
Location : UK South
Subscription (move) : Windows Azure Cloud Essentials
Subscription ID : d4b95917-fc3f-4e1b-9482-de2b96153db2
Tags (edit) : Click here to add tags
Policy state : Enabled
Policy mode : Detection
Associated application g... : 1
Policy settings
Configure multiple settings that apply to all rules within the policy.
Managed rules
Configure Azure managed rule sets that protect your web application from common threats.
Associated application gateways
Associate this WAF policy to one or more application gateways for protection.
Custom rules
Author custom rules to protect your web application from specifically identified threats.

Figure 9.13 – Policy overview

Application Gateway is a powerful tool for preventing malicious attacks, and the OWASP ruleset protects you against common code issues.

As mentioned previously, Azure Front Door works similarly but is better suited as an entry point for multi-region applications whereby your users may be coming from anywhere in the world! The setup is also similar, although slightly simpler, but again, you must remember to also lock down your web app so that you cannot bypass the WAF.

Securing keys and secrets

Securing your applications against network-level attacks is often the first line of defense. With the additional capabilities of a WAF, you can readily protect against many forms of attacks and issues in your code.

But not all attacks come head-on. Most attacks are usually through insider threats – either maliciously or accidentally. What we mean by this is that, if we think back to Mark Twain's proverb, it's those that seek to get around the front door that often cause the biggest problems.

Therefore, we need to think about attack vectors other than a direct one over the network. If our network is secure, and our frontend is secure, we must consider how the backend can be exploited.

From a network point of view, this is quite difficult; if an attacker has found a way around these controls, the next level of protection is usually authentication. Because the asset we want to protect is our data, we must consider strategies to protect the keys that provide access.

Simply put, the connection strings to the databases are the next area we need to secure. This applies to several different resources in Azure, including Azure SQL, but also includes storage accounts, Cosmos DB, and many services such as Azure Service Bus – they all use the concept of complex keys and connection strings to gain access.

Traditionally, these keys are recorded in a configuration file that the application uses when trying to connect. In the example code we looked at earlier, we used `ItemsContent` in the `appsettings.json` file.

This has two potential issues. The first is that we must provide the connection strings to our developers, and this, in turn, gives them access to the database itself. This is fine in a development environment, but when we go to production, we don't want to give our developers keys to potentially sensitive data.

The other issue is that if the key is stored in a configuration file that will most likely be stored in a Git code repository such as Azure DevOps or GitHub. This increases the population that can see that file, and, depending on how your repositories are configured, it could even be made public.

Another area to consider is management. Solutions often need to communicate with lots of different services and data stores and keeping track of all those keys can become problematic. In addition, you will have different keys for different environments – dev, test, and production.

To recap and summarize, what we need is a solution that can help us manage our keys and secrets, and ideally help us secure them so that we can tightly control who or what has access.

Azure Key Vault was designed to address these issues and more. It is a security service that helps you protect and manage sensitive information such as encryption keys, secrets, and certificates. By using Azure Key Vault, you can securely store and manage these assets, and then configure your applications to access them from Key Vault. This helps minimize the risk of unauthorized access to your application secrets because your developers never see or have direct access to the keys themselves, and because they are never stored in the code or configuration files, there is never any danger of inadvertently distributing them. Finally, Azure Key Vault simplifies the process of managing cryptographic keys and secrets by providing a central location to store them.

The best way to see the benefits of Azure Key Vault is to implement it.

Creating a Key Vault

The first step is to create a new Key Vault in Azure. To do this, perform the following steps:

1. Go back to the Azure portal home page and + **Create Resource**. Then, from **Popular Azure Services**, select **Key Vault**.
2. Enter the following details:
   - **Subscription**: Select your subscription
   - **Resource group**: Select the resource group we created earlier, `SecurityInDepth`
   - **Key vault name**: `PacktKeyVault`
   - **Region**: **East US**
   - **Pricing Tier**: **Standard**
   - **Days to retain deleted vaults**: `90`
   - **Purge protection**: **Disable**
3. Click **Review + create**.
4. Click **Create**.

Once the new Key Vault has been created, we need to configure it to allow our web app to access it. This is like how we granted access to our SQL Server from our web app – that is, we have to allow network access but also tell Key Vault to allow access from our web app.

Our web app is already connected to a VNet, which makes the networking process easier, but first, we need to set up an identity for it – we'll look at what this means in the next section:

1. In the Azure portal, find `PacktSecureWebApp` and, from the left-hand menu, under **Settings**, click **Identity**.
2. On this page, we will be taken to a **System assigned** tab, where **Status** is set to **Off**. Simply change this to **On** and click **Save**.

   Next, we need to create an access policy. An access policy is similar to assigning role-based access to a user, but this is very specific and granular to providing access to our Key Vaults.
3. Still in the Azure portal, find and select the new Key Vault you just created. Then, from the left-hand menu, under **Settings**, click **Access policies**.
4. From the left-hand menu, click **Access policies**, then click + **Create**.
5. We can store three types of security information in Key Vaults – keys, secrets, and certificates. A connection string is a secret, so on the permissions page, click the **Get** and **List** options under **Secret permissions**.

> **Note**
>
> When you created the Key Vault, an access policy will have been created automatically to give you full access. However, if anybody else wants to be able to manage keys, secrets, and certificates in the Key Vault, they will need their own access policy to be set up.

6. Click **Next**, then search for and select `PacktSecureWebApp` – this is only available because we enabled the system identity on it. Click **Next** again.
7. Click **Next** again then **Create** to create your new policy. Your screen should look like *Figure 9.14*:

PacktKeyVault | Access policies
Key vault
Search
Create
Refresh
Delete
Edit
Overview
Activity log
Access control (IAM)
Tags
Diagnose and solve problems
Access policies
Events
Objects
Keys
Access policies enable you to have fine grained control over access to vault items. Learn more
Search
Permissions : All
Type : All
Showing 1 to 2 of 2 records.
Name
Email
Key Permissions
Secret Permissions
Certificate Permissions
APPLICATION
PacktSecureWebApp
Get, List
USER
Brett Hargreaves
Get, List, Update, Create, Import, ...
Get, List, Set, Delete, Recover, Bac...
Get, List, Update, Create, Import, ...

Figure 9.14 – Configuring Key Vault access policies

8. Now, we need to create a secret and copy our connection string. Go to your code in Visual Studio and copy `ItemsContext` under `ConnectionStrings` from the `appsettings.json` file.
9. Back in the Azure portal, still viewing the Key Vault, click **Secrets** under **Objects** from the left-hand menu. Click + **Generate/Import** and fill in the following details:
   - **Upload options: Manual**
   - **Name**: `ItemsContextConnectionString`
   - **Secret value**: Paste in the connection string you copied from your code
   - **Set expiration date**: Untick this
10. Everything else can be left as-is, so click **Create**.
11. With that, a new secret has been created. Now, we need to get a URL that references it. So, click on the secret; you will be presented with a list of versions – there should only be one (as we've only just created it!), so click on that.

12. Finally, you will see **Secret Identifier**, as shown in *Figure 9.15*. Copy that value somewhere safe; we'll need it in a minute:

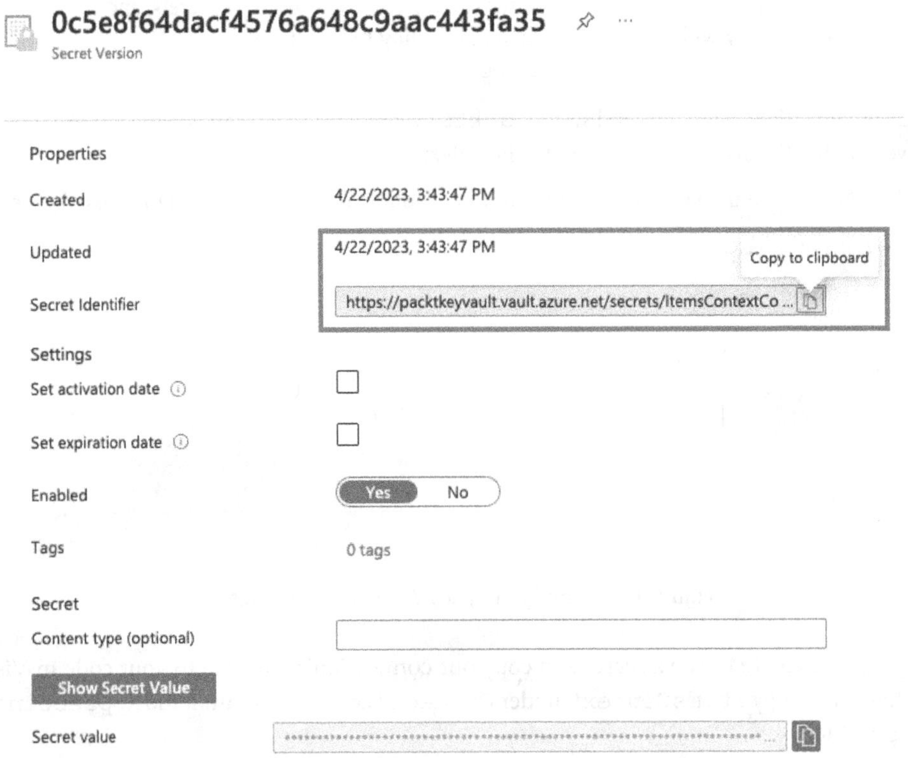
0c5e8f64dacf4576a648c9aac443fa35
Secret Version
Properties
Created
4/22/2023, 3:43:47 PM
Updated
4/22/2023, 3:43:47 PM
Copy to clipboard
Secret Identifier
https://packtkeyvault.vault.azure.net/secrets/ItemsContextCo ...
Settings
Set activation date
Set expiration date
Enabled
Yes
No
Tags
0 tags
Secret
Content type (optional)
Show Secret Value
Secret value

Figure 9.15 – Getting a secret identifier

We can now update our connection string in the configuration of the web app in Azure. By doing this, we don't need to make any code changes.

13. In the Azure portal, navigate back to your web app and click **Configuration** under **Settings**.

    Here, you can set applications settings and connection strings – so long as your code is built to use the .NET `ConfigurationManager` class. For example, since our code does this in the `Program.cs` file with the `builder.Configuration.GetConnectionString("ItemsContext")` line, we can override the `appSettings.json` file.

14. On the **Configuration** page, under **Connection strings**, click + **New connection string**. This will bring up a dialogue for adding/editing a connection string, as shown in *Figure 9.16*:

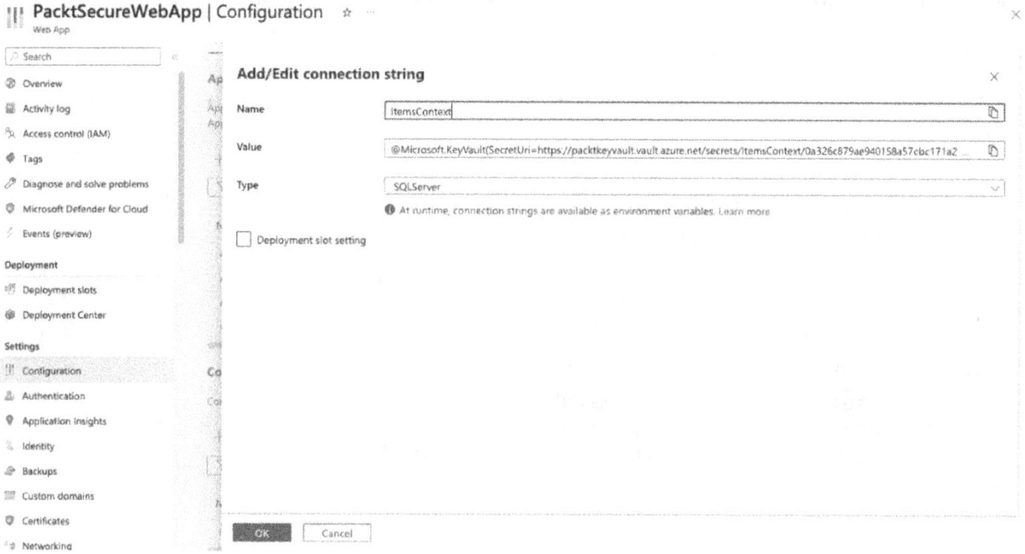
PacktSecureWebApp | Configuration
Web App
Search
Overview
Activity log
Access control (IAM)
Tags
Diagnose and solve problems
Microsoft Defender for Cloud
Events (preview)
Deployment
Deployment slots
Deployment Center
Settings
Configuration
Authentication
Application Insights
Identity
Backups
Custom domains
Certificates
Networking
Add/Edit connection string
Name
ItemsContext
Value
@Microsoft.KeyVault(SecretUri=https://packtkeyvault.vault.azure.net/secrets/ItemsContext/0a326c879ae940158a57cbc171a2
Type
SQLServer
At runtime, connection strings are available as environment variables. Learn more
Deployment slot setting
OK
Cancel

Figure 9.15 – Setting web app connection strings

Name needs to be the same as the connection string name in your appSettings.json file. **Value** needs to be in the @Microsoft.KeyVault(SecretUri=<Your key vault reference>) format, where <Your key vault reference> is the secret identifier you just copied from the Key Vault. @Microsoft.KeyVault() tells the configuration manager that the actual connection string we want is in the Key Vault specified.

Here is an example value:

```
@Microsoft.KeyVault(SecretUri=https://packtkeyvault.vault.azure.
net/secrets/ItemsContext/0a326c879ae940158a57cbc171a246a5)
```

15. Finally, set **Type** to **SQLAzure**.
16. Once you've finished, click **OK**, then click **Save** at the top of the **Configuration** page.

As a proper test, you could edit the appSettings.json file in your code and either remove the connection string (leave the name, just blank out the value) or, better still, set it to a local SQL Server instance so that developers can run the program locally against a local database.

For example, it may now look like this:

```
{
  "Logging": {
    "LogLevel": {
      "Default": "Information",
      "Microsoft.AspNetCore": "Warning"
    }
  },
  "AllowedHosts": "*",
  "ConnectionStrings": {
    "ItemsContext": "Server=(localdb)\\
mssqllocaldb;Database=ItemsDatabase;Trusted_
Connection=True;MultipleActiveResultSets=true"
  }
}
```

Once you've redeployed your application, browse to it – the connection to the database should still work, even though we've blanked it out in our configuration file.

> **Note**
> Sometimes, you might get an error when running your app. This will probably be because you have not set up the Key Vault connection policy correctly. Alternatively, it sometimes takes around 15 minutes for Azure to reconfigure itself in the background. So, if you do get an error, leave it for a few minutes and try again.

This walkthrough has demonstrated how we can use Azure Key Vault to store our connection strings, which makes them far safer and reduces the audience who need direct access to them.

We could further lock down access to the Key Vault at the network level. By default, Key Vaults are set to allow access from all public endpoints, but we can change this to set IP addresses or VNets.

To do this, find and select your Key Vault in the Azure portal and click **Networking** under **Settings**. You will see that the vault is set to **Allow public access from all networks**. We want to lock this down to just our web app, so change this to **Allow public access from the specific virtual networks and IP addresses**.

Next, under **Virtual Networks**, click **Add a virtual network | Add existing virtual networks**. Then, select the SecureVNET VNet we created, as well as the default subnet we attached our web app to.

Click **Enable**, then click **Add**. Finally, click **Apply**. See *Figure 9.17* for an example:

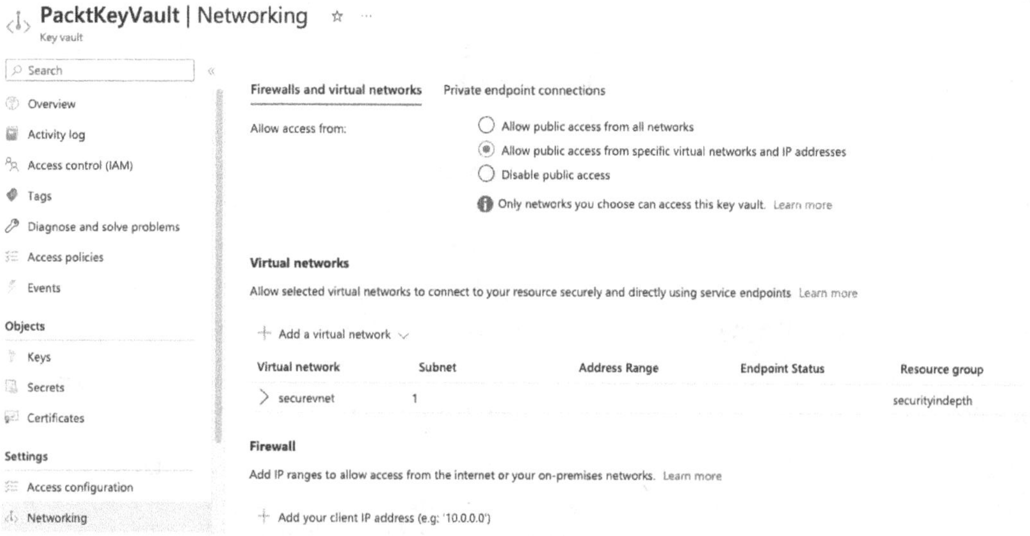
PacktKeyVault | Networking
Key vault
Search
Overview
Activity log
Access control (IAM)
Tags
Diagnose and solve problems
Access policies
Events
Objects
Keys
Secrets
Certificates
Settings
Access configuration
Networking
Firewalls and virtual networks
Private endpoint connections
Allow access from:
Allow public access from all networks
Allow public access from specific virtual networks and IP addresses
Disable public access
Only networks you choose can access this key vault. Learn more
Virtual networks
Allow selected virtual networks to connect to your resource securely and directly using service endpoints Learn more
Add a virtual network
Virtual network
Subnet
Address Range
Endpoint Status
Resource group
securevnet
1
securityindepth
Firewall
Add IP ranges to allow access from the internet or your on-premises networks. Learn more
Add your client IP address (e.g: '10.0.0.0')

Figure 9.17 – Restricting Key Vault access at the network level

For this to work, our web app needs to have VNet integration enabled, which we did earlier to allow access to SQL Server. Also, be aware that locking the vault down at the network level will also prevent you from creating and amending access policies as you will have also locked out your computer!

To get around this, you can also add your IP address under the **Firewall** section, and that has the same effect. The best practice would be to do this as required and always remove your public IP when you're finished.

When we created the vault, we also enabled our web app with a system-managed identity, which we then used to grant access to our vault from the Web App. This is called a **managed identity**, and you may be wondering why we can't just do the same thing to provide access to SQL Server itself.

Actually, we can! We'll explore that next.

Using managed identities

Many resources in Azure can have identities attached to them, and we can use those identities to provide access to other Azure services without the need for passwords and usernames.

Managed identities are a special type of account called a **service principal**, and as they are managed, it means we never have to see the details – Azure just manages them for us.

Managed identities are arguably the most secure method of providing access between components as we never have to record sensitive passwords anywhere.

When we created our SQL Server, we chose **Use both SQL and Azure AD Authentication**. The AD part is important when using managed identities because, under the hood, they are just Active Directory accounts.

We also need to enable our web app with an identity, which we did in the previous section when we set up the Key Vault. Therefore, all we need to do now is grant access to that account to SQL Server. We must do that by using T-SQL commands on the database.

Because SQL Server is locked down at the network level, the first thing we need to do is grant access to our IP address – even though we will be using the Azure portal to issue the commands.

In the Azure portal, search for and select your SQL database. From the **Overview** view, click **Set server firewall**, just like we did earlier when we granted access to the VNet.

This time, under **Firewall rules**, click **Add your client Ipv4 address**. Usually, the portal will detect what your IP is, but if it hasn't, simply open another browser window and, using your search engine of choice, type `what is my ip`. Get the returned IP address and paste it into the **Start IPv4 address** and **End IPv4 address** boxes, and then set **Rule name** to `MyIPAddress`, for example.

Your screen should look something like *Figure 9.18*:

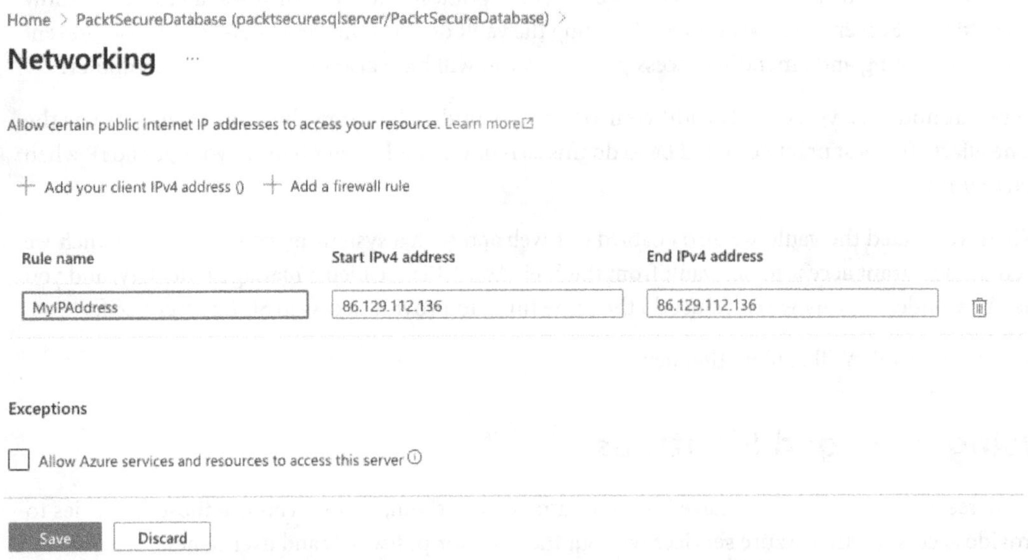
Home > PacktSecureDatabase (packtsecuresqlserver/PacktSecureDatabase) >
Networking
Allow certain public internet IP addresses to access your resource. Learn more
Add your client IPv4 address ()
Add a firewall rule
Rule name
Start IPv4 address
End IPv4 address
MyIPAddress
86.129.112.136
86.129.112.136
Exceptions
Allow Azure services and resources to access this server
Save
Discard

Figure 9.18 – Allowing your IP address in SQL Server

Now, click **Save**. Once you've finished, go back to the **Database Overview** page and, from the left-hand menu, click **Query editor**.

You will see a screen like *Figure 9.19*. Here, you must sign in using your AD account, so click **Continue as <your username>** under **Active Directory authentication**:

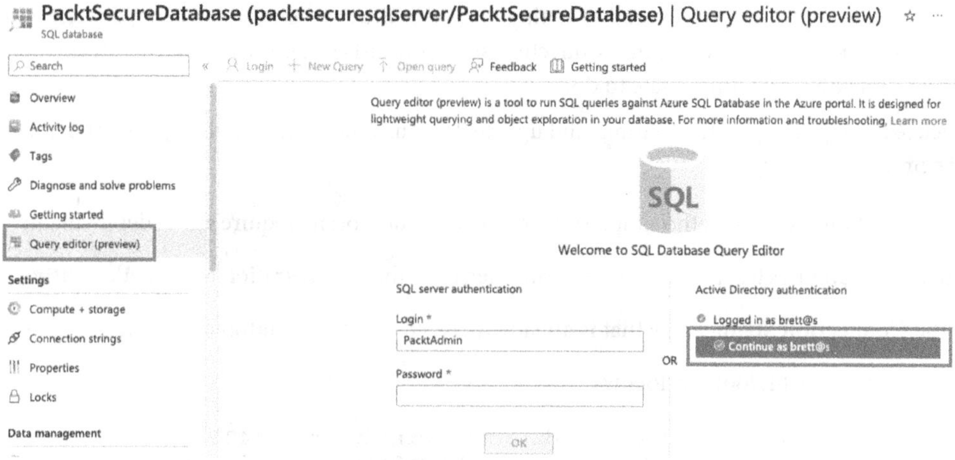
PacktSecureDatabase (packtsecuresqlserver/PacktSecureDatabase) | Query editor (preview)
SQL database
Search
Login
New Query
Open query
Feedback
Getting started
Overview
Activity log
Tags
Diagnose and solve problems
Getting started
Query editor (preview)
Settings
Compute + storage
Connection strings
Properties
Locks
Data management
Query editor (preview) is a tool to run SQL queries against Azure SQL Database in the Azure portal. It is designed for lightweight querying and object exploration in your database. For more information and troubleshooting, Learn more
SQL
Welcome to SQL Database Query Editor
SQL server authentication
Login *
PacktAdmin
Password *
OK
OR
Active Directory authentication
Logged in as brett@s
Continue as brett@s

Figure 9.19 – Logging into Azure SQL through the portal

You will be taken to a view showing your tables, views, and stored procedures. To the right of this view is a query editor. Paste the following into that window:

```
CREATE USER [PacktSecureWebApp] FROM EXTERNAL PROVIDER;
ALTER ROLE db_datareader ADD MEMBER [PacktSecureWebApp];
ALTER ROLE db_datawriter ADD MEMBER [PacktSecureWebApp];
```

Here, `[PacktSecureWebApp]` is the name of the web app that you created. Then, click the **Run** button, as shown in *Figure 9.20*. You should see a message that says `Query succeeded: Affected rows: 0`:

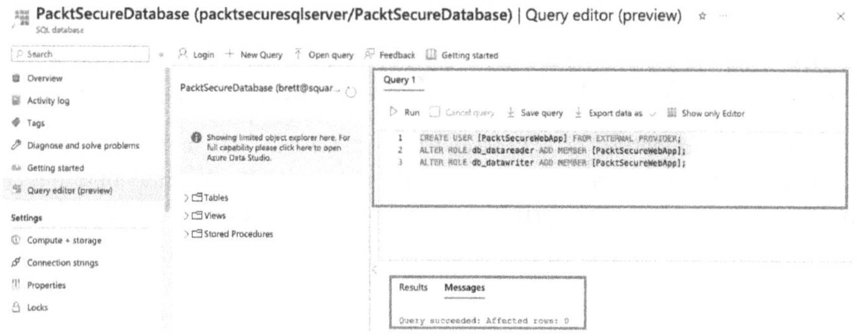
PacktSecureDatabase (packtsecuresqlserver/PacktSecureDatabase) | Query editor (preview)
Query 1
Run
Save query
Export data as
Show only Editor
Tables
Views
Stored Procedures
Results
Messages

Figure 9.20 – Granting access to your web app

Finally, we need to change the connection string so that it uses an Active Directory user. We have a few options:

1. Continue using the Key Vault. We would need to update the Key Vault key by creating a new version and then updating the Key Vault reference in the app configuration in Azure.
2. Continue using the app configuration settings but enter the new connection string directly and bypass the Key Vault. Because the connection string doesn't contain any sensitive information (such as a password), this is safe to do.
3. Delete the app configuration settings and update the connection string in the `appSettings.json` file in our app.

I'm going for option 2 since it is the simplest, is still secure, and doesn't require any redeployments.

Search for and find your web app in the Azure portal, then go to the **Configuration** menu under **Settings**.

Next to the **Connection String** entry that is already there, click the edit button on the right.

Update the value so that it looks as follows:

```
Server=tcp:packtsecuresqlserver.database.windows.net,1433;Initial Cat-
alog=PacktSecureDatabase;Encrypt=True;TrustServerCertificate=False;-
Connection Timeout=30;Authentication="Active Directory Default";
```

Here, `packtsecuresqlserver` is the name of your SQL Server.

Click **OK**, then click **Save**.

> **Note**
>
> If you are unsure of the connection string, you can go to the SQL database in the Azure portal and select **Connection Strings** from the left-hand menu. Various examples are available; for this exercise, we need the one labeled **ADO.NET (Active Directory passwordless authentication)**.

In this example, we learned how to secure our connection strings using the built-in capabilities of Azure Active Directory and think differently about how we store sensitive information.

Managed identities and Azure Key Vault are two great ways to separate the protection of data from the development of our platforms, which is a key consideration in securing those applications.

Whether you use managed identities or Azure Key Vault, or even both, is a matter of personal choice, or at least a decision that is dependent on your particular use case. However, either option is far more secure than the traditional method of storing information such as this in configuration files.

Summary

In this chapter, we looked at why we need to think about our application architecture in terms of securing access, and which tools can help us achieve this.

We looked at using a layered network approach to securely lock down our backend data by employing tools such as VNet integration, SQL firewalls, Azure Firewall, Azure Application Gateway, and Azure Front Door.

We then looked at securing the connection strings to our databases, why we need to do this, and, again, the different options, such as Azure Key Vault and managed identities.

In the next chapter, we will look at how to control what users can do within Azure using its built-in governance tools.

Part 3 – Making the Most of Infrastructure-as-Code for Azure

Infrastructure-as-code enables organizations to define and manage their infrastructure and application resources in a programmatic and automated manner. By using configuration files or scripts, developers and operations teams can describe their desired Azure infrastructure and deploy it consistently and reliably.

This part will discuss how infrastructure-as-code in Azure empowers organizations to optimize their cloud infrastructure, increase productivity, and drive successful digital transformations.

This part has the following chapters:

10

Governance in Azure – Components and Services

Cloud computing governance refers to the policies, procedures, and standards organizations use to manage cloud services. This includes ensuring compliance with regulations, managing security and risk, and ensuring the proper use of resources.

Effective cloud governance also involves continuous monitoring and reporting to track usage, costs, and performance. Put simply, a clear and well-defined cloud governance strategy is key to effectively managing the use of cloud services in an organization.

In this chapter, we will analyze how a comprehensive cloud governance strategy can help organizations effectively manage their cloud services, while ensuring compliance with regulations, security, and risk and getting the most value from their investment.

In this chapter, we'll cover the following main topics:

- Azure governance
- Azure governance – components and services
- Microsoft cost management

Let's get started!

Planning a comprehensive cloud governance strategy

What does it mean for you to adopt a comprehensive cloud governance strategy? How can we measure the impact of adopting a cloud governance strategy in an organization? Let's start with an example of an ideal approach by seeing how SpringToys adopted a comprehensive cloud governance strategy.

Let's imagine SpringToys has a team of IT professionals responsible for managing the company's use of cloud services. The company has established policies and procedures to follow when using cloud services. These policies cover areas such as security, compliance, and resource management.

SpringToys has also implemented a system for monitoring and reporting the use of cloud services. This includes tracking usage and costs, as well as monitoring the performance of the services. This information is used to identify areas where the company can improve its use of cloud services and ensure that it gets the most value from its investment.

In addition, SpringToys has set up a process to approve and manage the use of new cloud services, including a thorough assessment of the services to ensure that they meet the company's security and compliance requirements, as well as a review of the service's cost and performance. The company also has a process to decommission cloud services that are no longer needed.

How did SpringToys achieve this? First, it is necessary to agree on the terminology related to governance.

Understanding Azure governance

Azure governance is a way to ensure that an organization's use of the Azure cloud platform is secure, compliant, and efficient. Think of Azure governance as a series of guidelines to help organizations keep track of their Azure subscriptions, resources, and policies, ensuring that only authorized users have access to specific resources and that compliance and security requirements are met.

Managing and controlling cloud resources is a continuous process. Cloud governance should work in conjunction with policies that organizations already have for procedures to govern their on-premises IT environment. The level of integration between on-premises and cloud policies varies, depending on an organization's level of cloud governance maturity and the digital assets they have in the cloud. As the organization's cloud environment evolves, so must its cloud governance processes and policies to adapt to changes.

To establish a solid foundation for your cloud governance, you can start by following these steps:

- **Define your approach**: Understand the methodology used in the Cloud Adoption Framework to guide your decision-making and develop a clear vision for your end goal
- **Evaluate your current state**: Use the governance benchmark tool to assess where you currently stand and where you want to be in the future
- **Start small**: Implement a basic set of governance tools known as a **minimum viable product (MVP)** to begin your governance journey
- **Continuously improve**: As you progress through your cloud adoption plan, regularly add governance controls to address specific risks and work toward your end goal

Azure governance and Microsoft Cloud Adoption Framework for Azure are closely related, as they provide guidance and best practices to manage and control resources on the Azure platform. The Cloud Adoption Framework for Azure is a methodology that helps organizations plan, design, and execute their cloud adoption journey. It provides a set of best practices and guidance on assessing the current environment, identifying the goals and objectives, designing the target architecture, and executing the migration plan. It also guides how to operate and optimize the cloud environment.

Azure governance also provides the tools and features to manage and govern Azure resources. It includes features such as Azure Blueprints, Azure Policy, Azure management groups, and Azure locks, designed to help organizations manage their Azure environment.

Together, these two frameworks provide a comprehensive solution for organizations looking to adopt Azure, by helping them to understand the best way to approach their migration and then providing the means to govern and manage the resources once they are in the cloud.

Azure governance – components and services

Azure governance refers to the set of practices, policies, and technologies that organizations use to manage their Azure resources and ensure compliance with regulatory requirements. It includes several components and services that provide centralized management, control, and monitoring of Azure resources. The key components of Azure Governance include Azure Policy, Azure Blueprints, Azure management groups, and Azure Resource Manager. Let's understand each of these components.

Management groups

Let's imagine SpringToys has multiple departments, each with its own Azure subscription. In this case, SpringToys' IT Team can leverage Azure management groups to manage and organize various subscriptions in a hierarchical structure, providing centralized control over access, policies, and compliance.

Every Azure AD tenant is allocated a sole top-level management group known as the root management group. This group is an integral part of the subscription hierarchy, encompassing all other management groups and subscriptions. It serves as the central point to implement global policies and Azure role assignments at the directory level.

The root management group is automatically established when you access the Azure portal and select **Management Groups**, when you create a management group using an API call, or when you create a management group using PowerShell.

SpringToys' IT Team can create management groups for each department in the organization and assign the appropriate access controls and policies.

Once each department's subscription is assigned to the corresponding management group, a particular team or business unit can manage its own resources while still following the organization's policies and standards.

Note that SpringToys' IT Team can use the *auditing and reporting* features of Azure management groups to monitor compliance with policies and standards across all business units and track changes to resources.

Azure management groups enable centralized management of multiple Azure subscriptions, making it easier to manage large-scale Azure deployments, as they provide a hierarchical structure to organize and manage access, policies, and compliance for Azure resources.

The following figure shows the hierarchy of management groups and subscriptions.

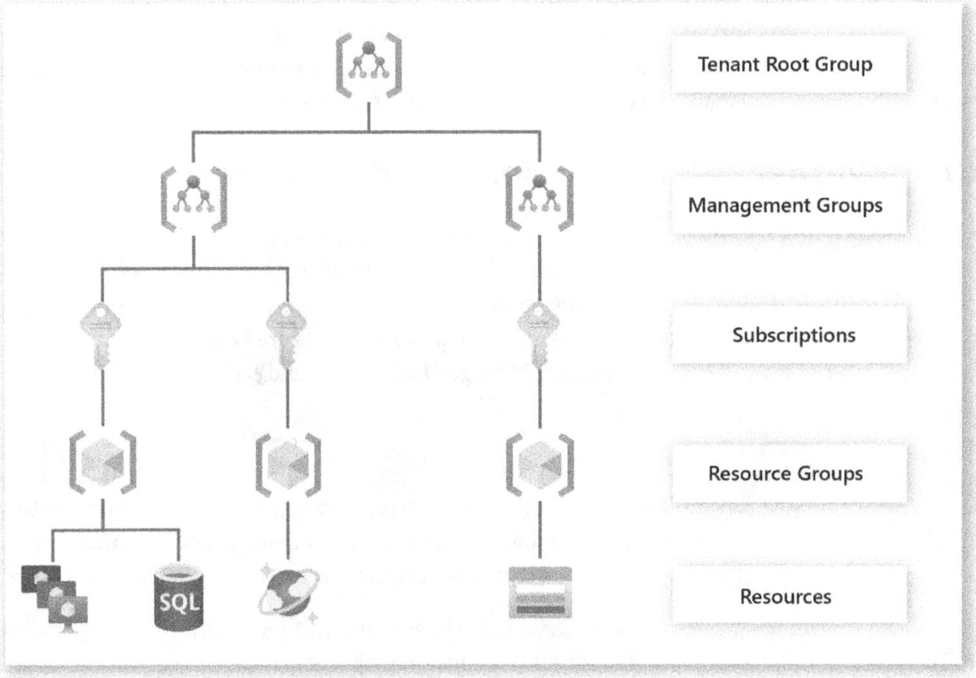
Tenant Root Group
Management Groups
Subscriptions
Resource Groups
Resources
SQL

Figure 10.1 – The hierarchy of management groups and subscriptions

You can create a hierarchy using management groups and apply specific policies across your organization. For example, you could create a policy to limit the locations where virtual machines can be provisioned, or establish a policy to allow only Linux OSs for the virtual machines. We will discuss how to work with policies in the following section.

Before you can leverage management groups, ensure you have an active Azure subscription, an Azure AD tenant, and a user account with the `Global Administrator` role. Then, you can proceed to work with management groups.

Once you have the correct user account, go to the Azure portal, and in the **Properties** section of the **Azure Active Directory** page, ensure that the **Access management for Azure resources** option is set to **Yes**, as shown here:

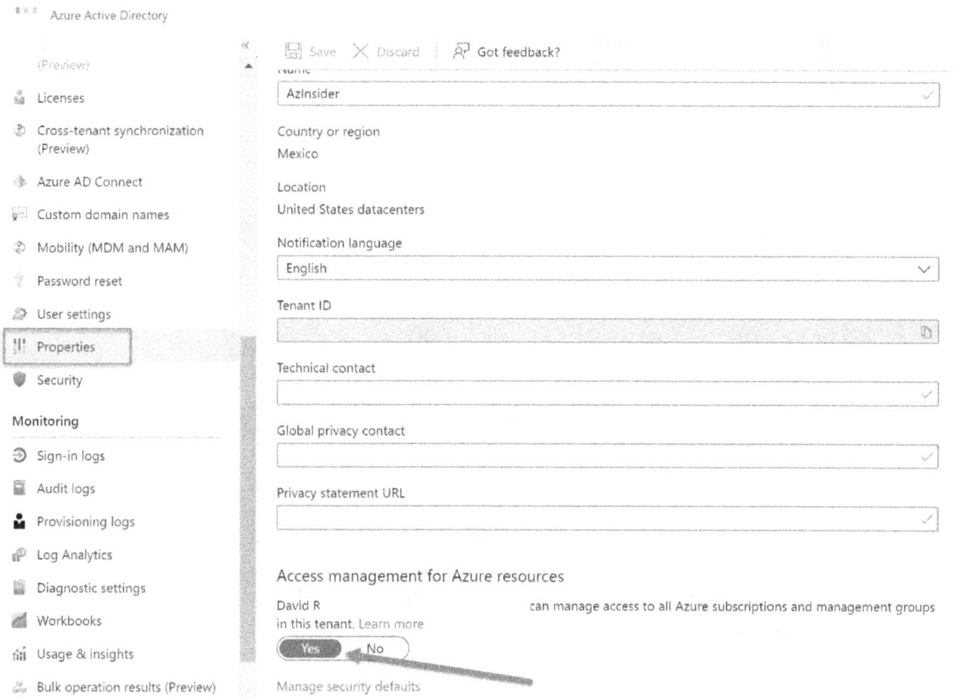
Azure Active Directory
Save
Discard
Got feedback?
(Preview)
Licenses
Cross-tenant synchronization (Preview)
Azure AD Connect
Custom domain names
Mobility (MDM and MAM)
Password reset
User settings
Properties
Security
Monitoring
Sign-in logs
Audit logs
Provisioning logs
Log Analytics
Diagnostic settings
Workbooks
Usage & insights
Bulk operation results (Preview)
AzInsider
Country or region
Mexico
Location
United States datacenters
Notification language
English
Tenant ID
Technical contact
Global privacy contact
Privacy statement URL
Access management for Azure resources
David R
can manage access to all Azure subscriptions and management groups
in this tenant. Learn more
Yes
No
Manage security defaults

Figure 10.2 – Access management for Azure resources

Once you enable the option shown in the preceding figure, you can start working with management groups. Traditionally, an Azure hierarchy includes four levels – management groups, subscriptions, resource groups, and resources.

A tenant root group contains all the management groups and subscriptions.

The following figure shows an example of a typical Azure hierarchy:

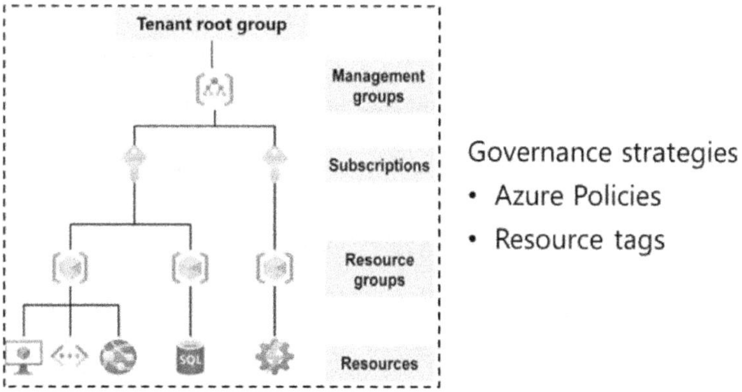
Tenant root group
Management groups
Subscriptions
Resource groups
Resources
SQL
Governance strategies
• Azure Policies
• Resource tags

Figure 10.3 – An example of an Azure hierarchy

Now, let's look at how you can create a management group using the Azure portal:

1. Log in to the Azure portal, go to the search bar, and look for `management groups`, as shown here:

management groups
All
Services (23)
Azure Active Directory (0)
Services
Management groups
API Management services

Figure 10.4 – The Azure portal – search for management groups

2. Then, you can add management groups and assign a subscription under the management group created, as shown here:

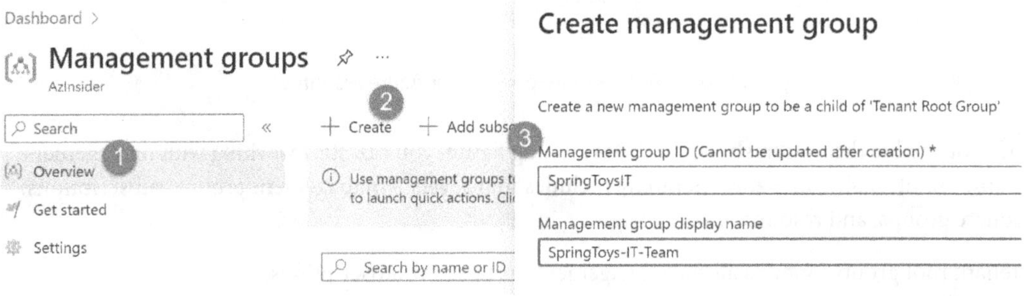
Dashboard >
Management groups
AzInsider
Search
Overview
Get started
Settings
Create
Search by name or ID
Create management group
Create a new management group to be a child of 'Tenant Root Group'
Management group ID (Cannot be updated after creation) *
SpringToysIT
Management group display name
SpringToys-IT-Team

Figure 10.5 – Creating a management group

Once a management group is created, its name cannot be changed.

3. Once the management group is created, you can select it and add a subscription. In this case, we will add an existing subscription called **SpringToys**, as shown here:

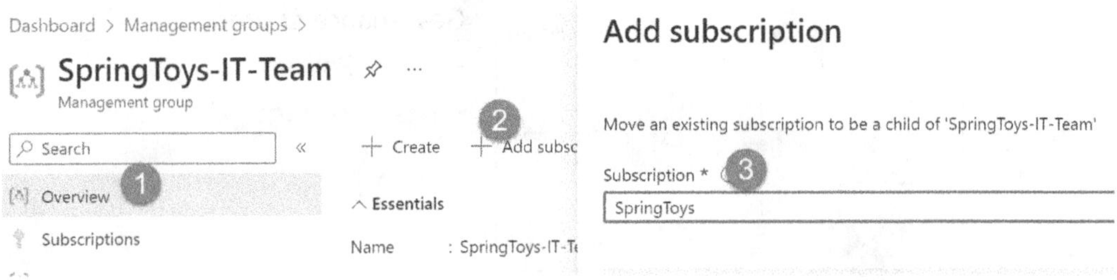
Dashboard > Management groups >
SpringToys-IT-Team
Management group
Search
Overview
Subscriptions
Create
Essentials
Add subscription
Move an existing subscription to be a child of 'SpringToys-IT-Team'
Subscription *
SpringToys

Figure 10.6 – Adding a subscription to a management group

After adding the subscription, you should see the subscription under the **SpringToys-IT-Team** management group, as shown here.

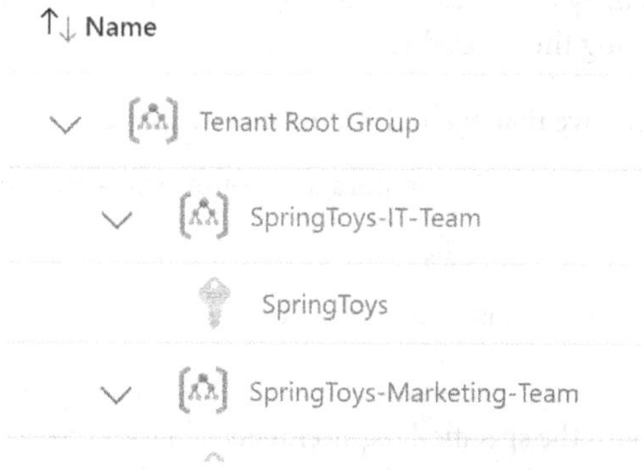
Name
Tenant Root Group
SpringToys-IT-Team
SpringToys
SpringToys-Marketing-Team

Figure 10.7 – The subscription and management group

With management groups, we can enable administrators to define and enforce policies across multiple subscriptions and resources, ensuring consistency and compliance. Now, let's understand how we can leverage Azure Policy to enforce compliance with standards across all business units in an organization.

Azure Policy

Azure Policy is a powerful service that enables the enforcement of organizational standards and facilitates comprehensive compliance assessment on a large scale. Its compliance dashboard offers an aggregated view to assess the overall state of an environment, allowing users to drill down to the granular level of each resource and policy. Azure Policy aids in bringing resources into compliance by offering bulk remediation for existing resources and automatic remediation for new resources.

Azure Policy addresses various critical aspects such as resource consistency, regulatory compliance, security, cost, and management. Specifically, Azure Policy empowers you to enforce essential governance actions, including the following:

- Restricting Azure resource deployment to approved regions only
- Ensuring consistent application of taxonomic tags across resources
- Mandating the sending of diagnostic logs from resources to a designated Log Analytics workspace

All Azure Policy data and objects are encrypted at rest, providing robust security measures.

Azure Policy allows you to set and enforce resource rules and standards within your Azure subscription. For example, if SpringToys wants to ensure that all virtual machines created in Azure have certain specific configurations for security purposes, then SpringToys could define an Azure policy to enforce these requirements by specifying the desired values for the following:

- Virtual machines must have managed disks, not classic storage
- Virtual machines must use a particular storage account type – for example, premium storage
- Virtual machines must have a specified size – for instance, `Standard_B2s`
- Virtual machines must use the latest version of the Windows OS

By creating and assigning this policy, SpringToys can ensure that all virtual machines created in their Azure environment comply with the specified requirements. If a user tries to create a virtual machine that doesn't meet the policy, Azure will prevent the deployment and return an error message explaining why it cannot be created.

Azure Policy helps to ensure compliance with an organization's standards and regulations and control resource configuration and deployment.

It is essential to understand the most relevant components of Azure Policy:

- **Policy definitions**: This is the basic building block of Azure Policy, where you define the policy rules and conditions that need to be enforced in the form of a JSON-defined object. A policy definition consists of a policy rule and an associated effect, which can be used either to enforce compliance or deny deployment.
- **Policy assignments**: Policy assignments are JSON-defined objects used to apply policy definitions to specific Azure resources, resource groups, or subscriptions. Multiple policy assignments can be created for a single policy definition.
- **Policy effects**: In Azure Policy, each policy definition is associated with a single effect that determines the outcome when the policy rule is evaluated for a match. The behavior of these effects varies, depending on whether they apply to a new resource, an updated resource, or an existing resource. The following effects are currently supported in a policy definition: `Append`, `Audit`, `AuditIfNotExists`, `Deny`, `DenyAction` (preview), `DeployIfNotExists`, `Disabled`, `Manual` (preview), and `Modify`.
- **Policy evaluation**: Azure Policy evaluates resources against assigned policy definitions whenever a change is made to a resource or on a defined schedule. The results of policy evaluations are displayed in the Azure portal and can be viewed by administrators.
- **Built-in policies**: Azure Policy includes several built-in policies, which cover common scenarios such as controlling the use of specific virtual machine sizes, enforcing specific disk types, or ensuring that virtual networks are correctly configured. You can review examples of built-in policies at the following GitHub repository: `https://github.com/Azure/azure-policy`.

- **Custom policies**: You can also create custom policies to meet your organization's specific requirements.
- **Initiative**: Think of an initiative as a collection of policy definition IDs. By using initiatives, you can centralize multiple policy definitions with a common goal.

By using Azure Policy, you can ensure that resources deployed in Azure meet your organization's standards, improve security and compliance, and reduce the likelihood of misconfigurations.

To create a custom Azure policy to enforce your desired standards and ensure that all resources deployed in your Azure environment comply with your organization's policies, follow these steps:

1. Go to the Azure portal and search for the Azure policy service, as shown here:

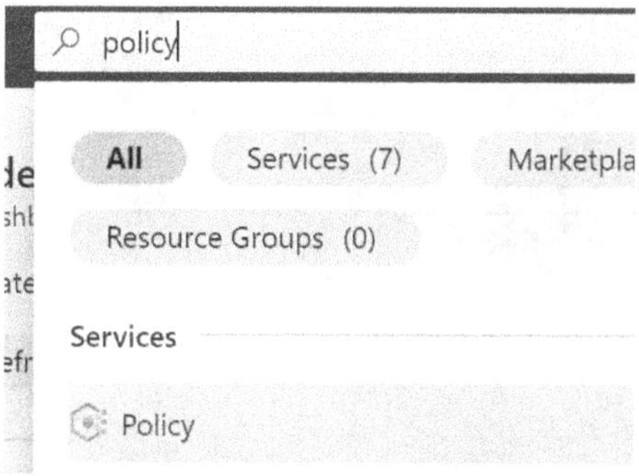
policy
All
Services (7)
Marketpla
Resource Groups (0)
Services
Policy

Figure 10.8 – Search for Azure Policy

2. Then, select the **Definitions** tab and then the + **Policy definition** option.

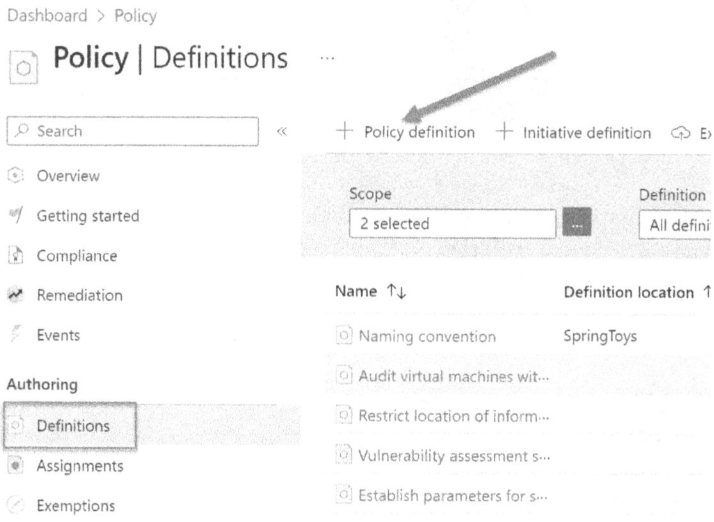
Dashboard > Policy
Policy | Definitions
Search
+ Policy definition
+ Initiative definition
Overview
Getting started
Compliance
Remediation
Events
Authoring
Definitions
Assignments
Exemptions
Scope
2 selected
Definition
All defini
Name
Definition location
Naming convention
SpringToys
Audit virtual machines wit...
Restrict location of inform...
Vulnerability assessment s...
Establish parameters for s...

Figure 10.9 – Creating an Azure policy

3. Then, provide the parameters as shown in the following figure:

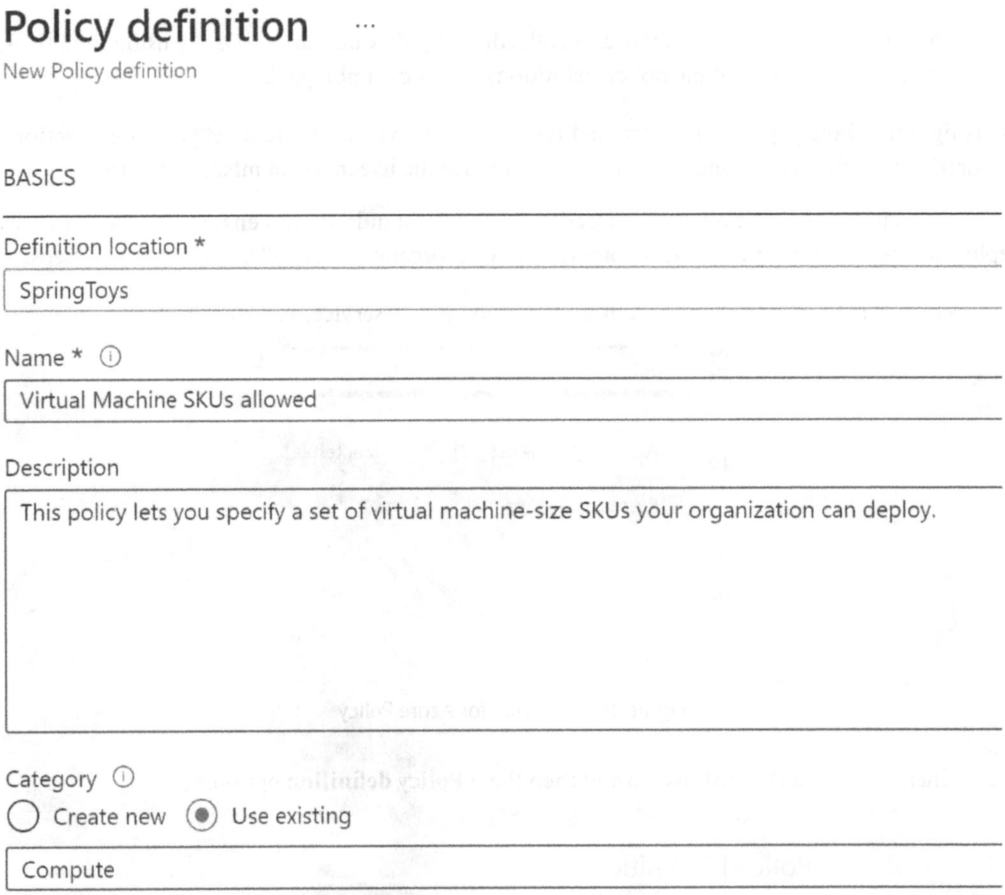
Policy definition ...
New Policy definition
BASICS
Definition location *
SpringToys
Name *
Virtual Machine SKUs allowed
Description
This policy lets you specify a set of virtual machine-size SKUs your organization can deploy.
Category
Create new
Use existing
Compute

Figure 10.10 – Azure Policy parameters

4. Next, we will define the policy and choose the effect of the policy to determine what will happen if a resource doesn't meet the conditions specified in the policy rule, using the following JSON definition:

```
{
  "properties": {
    "displayName": "Allowed virtual machine size SKUs",
    "policyType": "BuiltIn",
    "mode": "Indexed",
    "description": "This policy enables you to specify a set of
virtual machine size SKUs that your organization can deploy.",
    "metadata": {
```

```
        "version": "1.0.1",
        "category": "Compute"
      },
      "version": "1.0.1",
      "parameters": {
        "listOfAllowedSKUs": {
          "type": "Array",
          "metadata": {
            "description": "The list of size SKUs that can be
specified for virtual machines.",
            "displayName": "Allowed Size SKUs",
            "strongType": "VMSKUs"
          }
        }
      },
      "policyRule": {
        "if": {
          "allOf": [
            {
              "field": "type",
              "equals": "Microsoft.Compute/virtualMachines"
            },
            {
              "not": {
                "field": "Microsoft.Compute/virtualMachines/sku.
name",
                "in": "[parameters('listOfAllowedSKUs')]"
              }
            }
          ]
        },
        "then": {
          "effect": "Deny"
        }
      }
    },
    "id": "/providers/Microsoft.Authorization/policyDefinitions/
cccc23c7-8427-4f53-ad12-b6a63eb452b3",
    "name": "cccc23c7-8427-4f53-ad12-b6a63eb452b3"
}
```

You can find the preceding code in the following GitHub repository:

https://github.com/PacktPublishing/Azure-Architecture-Explained/blob/main/chapters/chapter-10/azure-policy-sample.json

You can use the preceding code and place it in the policy definition section, as shown here:

POLICY RULE
Import sample policy definition from GitHub
Learn more about policy definition structure
"properties": {
"displayName": "Allowed virtual machine size SKUs",
"policyType": "BuiltIn",
"mode": "Indexed",
"description": "This policy enables you to specify a set of virtual machine size SKUs th
"metadata": {
"version": "1.0.1",
"category": "Compute"
},
"version": "1.0.1",
"parameters": {
"listOfAllowedSKUs": {
"type": "Array",
"metadata": {
"description": "The list of size SKUs that can be specified for virtual machines."
"displayName": "Allowed Size SKUs",
"strongType": "VMSKUs"
}
}
},
"policyRule": {
"if": {
"allOf": [
{
"field": "type",
"equals": "Microsoft.Compute/virtualMachines"
},
{
"not": {
"field": "Microsoft.Compute/virtualMachines/sku.name",
"in": "[parameters('listOfAllowedSKUs')]"
}
}
]
},
"then": {
"effect": "Deny"
}
}
Save
Cancel

Figure 10.11 – An Azure Policy effect

5. Then, proceed to create the Azure policy. Once created, the next step is to create an assignment. Select the **Assign** option, as shown here:

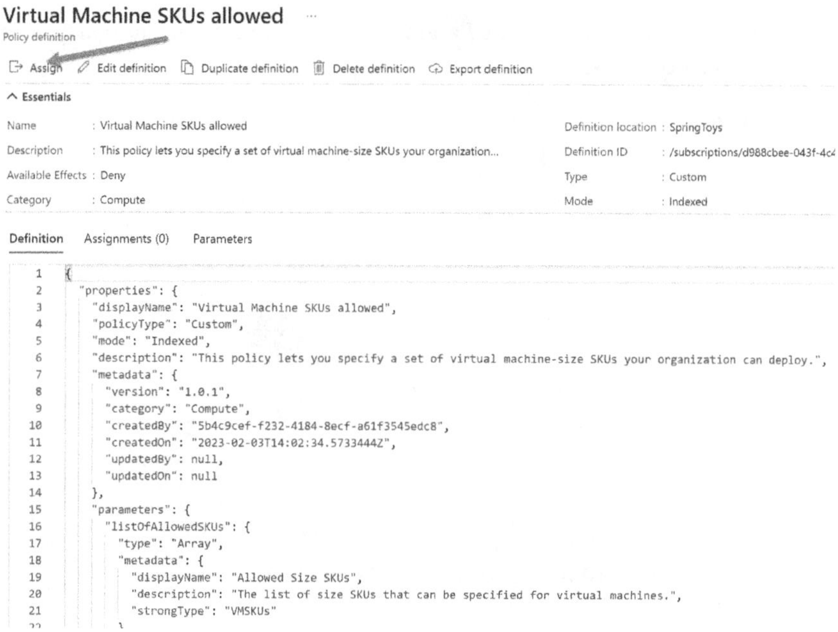
Virtual Machine SKUs allowed
Policy definition
Assign
Edit definition
Duplicate definition
Delete definition
Export definition
Essentials
Name : Virtual Machine SKUs allowed
Description : This policy lets you specify a set of virtual machine-size SKUs your organization...
Available Effects : Deny
Category : Compute
Definition location : SpringToys
Type : Custom
Mode : Indexed
Definition
Assignments (0)
Parameters
{
"properties": {
"displayName": "Virtual Machine SKUs allowed",
"policyType": "Custom",
"mode": "Indexed",
"description": "This policy lets you specify a set of virtual machine-size SKUs your organization can deploy.",
"metadata": {
"version": "1.0.1",
"category": "Compute",
"createdBy": "5b4c9cef-f232-4184-8ecf-a61f3545edc8",
"createdOn": "2023-02-03T14:02:34.5733444Z",
"updatedBy": null,
"updatedOn": null
},
"parameters": {
"listOfAllowedSKUs": {
"type": "Array",
"metadata": {
"displayName": "Allowed Size SKUs",
"description": "The list of size SKUs that can be specified for virtual machines.",
"strongType": "VMSKUs"

Figure 10.12 – Azure Policy – assignment

6. We can assign the scope to a resource group, subscription, or Azure management group. In this case, we will select the **SpringToys** subscription.

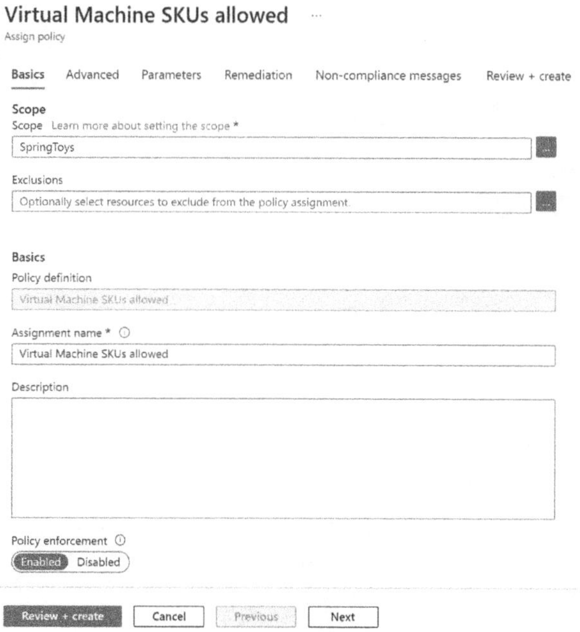
Virtual Machine SKUs allowed
Assign policy
Basics
Advanced
Parameters
Remediation
Non-compliance messages
Review + create
Scope
Scope Learn more about setting the scope *
SpringToys
Exclusions
Optionally select resources to exclude from the policy assignment
Basics
Policy definition
Virtual Machine SKUs allowed
Assignment name *
Virtual Machine SKUs allowed
Description
Policy enforcement
Enabled
Disabled
Review + create
Cancel
Previous
Next

Figure 10.13 – Assign policy

7. We can define the allowed SKUs in the **Parameters** tab for this particular policy. In this case, we will select two SKUs, as shown here:

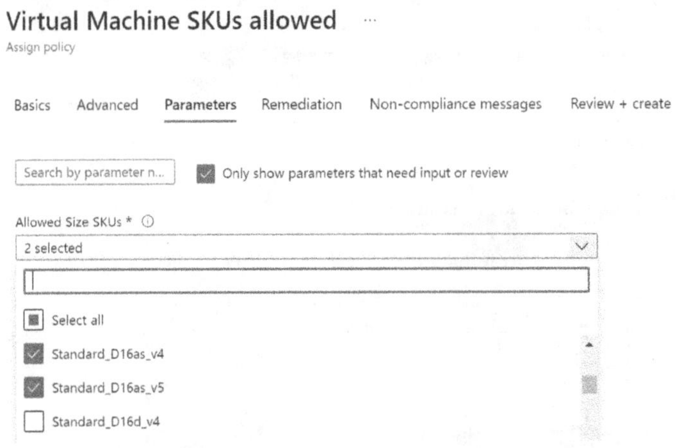
Virtual Machine SKUs allowed
Assign policy
Basics
Advanced
Parameters
Remediation
Non-compliance messages
Review + create
Search by parameter n...
Only show parameters that need input or review
Allowed Size SKUs *
2 selected
Select all
Standard_D16as_v4
Standard_D16as_v5
Standard_D16d_v4

Figure 10.14 – Assign policy – the Parameters tab

8. Then, proceed to create the policy assignment.

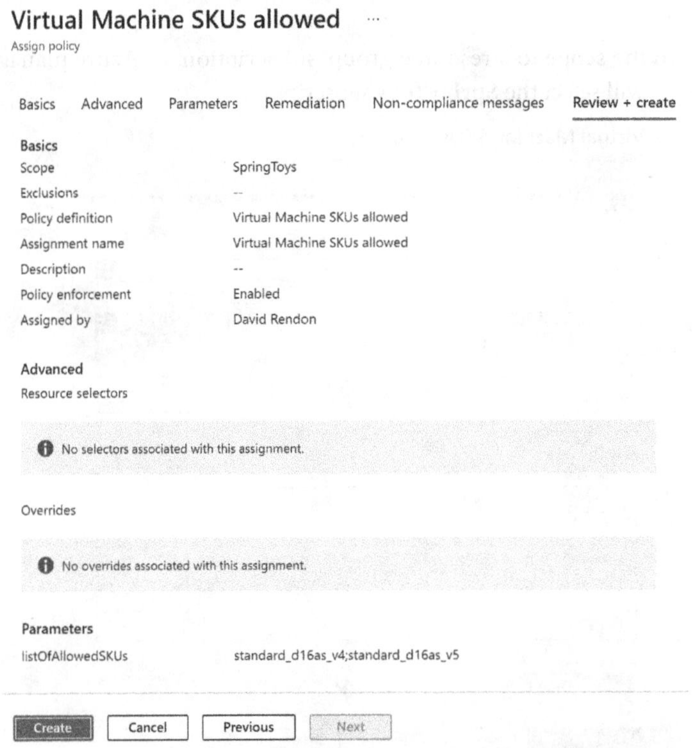
Virtual Machine SKUs allowed
Assign policy
Basics
Advanced
Parameters
Remediation
Non-compliance messages
Review + create
Basics
Scope
SpringToys
Exclusions
--
Policy definition
Virtual Machine SKUs allowed
Assignment name
Virtual Machine SKUs allowed
Description
--
Policy enforcement
Enabled
Assigned by
David Rendon
Advanced
Resource selectors
No selectors associated with this assignment.
Overrides
No overrides associated with this assignment.
Parameters
listOfAllowedSKUs
standard_d16as_v4;standard_d16as_v5
Create
Cancel
Previous
Next

Figure 10.15 – Assign policy – the Review + create tab

9. Once the policies and assignments are configured, you can monitor the compliance of resources in the Azure portal by using the **Compliance** tab. After assigning the policy, it might take several minutes to get the compliance data, depending upon the condition that needs to be evaluated and the number of resources to be evaluated. This page will show a list of resources and their compliance status with the policies that have been assigned.

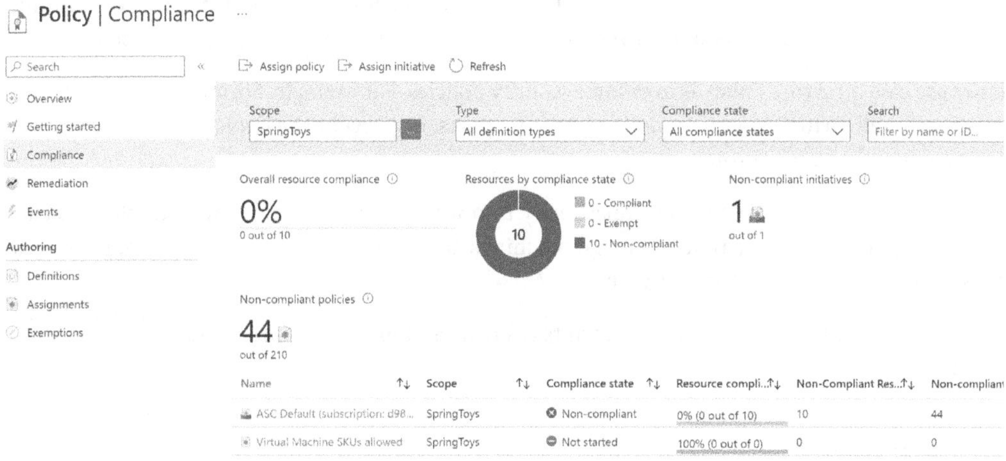
Policy | Compliance
Search
Overview
Getting started
Compliance
Remediation
Events
Authoring
Definitions
Assignments
Exemptions
Assign policy
Assign initiative
Refresh
Scope
SpringToys
Type
All definition types
Compliance state
All compliance states
Search
Filter by name or ID...
Overall resource compliance
0%
0 out of 10
Resources by compliance state
10
0 - Compliant
0 - Exempt
10 - Non-compliant
Non-compliant initiatives
1
out of 1
Non-compliant policies
44
out of 210
Name
Scope
Compliance state
Resource compli...
Non-Compliant Res...
Non-complian
ASC Default (subscription: d98...
SpringToys
Non-compliant
0% (0 out of 10)
10
44
Virtual Machine SKUs allowed
SpringToys
Not started
100% (0 out of 0)
0
0

Figure 10.16 – Azure Policy – compliance tab

Remediation is a powerful feature in Azure Policy that brings non-compliant resources into a compliant state, particularly for policies with `DeployIfNotExists` or Modify effects. By utilizing remediation tasks, you can deploy the deployIfNotExists template or execute modify operations on your existing resources and subscriptions. This applies to policy assignments at the management group, subscription, resource group, or individual resource level. To understand and effectively utilize remediation with Azure Policy, refer to the steps outlined in this article.

In terms of access control, Azure Policy employs a managed identity associated with the policy assignment when initiating template deployments for deployIfNotExists policies or modifying resources for Modify policies. Policy assignments leverage managed identities for Azure resource authorization. You can use either a system-assigned managed identity provided by the policy service or a user-assigned identity of your choice.

The managed identity must be assigned the necessary minimum **Role-Based Access Control** (**RBAC**) roles required for resource remediation. If the managed identity lacks these roles, the portal will display an error during policy or initiative assignments. When using the portal, Azure Policy automatically grants the listed roles to the managed identity upon assignment commencement. However, if you use an Azure **Software Development Kit** (**SDK**), you must manually assign the roles to the managed identity. It is important to note that the location of the managed identity does not affect its functionality with Azure Policy.

Consider using Azure Policy when your organization wants to enforce specific standards and comply with regulations for its Azure resources.

For example, SpringToys could leverage Azure policy to meet regulatory and compliance requirements such as PCI-DSS, HIPAA, and GDPR by enforcing policies that ensure resources are correctly configured.

On the other hand, Azure Policy can be utilized to enforce policies to help control costs, by limiting the use of premium storage accounts or restricting the use of expensive virtual machine sizes.

Another use case for Azure Policy is to enforce security policies. For example, SpringToys could require encryption for all virtual machines, control specific ports, or prevent the deployment of resources without a network security group.

Lastly, Azure Policy can also enforce consistency in resource deployments. For example, you can implement a policy to ensure that all virtual machines are deployed in a specific region, resources group, or within a specific virtual network.

Now, we will review how you can use Blueprints to create and manage Azure resource collections consistently.

Azure Blueprints

With Azure Blueprints, organizations can establish a consistent and repeatable set of Azure resources that align with their standards, patterns, and requirements. This empowers development teams to rapidly create and launch new environments while ensuring compliance with organizational policies.

By leveraging Azure Blueprints, which is in preview at the time of writing, development teams gain the confidence that they can build within the confines of organizational compliance, thanks to a comprehensive set of built-in components. These components, such as networking, accelerate development, and delivery processes.

Blueprints offer a declarative approach to orchestrate the deployment of various resource templates and artifacts, including role assignments, policy assignments, **Azure Resource Manager** (**ARM**) templates, and resource groups.

Imagine SpringToys wants to deploy a new online store that is scalable, secure, and compliant with industry standards. Instead of creating each resource for a specific environment, SpringToys' IT team creates an Azure blueprint with the following components:

- **Web servers**: Virtual machines to host the frontend
- **Database servers**: Azure SQL databases to handle customer information and transaction data
- **Load balancers**: To distribute incoming traffic across the web servers and ensure that the online store is highly available

- **Policies:** The blueprint includes policies to enforce security best practices and ensure that resources are deployed in a consistent and compliant manner, such as requiring encrypted storage for sensitive data

Azure Blueprints allows you to create and manage collections of Azure resources quickly and consistently. You can define the infrastructure and policy part of your environment and ensure a consistent deployment every time.

Once you create a blueprint, you can apply it to a scope, assign it to an environment, and quickly deploy the resources without worrying about manual configuration errors. Then, you can track where blueprints have been applied and share them across your organization.

In this scenario, SpringToys IT Team can create a new blueprint, then assign it to the development environment, and streamline the deployment process, reducing the time and effort needed to deploy the online store.

Azure Blueprints can help SpringToys improve the speed, consistency, and security of the online store deployment, reducing the risk of configuration errors and security breaches. A blueprint packages different artifact types, such as resource groups, policies, role assignments, and ARM templates.

These blueprint packages can be versioned and utilized through a continuous integration and continuous delivery pipeline. Blueprints are assigned to a subscription and can be audited and tracked.

The following figure shows a high-level overview of Azure Blueprints:

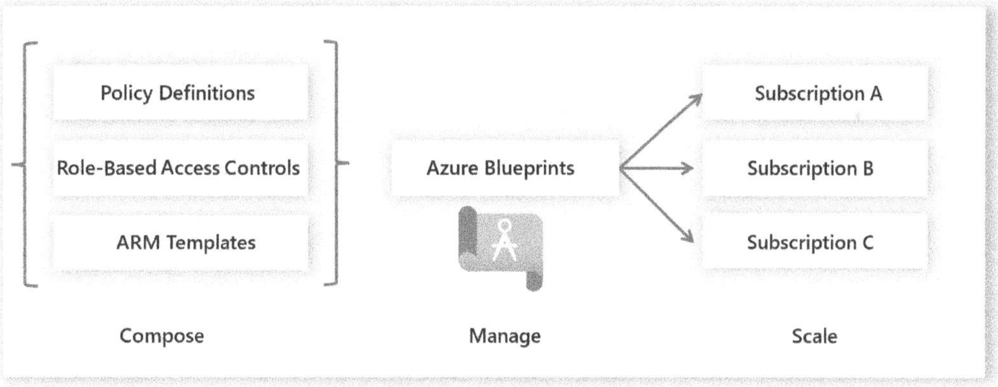
Policy Definitions
Role-Based Access Controls
ARM Templates
Azure Blueprints
Subscription A
Subscription B
Subscription C
Compose
Manage
Scale

Figure 10.17 – Azure Blueprints

Azure Blueprints comprise various artifacts, including policy definitions, role assignments, ARM templates, and resource groups. To create an Azure blueprint, you can use the Azure portal:

1. In the Azure portal, search for `blueprints`, as shown here:

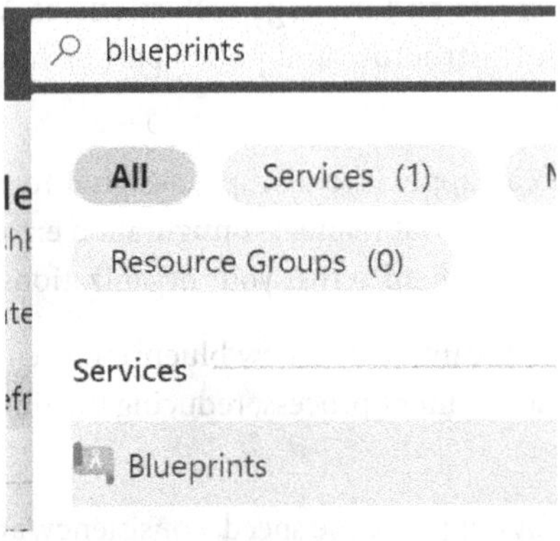
blueprints
All
Services (1)
Resource Groups (0)
Services
Blueprints

Figure 10.18 – Searching for blueprints

2. Now, proceed to create a new instance of this resource type. When you create an Azure blueprint, you can use a blank template or use an existing blueprint template, as shown here:

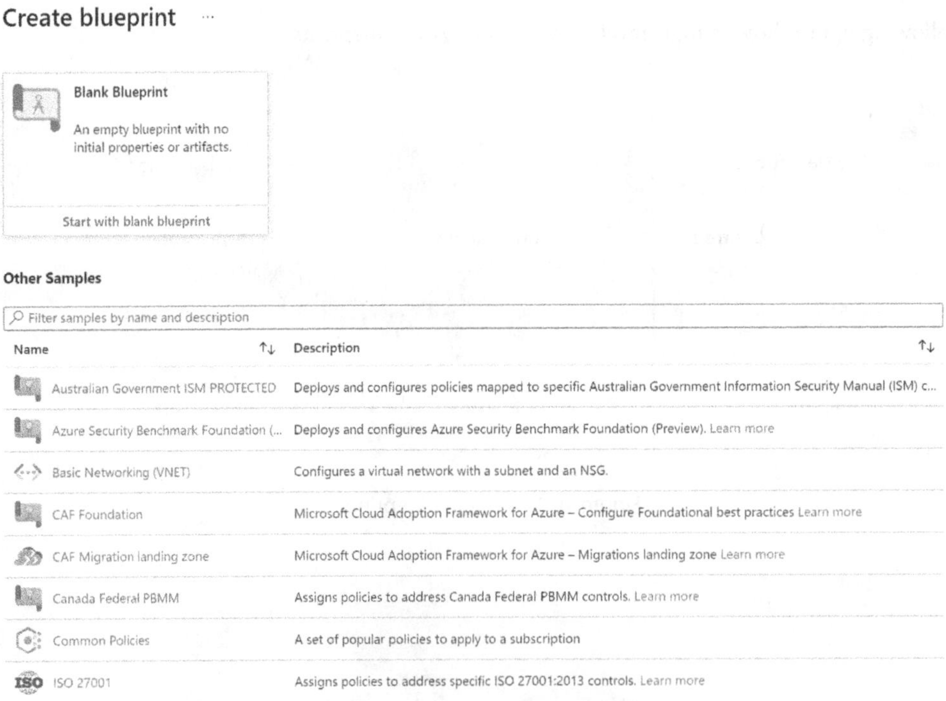
Create blueprint
Blank Blueprint
An empty blueprint with no initial properties or artifacts.
Start with blank blueprint
Other Samples
Filter samples by name and description
Name
Description
Australian Government ISM PROTECTED
Deploys and configures policies mapped to specific Australian Government Information Security Manual (ISM) c...
Azure Security Benchmark Foundation (...
Deploys and configures Azure Security Benchmark Foundation (Preview). Learn more
Basic Networking (VNET)
Configures a virtual network with a subnet and an NSG.
CAF Foundation
Microsoft Cloud Adoption Framework for Azure – Configure Foundational best practices Learn more
CAF Migration landing zone
Microsoft Cloud Adoption Framework for Azure – Migrations landing zone Learn more
Canada Federal PBMM
Assigns policies to address Canada Federal PBMM controls. Learn more
Common Policies
A set of popular policies to apply to a subscription
ISO 27001
Assigns policies to address specific ISO 27001:2013 controls. Learn more

Figure 10.19 – Creating a new blueprint

3. In this example, we will select the **CAF Foundation** blueprint. Then, we will provide the parameters for the **Basics** tab, as shown here:

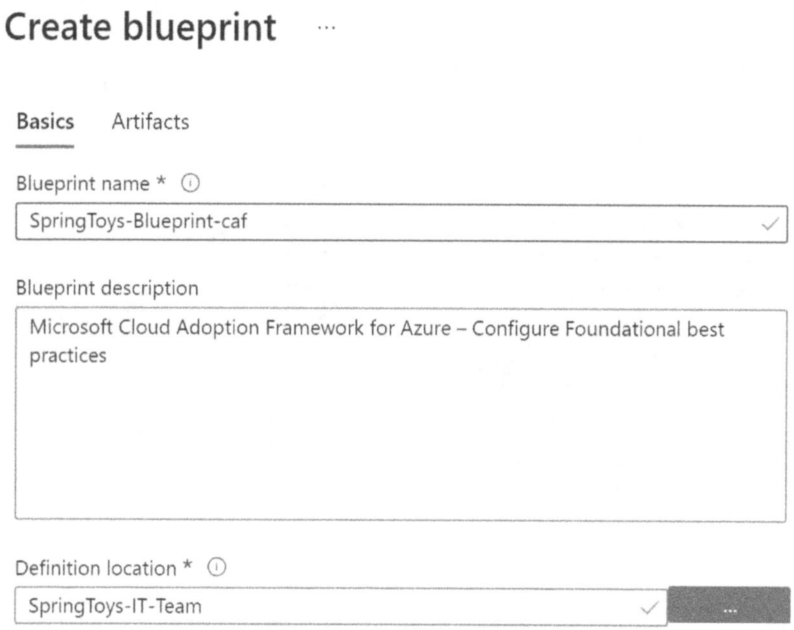
Create blueprint
Basics
Artifacts
Blueprint name *
SpringToys-Blueprint-caf
Blueprint description
Microsoft Cloud Adoption Framework for Azure – Configure Foundational best practices
Definition location *
SpringToys-IT-Team

Figure 10.20 – Creating a new blueprint – the Basics tab

> **Note**
>
> We provide the name of the blueprint, a brief description, and specify the location of the blueprint. The location can be a management group or subscription where you want to save the blueprint.

4. Next, we will add the artifacts to the blueprint. For the **CAF Foundation** blueprint, there are already a few artifacts, including policies, resource groups, and ARM templates, as shown here:

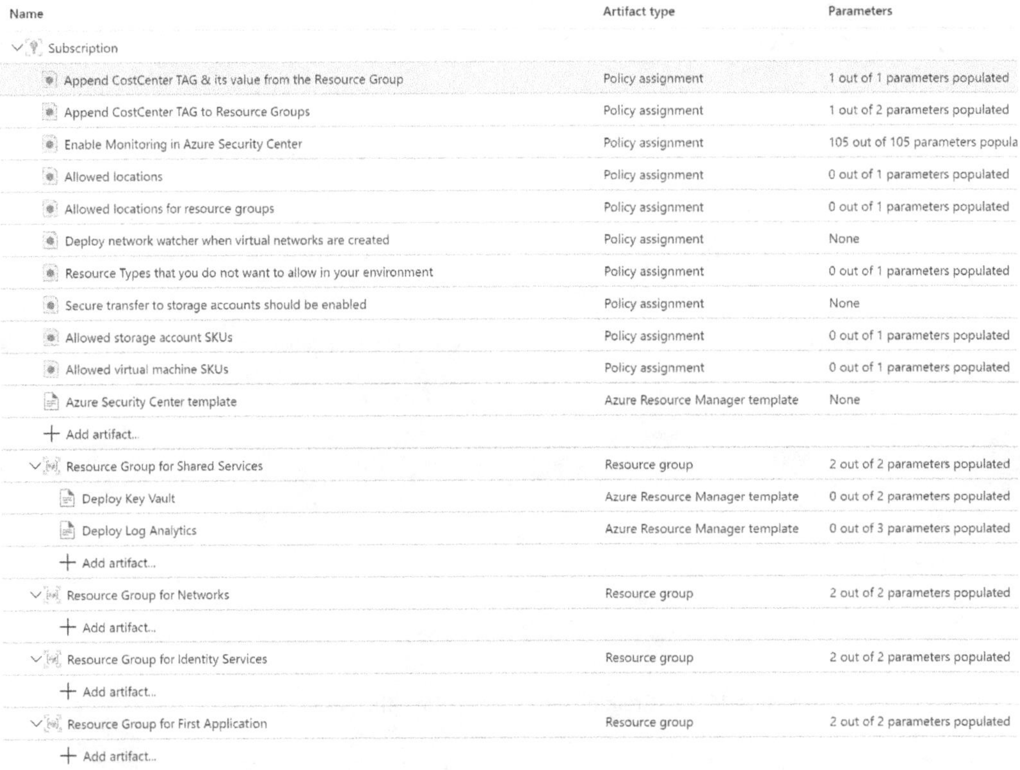
Name
Artifact type
Parameters
Subscription
Append CostCenter TAG & its value from the Resource Group
Policy assignment
1 out of 1 parameters populated
Append CostCenter TAG to Resource Groups
Policy assignment
1 out of 2 parameters populated
Enable Monitoring in Azure Security Center
Policy assignment
105 out of 105 parameters popula
Allowed locations
Policy assignment
0 out of 1 parameters populated
Allowed locations for resource groups
Policy assignment
0 out of 1 parameters populated
Deploy network watcher when virtual networks are created
Policy assignment
None
Resource Types that you do not want to allow in your environment
Policy assignment
0 out of 1 parameters populated
Secure transfer to storage accounts should be enabled
Policy assignment
None
Allowed storage account SKUs
Policy assignment
0 out of 1 parameters populated
Allowed virtual machine SKUs
Policy assignment
0 out of 1 parameters populated
Azure Security Center template
Azure Resource Manager template
None
Add artifact...
Resource Group for Shared Services
Resource group
2 out of 2 parameters populated
Deploy Key Vault
Azure Resource Manager template
0 out of 2 parameters populated
Deploy Log Analytics
Azure Resource Manager template
0 out of 3 parameters populated
Add artifact...
Resource Group for Networks
Resource group
2 out of 2 parameters populated
Add artifact...
Resource Group for Identity Services
Resource group
2 out of 2 parameters populated
Add artifact...
Resource Group for First Application
Resource group
2 out of 2 parameters populated
Add artifact...

Figure 10.21 – Creating a new blueprint – the Artifacts tab

5. Once you save this draft, you will see it in the **Blueprints definitions** section, as shown here:

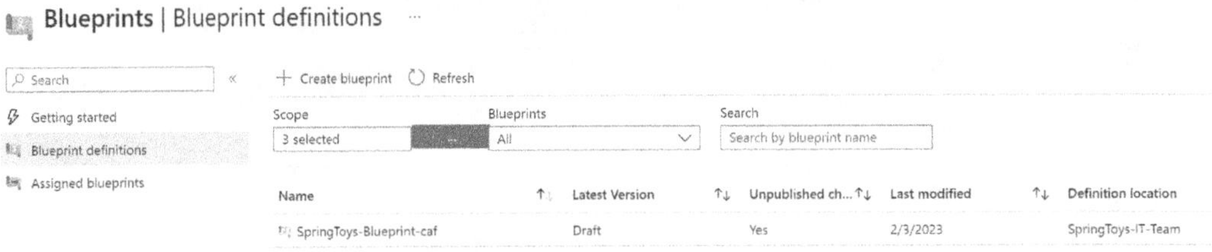
Blueprints | Blueprint definitions
Search
Getting started
Blueprint definitions
Assigned blueprints
Create blueprint
Refresh
Scope
3 selected
Blueprints
All
Search
Search by blueprint name
Name
Latest Version
Unpublished ch...
Last modified
Definition location
SpringToys-Blueprint-caf
Draft
Yes
2/3/2023
SpringToys-IT-Team

Figure 10.22 – Blueprint definitions

6. The next step is to publish the blueprint, and you can select the blueprint and then publish it in the **Overview** tab. Note that you can specify the version of the blueprint and include change notes:

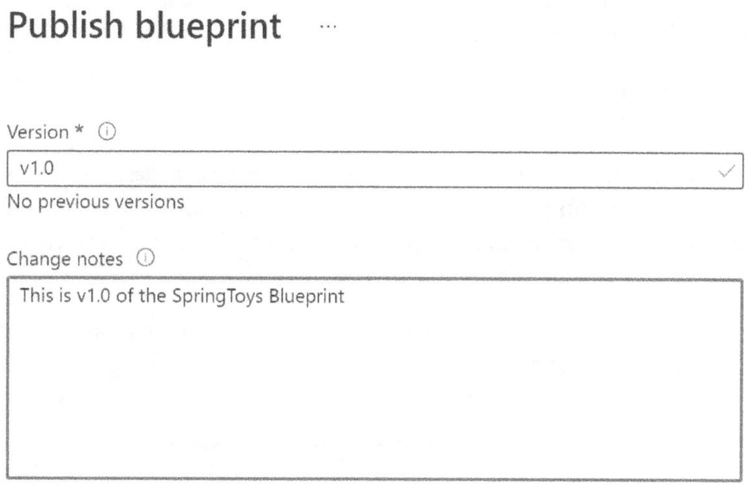
Publish blueprint
Version *
v1.0
No previous versions
Change notes
This is v1.0 of the SpringToys Blueprint

Figure 10.23 – Publish blueprint

In our scenario, SpringToys can leverage blueprints to ensure their online store deployment's consistency, security, and compliance.

This deployment process automation helps streamline the creation of complex environments, saving time and effort for the SpringToys IT team.

Now, what if SpringToys needs to obtain a complete inventory of the resources in the subscription, or enforce policies to improve its security posture?

Let's take a look at Azure Resource Graph.

Azure Resource Graph

Put simply, by using Azure Resource Graph, SpringToys can optimize its infrastructure management. SpringToys can leverage Azure Resource Graph to query and analyze the Azure resources, monitor resource utilization, identify underutilized resources, and make informed decisions about resource scaling and optimization.

For example, SpringToys could identify underutilized virtual machines and scale them as needed to reduce monthly consumption, or to automate policy enforcement and compliance checks to ensure the infrastructure meets security and compliance standards.

Using Azure Resource Graph's query language, you can query basic resource fields such as the resource name, ID, type, resource group, subscription, and location. This language is based on the **Kusto Query Language** (**KQL**) used by Azure Data Explorer. With Azure Resource Graph, you can streamline your resource management and gain valuable insights into your Azure environment.

You can easily access your resources' properties in a single query, eliminating the need for multiple API calls.

In addition, the resource change history feature allows you to view the changes made to your resources over the last seven days, including what properties were changed. This feature provides valuable insights into your Azure environment, helping you make informed decisions and ensuring efficient resource utilization.

To query Azure resources, you can use the Azure portal. In the search bar, look for `resource graph explorer`, as shown here:

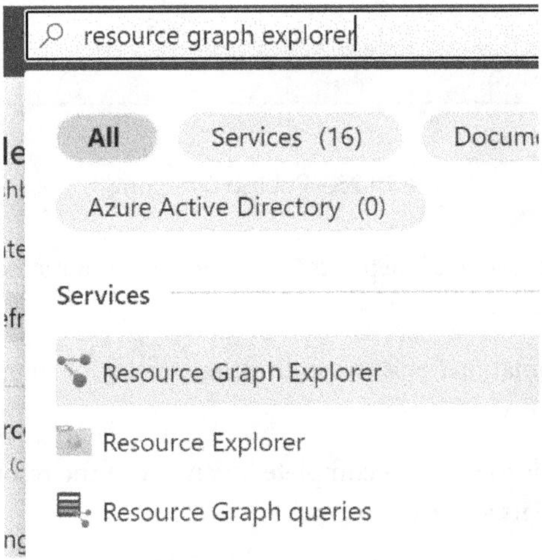
resource graph explorer
All
Services (16)
Azure Active Directory (0)
Services
Resource Graph Explorer
Resource Explorer
Resource Graph queries

Figure 10.24 – Search bar – resource graph explorer

Once in **Resource Graph Explorer**, you can customize and run your query. You can also leverage existing queries. The following figure shows a query that returns the number of Azure resources that exist in the subscription you have access to.

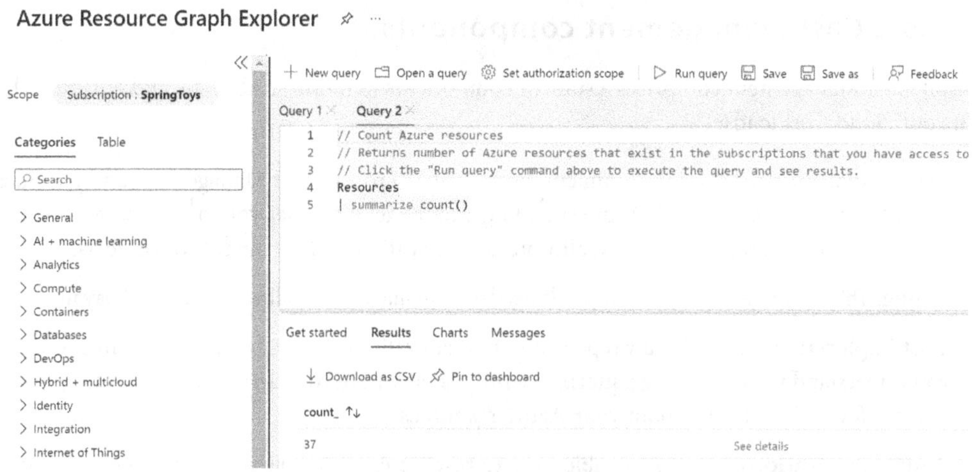
Azure Resource Graph Explorer
New query
Open a query
Set authorization scope
Run query
Save
Save as
Feedback
Scope
Subscription : SpringToys
Query 1
Query 2
// Count Azure resources
// Returns number of Azure resources that exist in the subscriptions that you have access to
// Click the "Run query" command above to execute the query and see results.
Resources
| summarize count()
Categories
Table
Search
General
AI + machine learning
Analytics
Compute
Containers
Databases
DevOps
Hybrid + multicloud
Identity
Integration
Internet of Things
Get started
Results
Charts
Messages
Download as CSV
Pin to dashboard
count_
37
See details

Figure 10.25 – Resource Graph Explorer

Your queries can be pinned to dashboards. Furthermore, you can import built-in dashboards. You can find a few examples in the following GitHub repository: `https://github.com/Azure-Samples/Governance/tree/master/src/resource-graph/portal-dashboards`.

As you can see, Azure Resource Graph can help SpringToys streamline its resource management and maintain a secure and compliant infrastructure. By providing a complete overview of resource configuration changes and properties, Azure Resource Graph empowers SpringToys to make informed decisions, optimize its resource utilization, and ultimately, drive its business forward.

Now, let's look at how SpringToys can gain insights into its Azure spending, track its usage, and create alerts and budgets to monitor and control its costs.

Microsoft Cost Management

Let's consider that SpringToys uses various Azure resources such as virtual machines to host its web servers, databases, and other applications. In addition to keeping track of those resources, the company has multiple business units such as marketing, IT, and operations that use Azure services. Each department has its own budget allocation and continuously manually adjusts its usage to reduce costs.

Using Microsoft Cost Management, SpringToys can monitor the cost of Azure resources and keep track of the spending of each business unit, ensuring each stays within its budget. Moreover, Microsoft Cost Management can be utilized to analyze Azure spending patterns over time to identify areas where an organization can optimize its usage and reduce costs.

Microsoft Cost Management components

Microsoft Cost Management comprises a suite of components to monitor, allocate, and optimize the cost of your cloud workloads:

- **Cost Management**: This is the main component of Microsoft Cost Management and provides insights into an organization's Azure spending and usage. It enables you to analyze your Azure bills, create alerts and budgets, and allocate costs to different departments or projects.
- **Billing**: This component provides the tools needed to manage your billing account and pay invoices.
- **Cost Explorer**: This tool visually represents an organization's Azure spending over time. It helps you understand your spending patterns, identify areas where you can optimize their usage, and make informed decisions about your Azure resources.
- **Azure Reservations**: This tool enables you to reserve Azure resources for a specified period, which can result in cost savings.
- **Azure Policy**: This tool can help you to enforce policies around your Azure resources, such as ensuring that virtual machines are shut down during off-peak hours to minimize costs.

Cost Management and Billing provide various capabilities to manage and monitor Azure costs. Some of the things you can do with them include the following:

- Generate cost reports and analyze them in the Azure portal, Microsoft 365 admin center, or by exporting data to an external system
- Stay ahead of costs with budget, anomaly, and scheduled alerts
- Assign shared costs with cost allocation rules
- Create and categorize subscriptions for personalized invoicing
- Set up payment options and process invoices
- Maintain and update billing information such as legal entity, tax information, and agreements

The following figure shows a high-level overview of Cost Management:

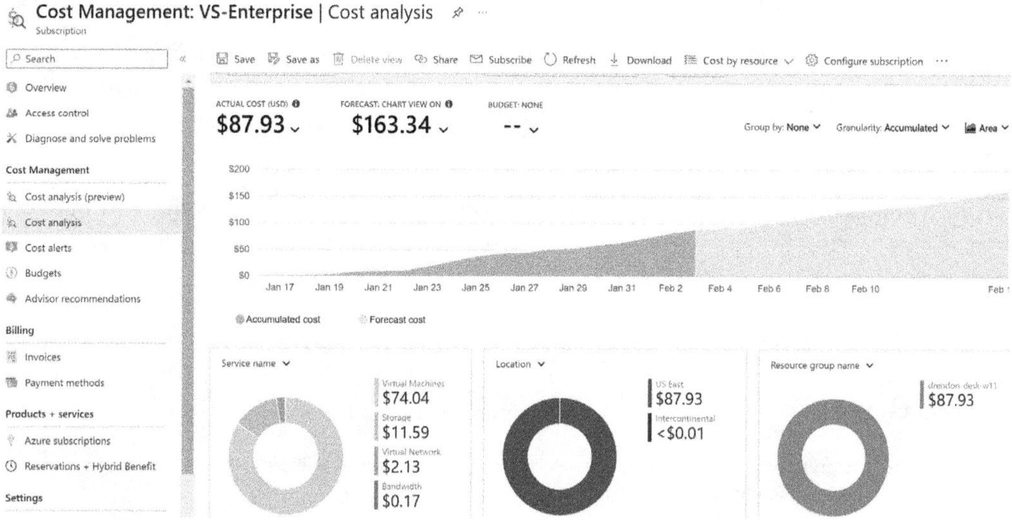

Cost Management: VS-Enterprise | Cost analysis
Subscription
Search
Save
Save as
Delete view
Share
Subscribe
Refresh
Download
Cost by resource
Configure subscription
Overview
Access control
Diagnose and solve problems
Cost Management
Cost analysis (preview)
Cost analysis
Cost alerts
Budgets
Advisor recommendations
Billing
Invoices
Payment methods
Products + services
Azure subscriptions
Reservations + Hybrid Benefit
Settings
ACTUAL COST (USD)
$87.93
FORECAST: CHART VIEW ON
$163.34
BUDGET: NONE
--
Group by: None
Granularity: Accumulated
Area
$200
$150
$100
$50
$0
Jan 17
Jan 19
Jan 21
Jan 23
Jan 25
Jan 27
Jan 29
Jan 31
Feb 2
Feb 4
Feb 6
Feb 8
Feb 10
Accumulated cost
Forecast cost
Service name
Virtual Machines
$74.04
Storage
$11.59
Virtual Network
$2.13
Bandwidth
$0.17
Location
US East
$87.93
Intercontinental
<$0.01
Resource group name
$87.93

Figure 10.26 – Cost Management

You can view forecast costs, which are based on historical resource consumption. This way, you can project the total of all your services every month. Then, you can break it down into details per service.

You can also review costs based on resources and use built-in views from Cost Management to view the cost by resource using filters, as shown in the following figure:

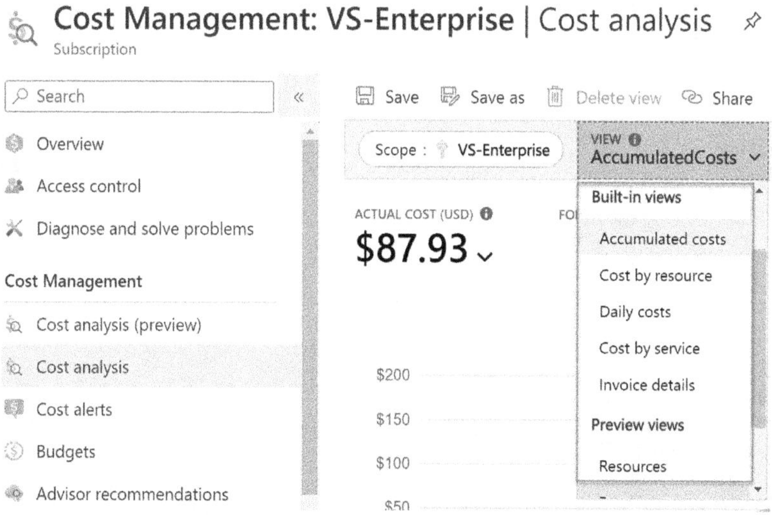

Cost Management: VS-Enterprise | Cost analysis
Subscription
Search
Save
Save as
Delete view
Share
Overview
Access control
Diagnose and solve problems
Cost Management
Cost analysis (preview)
Cost analysis
Cost alerts
Budgets
Advisor recommendations
Scope : VS-Enterprise
VIEW
AccumulatedCosts
ACTUAL COST (USD)
$87.93
Built-in views
Accumulated costs
Cost by resource
Daily costs
Cost by service
Invoice details
Preview views
Resources
$200
$150
$100

Figure 10.27 – Cost Management – views

Budgets in Cost Management are another useful tool to manage and monitor spending. Budgets allow you to plan your expenses, hold individuals accountable, and stay informed of your spending progress.

By setting alerts based on actual or predicted costs, you can ensure that your spending stays within your desired thresholds. The alerts serve as reminders and do not impact your resources or halt consumption when the budget threshold is exceeded. Budgets also provide a means to compare and monitor spending as you analyze costs.

To create a new budget, in the **Cost Management** section, go to **Budgets**, click the **New budget** button, and then select the appropriate options, such as the budget type, spending limit, and time period, as shown here:

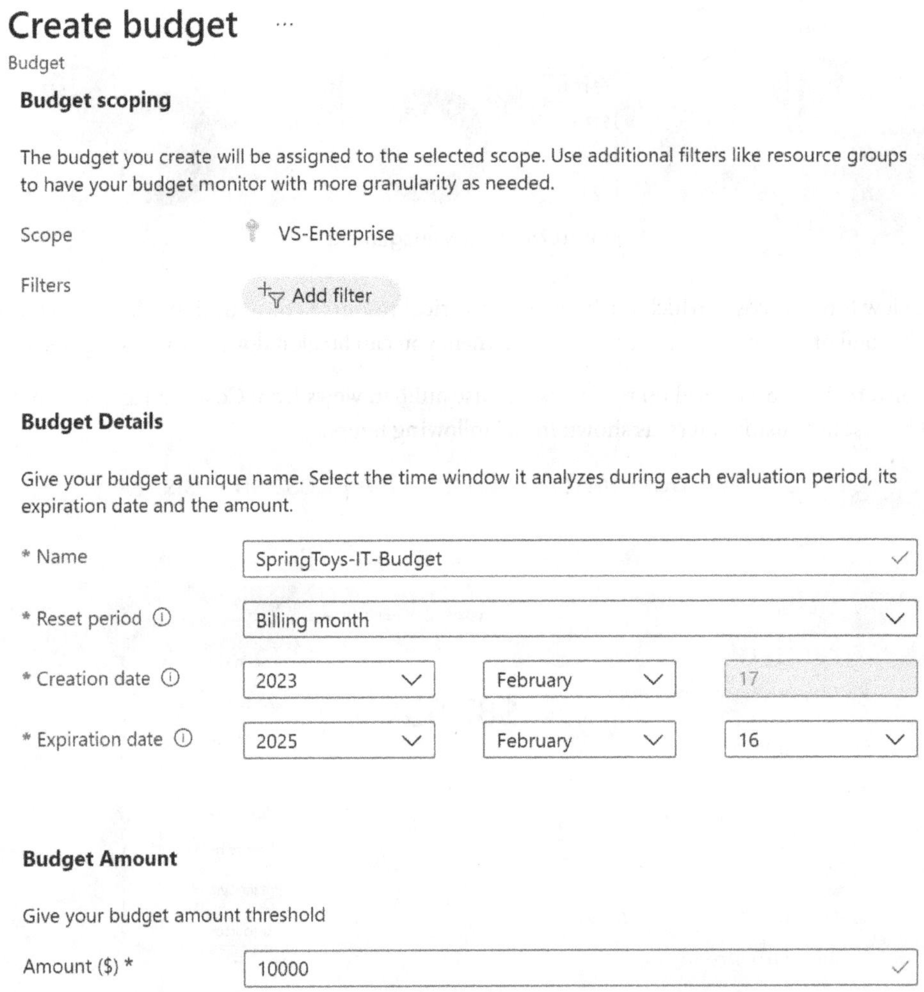
Create budget
Budget
Budget scoping
The budget you create will be assigned to the selected scope. Use additional filters like resource groups to have your budget monitor with more granularity as needed.
Scope
VS-Enterprise
Filters
Add filter
Budget Details
Give your budget a unique name. Select the time window it analyzes during each evaluation period, its expiration date and the amount.
* Name
SpringToys-IT-Budget
* Reset period
Billing month
* Creation date
2023
February
17
* Expiration date
2025
February
16
Budget Amount
Give your budget amount threshold
Amount ($) *
10000

Figure 10.28 – Cost Management – creating a budget

Note that you can specify the budget amount for the selected time period. The next step is configuring alerts to notify you when the budget threshold is exceeded.

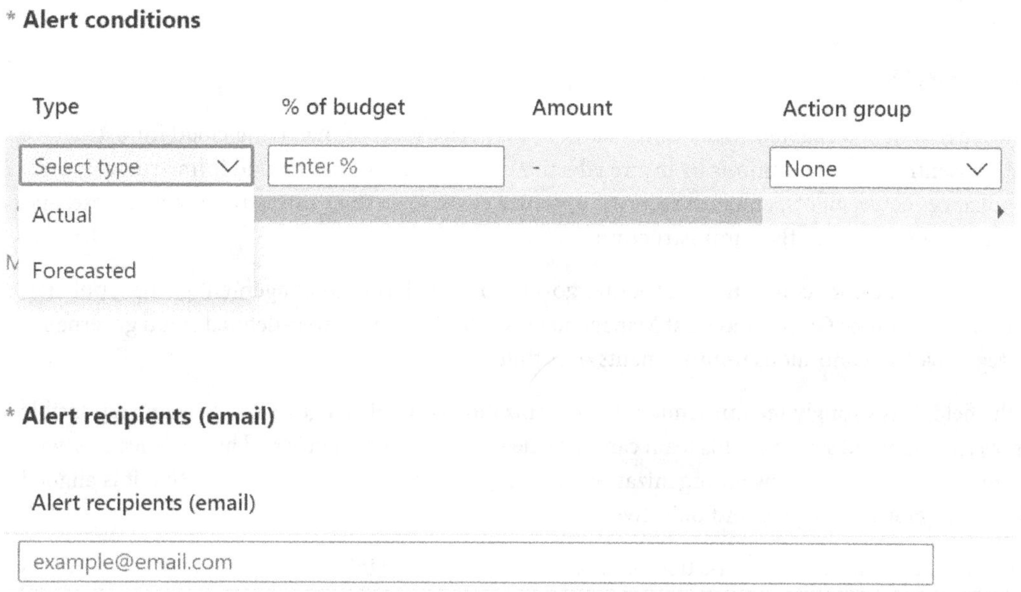
* Alert conditions
Type
% of budget
Amount
Action group
Select type
Enter %
-
None
Actual
Forecasted
* Alert recipients (email)
Alert recipients (email)
example@email.com

Figure 10.29 – Cost Management – Alert conditions

Once the budget has been created, you can view it on the **Budgets** page and monitor your spending against it. You can edit the budget anytime by clicking on its name and making the necessary changes.

When setting up or modifying a budget at the subscription or resource group level, you have the option to associate it with an action group. This action group enables you to define specific actions that will be triggered when your budget threshold is reached. By configuring Azure app push notifications within the action group, you can ensure that you receive mobile push notifications as soon as your budget threshold is met. This allows for timely and effective monitoring of your budget and financial management.

Azure Cost Management is a vital tool for organizations to monitor and control their Azure spending, optimize their resource usage, and ensure that their online store runs efficiently and effectively.

In addition to these controls, Azure landing zones, a set of pre-built, opinionated infrastructures and governance components, can enable organizations to quickly deploy and operationalize a secure and scalable multi-environment Azure architecture.

Azure landing zones help organizations quickly set up a secure and efficient Microsoft Azure environment, saving time and effort compared to setting it up from scratch.

You can find an Azure landing zone for governance at the following URL: `https://learn.microsoft.com/en-us/azure/cloud-adoption-framework/ready/landing-zone/design-area/governance`.

Summary

As discussed in this chapter, Azure governance is a crucial aspect of managing cloud infrastructure and is essential for organizations to ensure effective management of their cloud infrastructure, meet compliance requirements, enhance security, optimize costs, scale their infrastructure with ease, and maintain consistency in their infrastructure.

We reviewed the core components of Azure governance, including management groups, policies, blueprints, Resource Graph, and Cost Management. We learned that a well-defined cloud governance strategy involves continuous improvements over time.

In the field, it is strongly recommended that organizations establish a governance team responsible for overseeing cloud services. This team can be made up of IT professionals and business leaders, who can meet regularly to review an organization's cloud governance strategy to ensure that it is aligned with the organization's goals and objectives.

In the next chapter, we will discuss how organizations can leverage infrastructure as code using Azure's Bicep language to improve their infrastructure operations' reliability, consistency, and speed, while reducing the risk of manual errors.

11

Building Solutions in Azure Using the Bicep Language

As organizations increasingly move toward cloud-based infrastructure, the need for efficient and reliable deployment workflows has become critical. **Infrastructure as Code** (**IaC**) is a key approach to managing cloud resources, allowing teams to define their IaC and manage it through version control and automation tools. IaC is a paradigm shift that can enable organizations to improve their deployment workflows, reduce errors, and increase agility. In this chapter, we will explore the use of Bicep, a new **Domain-Specific Language** (**DSL**) for deploying Azure resources, as a tool for implementing IaC.

Bicep is a declarative language that enables developers and DevOps teams to define their IaC, making it easy to manage, automate, and deploy cloud resources. Bicep is an open source project that is optimized for use with Azure resources, providing a simple syntax for defining complex infrastructure topologies. In this chapter, we will provide an introduction to the key features of Bicep, including authoring Bicep files, working with parameters and modules, and using the what-if feature in Azure to preview deployment changes. By the end of this chapter, you will have a solid understanding of how Bicep can be used to improve your deployment workflows and enable the benefits of IaC.

In this chapter, we'll cover the following main topics:

- Authoring Bicep files
- Bicep file structure
- Working with parameters
- Bicep modules
- Previewing Azure deployment changes using what-if

Let's get started!

Unlocking the benefits of IaC with Azure Resource Manager

IaC is a state-of-the-art approach to managing and configuring IT infrastructure that uses a descriptive model. This method allows for the creation and management of infrastructure in a way that is automated, repeatable, and version-controlled. IaC has become increasingly popular in recent years due to its benefits, including improved efficiency, accuracy, and collaboration.

This technique has become essential for DevOps teams to automate their infrastructure deployment, management, and maintenance process. IaC provides consistency across multiple deployments and helps organizations increase efficiency, reduce operational costs, and improve time to market.

With cloud computing becoming more prevalent, cloud providers such as Microsoft Azure offer solutions such as **Azure Resource Manager** (**ARM**), which provides a consistent management layer to manage the life cycle of your environment through IaC.

IaC has become essential for developers, systems engineers, and DevOps professionals to ensure collaboration and take full advantage of IaC principles.

Think of IaC as a way of managing and provisioning IT infrastructure using code instead of manual configuration. This means that instead of manually setting up servers, networks, and other components, the infrastructure is defined in code and then automatically deployed and managed using software.

The code acts as a blueprint for the infrastructure, making it easier to manage, scale, and update. IaC helps ensure consistency, reduces errors, and makes it easier to manage and maintain the infrastructure over time. The result is a more efficient, reliable, and scalable IT environment.

And while transitioning to IaC may require a significant effort, it pays off in the end by enhancing the efficiency, consistency, and dependability of a company's IT operations. Additionally, it has the potential to cut costs and foster better collaboration.

The transition to IaC typically involves several phases, including the following:

1. **Assessing the current state**: The first step in the transition to IaC is to evaluate the current state of the infrastructure. This involves taking inventory of the existing resources, understanding how they are currently managed, and identifying areas that can be improved.
2. **Defining the desired target state**: The next step is to determine the target state of the infrastructure, including the desired architecture, configuration, and deployment processes. This can be done using IaC tools, such as Terraform or ARM templates, and should be based on best practices and industry standards.
3. **Developing the IaC code**: The IaC code is then developed using a scripting language, such as JSON or YAML, to define the desired state of the infrastructure. The IaC code should be modular, reusable, and version-controlled to make it easier to maintain and manage over time.

4. **Testing and validating the IaC code**: Before deploying, it's essential to thoroughly test the IaC code to ensure that it meets the desired requirements and is free of errors. This can be accomplished through the use of automated testing tools and the implementation of **continuous integration/continuous delivery** (**CI/CD**) pipelines. These practices help promote quality and consistency in infrastructure deployments, making it easier to identify and resolve any issues before they become a problem.
5. **Deploying and managing the infrastructure**: The final step is to deploy and manage the infrastructure using the IaC code. This can be done using tools such as Bicep or ARM templates, which automate the deployment and management of the infrastructure for Azure resources.

The adoption of IaC has been increasing in recent years as organizations look to take advantage of its benefits. By automating the deployment and management of IT infrastructure, IaC enables organizations to be more efficient, consistent, and reliable while also improving collaboration and reducing costs.

The impact of IaC has been significant as it enables organizations to improve the delivery of applications and services, reduce downtime, and increase the agility of their IT operations. IaC has become a key enabler for DevOps and cloud computing, providing organizations with a more streamlined and efficient way to manage their IT infrastructure.

And as organizations continue to increase the adoption of cloud computing, the adoption of IaC is likely to continue to grow exponentially as organizations look to take advantage of its many benefits, including the following:

- **Improved efficiency**: By automating the deployment of infrastructure resources, IaC reduces the time and effort required to set up, configure, and manage infrastructure. Automating routine tasks also frees up IT staff to focus on more strategic initiatives that can drive the business forward.
- **Better consistency**: IaC ensures that infrastructure resources are deployed consistently, reducing the risk of human error and configuration drift. This helps ensure that the infrastructure is always in a known and consistent state, making it easier to manage and troubleshoot. Consistent infrastructure also makes it easier to replicate and scale the online retail store as the business grows.
- **Improved reliability**: IaC enables version control, testing, and validation of infrastructure resources, helping ensure their reliability and stability. This is particularly important for critical systems, such as online retail stores, which must be highly available and performant. IaC can also help reduce the risk of downtime caused by misconfigured infrastructure, ensuring customers have a seamless experience when shopping online.
- **Improved collaboration**: IaC supports modular design and can be integrated into existing DevOps workflows, enabling teams to work more effectively and efficiently. This helps enhance the cooperation between groups, allowing them to focus on delivering new features and services to customers. Teams can also take advantage of the version control capabilities of IaC to manage changes to the infrastructure, ensuring that everyone is working with the same version of the code.

- **Better cost management**: IaC enables organizations to define and manage the infrastructure required to support their online retail store more cost-effectively. By automating the deployment of infrastructure resources, organizations can reduce the need for manual intervention, reducing the risk of overprovisioning or under-provisioning resources. This helps ensure that the online retail store is always cost-optimized, providing better value for money.

Imagine SpringToys need to automate the deployment of its online store infrastructure in Azure and ensure that it is always up-to-date and secure. Bicep makes it easier for the development team to maintain the infrastructure, allowing them to focus on building new features and improving the customer experience.

Azure Bicep Language is a **DSL** for deploying Azure resources. It uses a declarative syntax, making it easy for developers to specify what resources should be deployed and how they should be configured.

Think of Bicep as an abstraction on top of ARM templates for defining Azure resources using declarative IaC.

In this context, SpringToys can use Azure Bicep to automate the deployment of the resources part of the online store, such as the following:

- **Virtual machines**: These are used for its web and database servers. The VMs are configured with the necessary software and security settings to ensure the online store is always available and secure.
- **Storage accounts**: These are used for storing customer data, product information, and other critical business data. Bicep ensures that the storage accounts are deployed with the required performance, security, and redundancy settings.
- **Networking**: This configures the virtual network and subnets the online store uses to connect the VMs and storage accounts securely.
- **Load balancer**: This routes incoming traffic to the appropriate VM, ensuring high availability and scalability for the online store.

Instead of creating these resources manually in the Azure portal, you can leverage Azure Bicep, a high-level language designed specifically for building IaC solutions.

Now, let's look at how you can work with Azure Bicep to build your solutions in Azure.

Authoring Bicep files

Instead of using JSON schemas with ARM templates, Bicep can be utilized to create Azure resources. The JSON syntax for ARM templates can be lengthy and involve complex expressions, but Bicep simplifies this process and enhances the development experience.

Bicep is a transparent layer over the ARM template that retains all JSON templates' capabilities. When it comes to deployment, the Bicep **command-line interface** (**CLI**) transforms a Bicep file into an ARM JSON-based template.

The following figure highlights how Bicep transpiles into ARM templates for resource deployments in Azure:

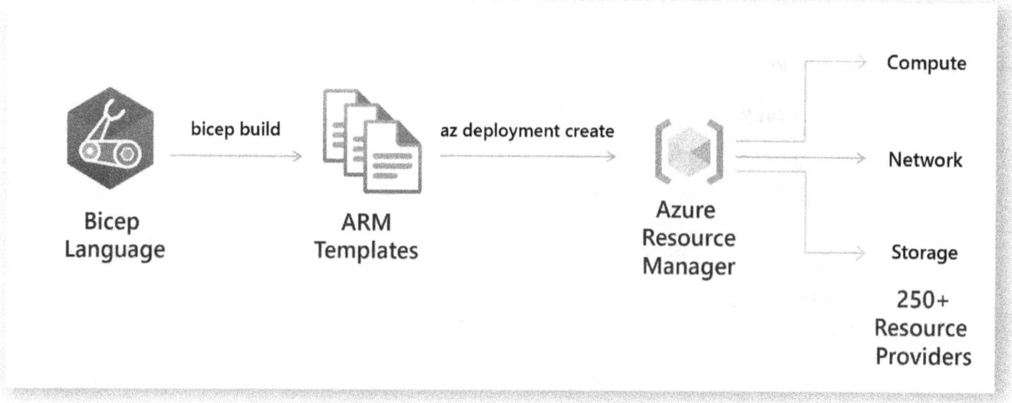
bicep build
az deployment create
Compute
Network
Storage
Bicep
Language
ARM
Templates
Azure
Resource
Manager
250+
Resource
Providers

Figure 11.1 – Bicep files overview

Before we start authoring our Bicep files, you must have the following tooling to help you build a solution in Azure more efficiently:

- An active Azure account. You can create an account for free at the following URL: `https://azure.microsoft.com/free/`.
- Azure Bicep installed on your local machine: `https://github.com/azure/bicep`.
- Azure PowerShell (`https://docs.microsoft.com/en-us/powershell/azure/install-az-ps`) or the Azure CLI (`https://docs.microsoft.com/en-us/cli/azure/install-azure-cli`).
- A resource group in your Azure subscription.
- Visual Studio Code with the Bicep extension: `https://marketplace.visualstudio.com/items?itemName=ms-azuretools.visualstudiobicep`.

Now, let's understand the definition of a Bicep file.

Bicep file structure

Bicep has its own syntax and structure for defining Azure resources in a simple and readable manner. The basic structure of an Azure Bicep file consists of the following:

- **Metadata**: A JSON object that defines the version of the Bicep language and any other metadata
- **Variables**: A section where you can declare variables to be used in the deployment
- **Parameters**: A section where you can define the input parameters required to deploy the resources
- **Resources**: A section where you can define the Azure resources you want to deploy
- **Outputs**: A section where you can define outputs that can be used to refer to the values of resources created during deployment

main.bicep > ...

```
//Parameters
param location string = 'eastus'
param storageAccountName string = 'springtoysstg'

//Variables
var storageSku = 'Premium_LRS'
var storageKind = 'StorageV2'

//Resources
resource storageAccount 'Microsoft.Storage/storageAccounts@2022-05-01' = {
  name: storageAccountName
  location: location
  kind: storageKind
  sku: {
    name: storageSku
  }
}

//Outputs
output stringOutput string = deployment().name
output endpoint string = storageAccount.properties.primaryEndpoints.blob
```

Figure 11.2 – A Bicep file that creates a storage account

Similar to ARM templates, you can define parameters, variables, resources, child resources, extensions, and even dependencies in your Bicep file. More advanced functionality, such as loops, conditions, and functions, can also be defined in your Bicep file.

Working with parameters

Defining parameters in Bicep works similarly to how it's done in ARM templates. However, it's important to keep in mind that there are some differences in the syntax between the two, which should be considered when using Bicep.

In Bicep, parameters serve as input values that can be defined within the template. These values can be supplied during deployment or referenced from a separate *parameters* file. To define parameters in Bicep, you must specify a name and type for each parameter within the template.

The following code demonstrates how to declare parameters in Bicep:

```
param location string
param size string
param adminUsername string
param adminPassword secureString
```

It's important to note that a parameter cannot share the same name as a variable, module, or resource in Bicep. Additionally, while parameters are useful for passing values that need to be different across deployments, it's best to avoid overusing them. Parameters are meant to be used sparingly and only when necessary.

As an example, when creating a storage account in Bicep, you are required to specify the SKU and storage account type. In this case, you can leverage parameters to dynamically pass in the desired values during deployment. This allows for greater flexibility and customization in your deployments.

When defining a virtual network and its associated subnets, you can use a parameter object to pass in the values for the virtual network and subnet definitions. This way, you can reuse the same code for multiple virtual network configurations.

It's always a good practice to provide descriptive information for your parameters in Bicep. Providing descriptions makes it easier for your team and colleagues to understand the purpose and usage of each parameter, which can help promote collaboration and reduce the risk of errors.

ARM resolves parameter values before starting the deployment operations. Now, let's take a quick look at the supported data types.

Parameter data types

Azure Bicep supports the following parameter data types:

- `array`
- `bool`
- `int`
- `object`

- `secureObject` and `secureString` — indicated by the `@secure()` decorator
- `string`

If you're wondering what decorators are, decorators in Bicep can be used to specify constraints and metadata for a parameter.

Here's a list of decorators available in Bicep:

- `allowed`
- `secure`
- `minLength` and `maxLength`
- `minValue` and `maxValue`
- `description`
- `metadata`

When should you use decorators?

Decorators can enforce constraints or provide metadata. By using decorators, you can indicate that a parameter should be treated as secure, define a set of allowed values, set specific minimum and maximum string lengths, establish minimum and maximum integer values, and provide a description for the parameter.

Decorators are a way to enhance or modify the behavior of resources and modules in your infrastructure code. They provide a mechanism to add additional features or configurations to the resources without directly modifying their underlying definition.

Decorators in Azure Bicep are defined using the @ symbol, followed by a specific decorator name, and they are placed directly above the resource or module they are intended to modify. These decorators accept parameters that allow you to specify the desired behavior or configuration.

For example, if you are deploying a virtual machine and want to specify the password for the username, you can use a parameter with the `@secure()` decorator. This way, the password will not be logged or stored in the deployment history, ensuring secure and compliant management of sensitive information.

The following code shows a few examples of parameters and the `@secure()` decorator:

```
// Username for the Virtual Machine.
param adminUsername string
// Type of authentication to use on the Virtual Machine. SSH key is
recommended.
param authenticationType string = 'sshPublicKey'
// SSH Key or password for the Virtual Machine. SSH key is
recommended.
@secure()
param adminPasswordOrKey string
```

In the preceding code, string-type parameters were defined, along with a third parameter for the password or key, which was designated with the `@secure()` decorator. This allows for the secure handling of sensitive information such as passwords or keys during deployment.

Now, let's look at an example of deploying a SQL database to understand this better.

Example – deploying SQL databases

The following Bicep template creates a SQL database and uses the `@secure()` decorator:

```
param serverName string = uniqueString('sql', resourceGroup().id)
param sqlDBName string = 'AzInsiderDb'
param location string = resourceGroup().location
param administratorLogin string

@secure()
param administratorLoginPassword string

resource server 'Microsoft.Sql/servers@2019-06-01-preview' = {
  name: serverName
  location: location
  properties: {
    administratorLogin: administratorLogin
    administratorLoginPassword: administratorLoginPassword
  }
}

resource sqlDB 'Microsoft.Sql/servers/databases@2020-08-01-preview' =
{
  name: '${server.name}/${sqlDBName}'
  location: location
  sku: {
    name: 'Standard'
    tier: 'Standard'
  }
}
```

In the preceding code, we defined a few parameters for the server name, the SQL database name, the location, and the admin login. Then, we used the `@secure()` decorator for the admin password.

Next, we will deploy the Bicep template to a resource group that we created previously with the following command:

```
New-AzResourceGroupDeployment -Name $deploymentName -ResourceGroupName
sql-database -TemplateFile .\main.bicep
```

During deployment time, we will provide the administrator login and the password, as shown here:

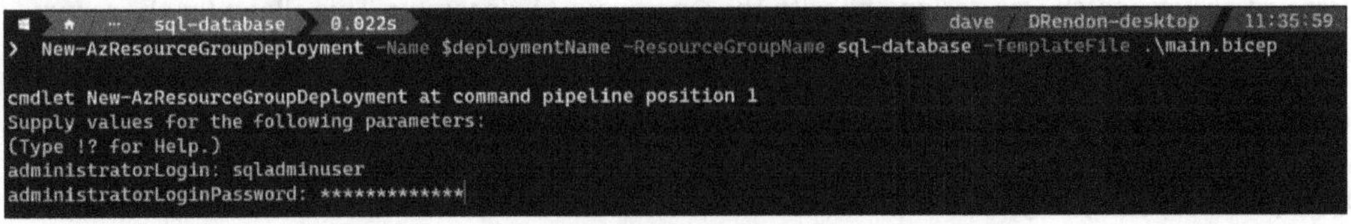
sql-database 0.022s
dave DRendon-desktop 11:35:59
> New-AzResourceGroupDeployment -Name $deploymentName -ResourceGroupName sql-database -TemplateFile .\main.bicep
cmdlet New-AzResourceGroupDeployment at command pipeline position 1
Supply values for the following parameters:
(Type !? for Help.)
administratorLogin: sqladminuser
administratorLoginPassword: *************

Figure 11.3 – Azure Bicep – deploying a SQL database

Similar to ARM templates, the actual value of the secret is never exposed. Using this type of parameter for your passwords and secrets is recommended. Sensitive data passed on as a secure parameter can't be read after resource deployment and isn't logged.

Now, let's look at the second option to pass on parameters: using a parameter file.

Deploying the Bicep file using a parameter file

As an alternative to specifying parameter values directly in the Bicep file, you can create a separate parameter file that contains the actual values. This parameter file is in JSON format and provides a convenient way to store the parameter values separately from the Bicep code.

Similar to ARM templates, Bicep templates can be organized with a main template file, typically named `main.bicep`, and a corresponding parameter file named `main.bicepparam`. This convention makes it clear and easy to understand the structure of the Bicep deployment.

The following code shows an example of a parameter file:

```
using './main.bicep'

param myString = 'test string'
param exampleInt = 2 + 2
param exampleBool = true
param exampleArray = [
  'value 1'
  'value 2'
]
param exampleObject = {
  property1: 'value 1'
  property2: 'value 2'
}
//
```

It is important to keep in mind that the data types specified in your parameter file must align with those in your Bicep file. This way, you can successfully deploy your Bicep template using either Azure PowerShell or the Azure CLI.

As you can see, the Bicep file structure allows you to define resources modularly and reuse variables, parameters, and outputs across multiple deployments.

As your cloud infrastructure grows in size and complexity, managing deployments can become increasingly challenging. To simplify this process and improve efficiency, Azure Bicep offers the concept of modules.

Bicep modules

Azure Bicep modules are a way to organize and reuse Bicep code across multiple deployments. Think of a module as a self-contained deployment unit that can define its own resources, variables, parameters, and outputs.

Modules in Bicep allow you to encapsulate logic and resources in a reusable unit and share it across multiple deployments. This enables you to manage your deployments in a more organized manner and reduces the amount of duplicated code.

To create a module, you need to define a Bicep file with the required resources, variables, parameters, and outputs. Then, you can use this module in other Bicep files by importing it using the `module` keyword. The imported module can then create instances of the defined resources in the current deployment.

Modules also allow you to define and pass parameters to the module, allowing it to be customized for each deployment. Additionally, modules can return outputs, which can be used in other parts of the deployment.

The following code shows the basic definition of a Bicep module:

```
module <symbolic-name> '<path-to-file>' = {
  name: '<linked-deployment-name>'
  params: {
    <parameter-names-and-values>
  }
}
```

We can use the symbolic name of the module to reference it in another part of our Bicep file.

Imagine you have two Bicep files: one that creates an app service called `appService.bicep` and another that creates an app service plan called `appServicePlan.bicep`.

Then, you can create a new `main.bicep` file and consume the two Bicep files as modules.

The following figure shows the definition of the app service:

appService.bicep > ...

```
param appServicePrefix string = 'appService'
param location string = 'eastus'
param appServicePlanId string

resource appService 'Microsoft.Web/sites@2021-01-15' = {
  name: '${appServicePrefix}site'
  location: location
  properties:{
    siteConfig:{
      linuxFxVersion: 'DOTNETCORE|3.0'
    }
    serverFarmId: appServicePlanId
  }
}
// Set an output which can be accessed by the module consumer
output siteURL string = appService.properties.hostNames[0]
```

Figure 11.4 – Azure Bicep – app service definition

The following figure shows the definition of the app service plan:

appServicePlan.bicep > ...

```
param appPlanPrefix string
param sku string = 'F1'
param location string = 'eastus'

resource appServicePlan 'Microsoft.Web/serverfarms@2021-01-15' = {
  //interpolate param
  name: '${appPlanPrefix}AppPlan'
  //pass on location param
  location: location
  kind: 'linux'
  sku: {
    //pass on sku param
    name: sku
  }
  properties:{
    reserved: true
  }

}
// Set an output which can be accessed by the module consumer
output appServicePlanId string = appServicePlan.id
```

Figure 11.5 – Azure Bicep – app service plan definition

When deploying modules in Bicep, they are executed simultaneously and you have the option to specify the location of the module. This location can be a local file or an external file, providing you with flexibility in how you want to organize and store your modules.

The following code shows the definition of the `main.bicep` file:

main.bicep > ...

```
//parameters
param location string = 'eastus'
param springToysPrefix string = 'springtoys'

//define target
targetScope = 'subscription'

//define new resoruce group
resource resourceGroup 'Microsoft.Resources/resourceGroups@2021-04-01' = {
  name: '${springToysPrefix}Rg'
  location: location
}

//consume appServicePlan as module
module appServicePlan 'appServicePlan.bicep' = {
  name:'appServicePlan'
  scope: resourceGroup
  params: {
    location: location
    appPlanPrefix: springToysPrefix
  }
}

//consume appService as module
module appService 'appService.bicep' = {
  name: 'appService'
  scope: resourceGroup
  params: {
    location: location
    appServicePlanId: appServicePlan.outputs.appServicePlanId
    appServicePrefix: springToysPrefix
  }
}
```

Figure 11.6 – Azure Bicep – main file

Now, we can define a target scope. In this case, we will target a subscription scope.

You can target your deployment at different scopes – the resource group, subscription, management group, and tenant levels – as shown here:

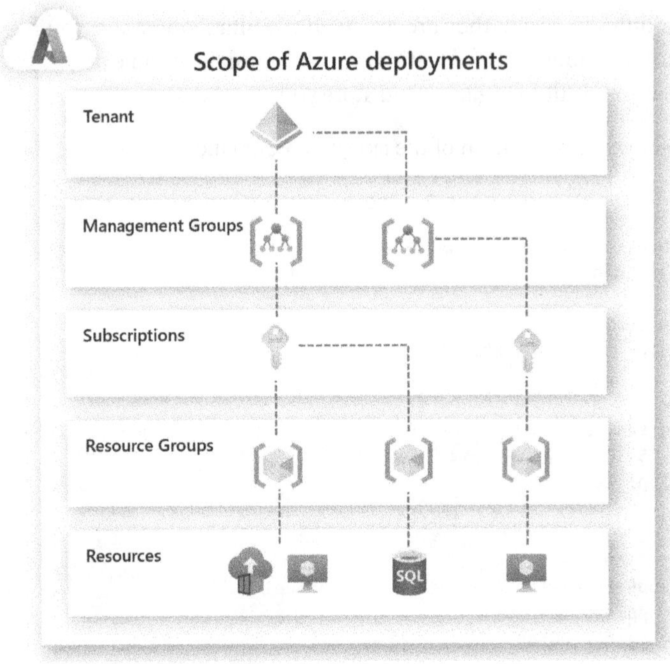
Scope of Azure deployments
Tenant
Management Groups
Subscriptions
Resource Groups
Resources
SQL

Figure 11.7 – Azure Bicep – deployment scopes

In addition, you can leverage the public Bicep registry modules available in the following GitHub repository: `https://github.com/azure/bicep-registry-modules`.

Before deploying this environment, we can leverage the "What-If" feature. The "What-If" feature allows you to preview the impact of changes to your infrastructure deployment without actually making those changes in your environment.

With "What-If," you can see how your infrastructure deployment would look after making changes to your Bicep code and verify whether the changes would result in any deployment errors.

This feature helps reduce the risk of unintended consequences while you're making changes to your deployment and enables you to make informed decisions before deploying the changes.

Previewing Azure deployment changes using what-if

In our working directory, we have three files – `appService.bicep`, `appServicePlan.bicep`, and `main.bicep` – as shown in the following figure:

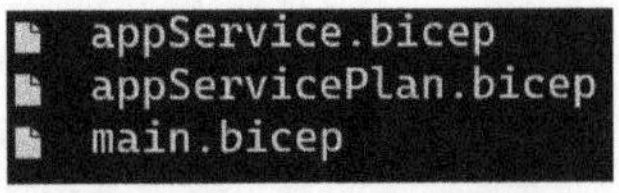
appService.bicep
appServicePlan.bicep
main.bicep

Figure 11.8 – Bicep files

We will target the deployment to the subscription level. We will use the following command to deploy the environment:

```
$date = Get-Date -Format "MM-dd-yyyy"
$rand = Get-Random -Maximum 1000
$deploymentName = "SpringToysDeployment-"+"$date"+"-"+"$rand"

New-AzDeployment -Name $deploymentName -TemplateFile .\main.bicep
-Location eastus -c
```

Note that we only refer to the `main.bicep` file; there's no need to reference all modules. We use the `-c` flag at the end to preview the changes and get prompted to continue with the deployment.

The following figure shows a preview of the deployment:

Note: The result may contain false positive predictions (noise).
You can help us improve the accuracy of the result by opening an issue here: https://aka.ms/WhatIfIssues.
Resource and property changes are indicated with this symbol:
+ Create
The deployment will update the following scopes:
Scope: /subscriptions/
+ resourceGroups/springtoysRg [2021-04-01]
apiVersion: "2021-04-01"
id: "/subscriptions/ /resourceGroups/springtoysRg"
location: "eastus"
name: "springtoysRg"
type: "Microsoft.Resources/resourceGroups"
Scope: /subscriptions/ /resourceGroups/springtoysRg
+ Microsoft.Web/serverfarms/springtoysAppPlan [2021-01-15]
apiVersion: "2021-01-15"
id:
"/subscriptions/ /resourceGroups/springtoysRg/providers/Microsoft.Web/serverfarms/springtoysAppPlan"
kind: "linux"
location: "eastus"
name: "springtoysAppPlan"
properties.reserved: true
sku.name: "F1"
type: "Microsoft.Web/serverfarms"
+ Microsoft.Web/sites/springtoyssite [2021-01-15]
apiVersion: "2021-01-15"
id:
"/subscriptions/ resourceGroups/springtoysRg/providers/Microsoft.Web/sites/springtoyssite"
location: "eastus"
name: "springtoyssite"
properties.serverFarmId:
"/subscriptions/ resourceGroups/springtoysRg/providers/Microsoft.Web/serverfarms/springtoysAppPlan"
properties.siteConfig: "*******"
type: "Microsoft.Web/sites"
Resource changes: 3 to create.
Are you sure you want to execute the deployment?
[Y] Yes [A] Yes to All [N] No [L] No to All [S] Suspend [?] Help (default is "Y"):

Figure 11.9 – Bicep deployment preview

Once you've validated the changes, you can proceed to execute the deployment.

Next, you should see the following output:

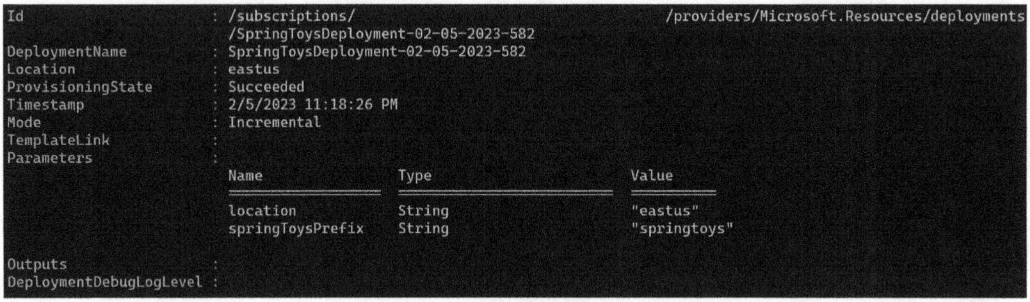
Id : /subscriptions/ /providers/Microsoft.Resources/deployments
/SpringToysDeployment-02-05-2023-582
DeploymentName : SpringToysDeployment-02-05-2023-582
Location : eastus
ProvisioningState : Succeeded
Timestamp : 2/5/2023 11:18:26 PM
Mode : Incremental
TemplateLink :
Parameters :
Name Type Value
location String "eastus"
springToysPrefix String "springtoys"
Outputs :
DeploymentDebugLogLevel :

Figure 11.10 – Bicep deployment output

You can go to the Azure portal and review the deployment history. Note that the deployment reflects a status of *Succeeded*, and we have two deployments – one for the app service and another for the app service plan:

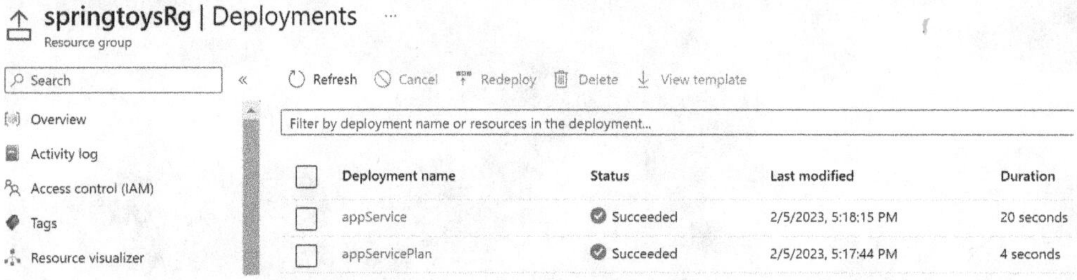
springtoysRg | Deployments
Resource group
Search
Overview
Activity log
Access control (IAM)
Tags
Resource visualizer
Refresh
Cancel
Redeploy
Delete
View template
Filter by deployment name or resources in the deployment...
Deployment name
Status
Last modified
Duration
appService
Succeeded
2/5/2023, 5:18:15 PM
20 seconds
appServicePlan
Succeeded
2/5/2023, 5:17:44 PM
4 seconds

Figure 11.11 – Deployments – the Azure portal

You can download the complete example for this code from the following GitHub repository: `https://github.com/PacktPublishing/Azure-Architecture-Explained/tree/main/Chapter11/bicep-modules`.

Additionally, you will find more advanced scenarios on how you can leverage IaC using Bicep in the following GitHub repository: `https://github.com/daveRendon/azinsider/tree/main/application-workloads`.

The preceding repository contains more than 60 examples of real scenarios that use Bicep.

Summary

Organizations using cloud services in Azure should consider Azure Bicep as it provides several benefits that can enhance their cloud infrastructure deployment and management processes.

Bicep simplifies the provisioning of Azure resources by allowing for the creation of templates that define the desired IaC. This results in consistent and repeatable deployments, reduces human error, and enables easier version control.

Bicep also integrates seamlessly with Azure DevOps and other Azure services, as will see in the following chapter, providing a unified experience for the entire application life cycle.

Additionally, Bicep provides a simplified syntax, making it easier for developers to write and understand templates, reducing the learning curve for new users. All these advantages make Azure Bicep an attractive option for organizations looking to streamline their cloud infrastructure deployment and management processes.

In the next chapter, we will review how you can work with Azure Pipelines, a tool that helps you automate the process of building and deploying your infrastructure in Azure.

12

Using Azure Pipelines to Build Your Infrastructure in Azure

As software development practices continue to evolve, so do the tools and technologies used to streamline the process. One such tool is Azure DevOps, a powerful suite of services for managing and delivering applications. In this chapter, we'll delve into some of the key features of Azure DevOps, focusing specifically on Azure Pipelines. We'll cover the configuration of Azure DevOps and Azure Repos, as well as the creation of both build and release pipelines. By the end of this chapter, you'll have a solid understanding of how to leverage Azure DevOps to enhance your software development workflow.

In this chapter, we'll cover the following main topics:

- Understanding the relationship between continuous integration, continuous delivery, and pipelines
- Understanding Azure Pipelines
- Configuring Azure DevOps
- Configuring Azure Repos
- Configuring a build pipeline in Azure DevOps
- Configuring a release pipeline in Azure DevOps
- Configuring Azure Pipelines with YAML

Let's get started!

Understanding the relationship between continuous integration, continuous delivery, and pipelines

As organizations adopt a cloud strategy, they realize the benefits of automating the build, test steps, and deployment steps. By automating the entire software delivery process, organizations can deliver software faster and with more confidence while reducing the risk of errors.

Continuous Integration (**CI**) and **Continuous Delivery** (**CD**) are practices that have evolved from Agile software development and DevOps culture that involve automatically building and testing code changes as soon as they are pushed to a source control repository, such as Git.

This helps catch and fix issues early in the development process, reducing the risk of more significant problems later on. They emerged in response to the need for faster and more reliable software delivery in a rapidly changing software development landscape.

As Agile and DevOps practices gained popularity, CI and CD became widely adopted and are now considered best practices in modern software development. They are supported by various tools and technologies, including CI/CD platforms such as Azure Pipelines, Jenkins, and Travis CI, making it easier for organizations to implement and benefit from these practices.

Let's stop for a moment to explain how CI and pipelines relate to one another. CI and pipelines are closely related concepts in the context of software development. Think of a pipeline as a series of automated steps to build, test, and deploy software.

In the context of CI, pipelines are used to automate the steps involved in CI. The pipeline takes the code changes, builds the code, runs automated tests, and deploys the code if all tests pass.

Put simply, pipelines are tools used to implement CI and make the process faster, more reliable, and less manual. The pipeline takes care of the build, test, and deploy steps involved in CI, helping organizations deliver software faster and more confidently.

By adopting CI/CD, organizations can deliver software faster, with improved quality and reliability, while also reducing the risk of errors and increasing efficiency. And there are several reasons why organizations adopt the CI/CD practice:

- **Faster software delivery**: CI/CD helps organizations deliver software faster by automating the build, test, and deploy steps. This reduces manual effort and the risk of errors, making it possible to deliver new features and bug fixes to customers faster.
- **Improved quality**: CI/CD helps catch and fix issues early in the development process, reducing the risk of more significant problems later on. Automated tests help ensure that software changes are thoroughly tested before they are deployed, reducing the risk of bugs and improving the quality of the software.
- **Increased collaboration**: CI/CD helps promote collaboration between development, testing, and operations teams, ensuring that everyone works together toward a common goal.
- **Better visibility**: CI/CD provides visibility into the software delivery process, making it easier to understand what is happening and identify areas for improvement.
- **Increased efficiency**: CI/CD helps improve the efficiency of the software delivery process by automating repetitive and time-consuming tasks. This reduces manual effort and the risk of errors, freeing up time for other essential tasks.

- **Improved security**: CI/CD helps enhance the security of the software delivery process by making it easier to detect and fix vulnerabilities early in the development process.
- **Better customer satisfaction**: By delivering new features and bug fixes faster, CI/CD helps improve customer satisfaction by providing a better user experience.

Implementing a CI/CD practice involves several steps, and the right strategy will depend on an organization's specific needs and constraints. However, here are some general best practices that can be followed to implement a successful CI/CD strategy:

- **Define the CI/CD pipeline**: Identify the stages involved in building, testing, and deploying software and define the steps in the pipeline. This includes the tools and technologies that will be used in each stage.
- **Automate testing**: Automate as many tests as possible to ensure software changes are thoroughly tested before deployment.
- **Use a source control repository**: Use a source control repository, such as Git, to store the source code and track changes. This makes it easier to manage the code and collaborate with other developers.
- **Integrate with other tools**: Integrate the CI/CD pipeline with tools and systems used in the development process, such as issue trackers and deployment environments.
- **Continuously monitor and optimize**: Monitor the CI/CD pipeline and make necessary improvements. This includes identifying bottlenecks and improving the speed and efficiency of the pipeline.

Imagine SpringToys intends to implement a CI/CD practice to develop and maintain its systems. SpringToys could adopt the following process:

- **CI**: Whenever a developer pushes code changes to the Git repository, the CI/CD pipeline is triggered. The pipeline automatically builds the application and runs automated tests to verify that the code changes won't break any existing functionality.
- **CD**: If the tests pass, the pipeline automatically deploys the code changes to a staging environment for further testing. If the tests in the staging environment are successful, the pipeline can deploy the code changes to production, making the new features and bug fixes available to its customers.

Azure Pipelines can help SpringToys deliver software faster and more confidently by automating the delivery process and improving team collaboration. Additionally, Azure Pipelines provides a centralized platform for monitoring and managing the delivery pipeline, making it easier for teams to identify and resolve issues.

Azure Pipelines, an essential component of Azure DevOps, offers a solution to this challenge by automating the entire software delivery process. By automating the build, test, and deployment steps,

Azure Pipelines can help organizations save time and reduce the risk of errors. Let's review how Azure Pipelines can help organizations automate their software delivery process.

Understanding Azure Pipelines

Azure Pipelines is one of the components of Azure DevOps, a CI/CD platform offered by Microsoft. It helps organizations automate the process of building, testing, and deploying software.

Here's how it works in simple terms:

1. Code is pushed to a source control repository, such as Git.
2. Azure Pipelines is triggered by this code change and performs a series of steps as defined in the pipeline.
3. **Build**: The code is compiled and turned into artifacts, such as executables or Docker images.
4. **Test**: Automated tests are run on the artifacts to verify that the application works as expected.
5. **Deploy**: If the tests pass, the artifacts are deployed to the desired environment, such as a production server or a staging environment, for further testing

Azure Pipelines helps automate the entire software delivery process, from building and testing code changes to deploying to production. This helps reduce manual errors, speed up delivery times, and ensure that the software is always in a deployable state.

In Azure DevOps, there are two main approaches to creating pipelines:

- **Classic Editor**: The Classic Editor is a graphical interface that allows users to create pipelines using a visual designer. It provides a drag-and-drop interface for defining various stages, tasks, and dependencies in the pipeline. Users can select from a wide range of pre-built tasks and configure them as needed. The Classic Editor is well suited to users who prefer a visual approach and do not have extensive knowledge of scripting or coding.
- **YAML Pipelines**: Azure DevOps supports defining pipelines using YAML files, which describe the entire pipeline in a declarative manner. YAML pipelines offer greater flexibility and version control capabilities compared to the Classic Editor. With YAML, you define the pipeline's stages, jobs, steps, and dependencies using plain text files that can be stored in source control. This approach is popular among developers who prefer to have their pipeline configurations as code and want to take advantage of features such as code reviews, branching, and versioning.

Both approaches have their advantages. The Classic Editor is more beginner-friendly and provides an intuitive visual representation of the pipeline. It is suitable for simple pipelines and users who prefer a graphical interface. On the other hand, YAML pipelines offer more control and transparency, as the pipeline configuration is stored as code. It allows for versioning, code reviews, and easier collaboration between team members. YAML pipelines are generally preferred for complex pipelines or scenarios that require more customization and advanced scripting capabilities.

Ultimately, the choice between the Classic Editor and YAML pipelines depends on the specific requirements of your project, the complexity of the pipeline, and the preferences and skill sets of your team members.

Now, let's set up our organization in Azure DevOps.

Configuring Azure DevOps

Let's explore how to set up Azure DevOps for your organization, including configuring Azure Repos and creating build and release pipelines. By the end of this section, you'll have a solid understanding of how to use Azure Pipelines to streamline your software delivery process and improve your overall software development workflow. Follow these steps:

1. The first step is to go to `https://dev.azure.com` and create a new project. Note you will need an organization as well. And we will provide the name of the project called, **SpringToysApp**, and set the visibility to **Public**, as shown here:

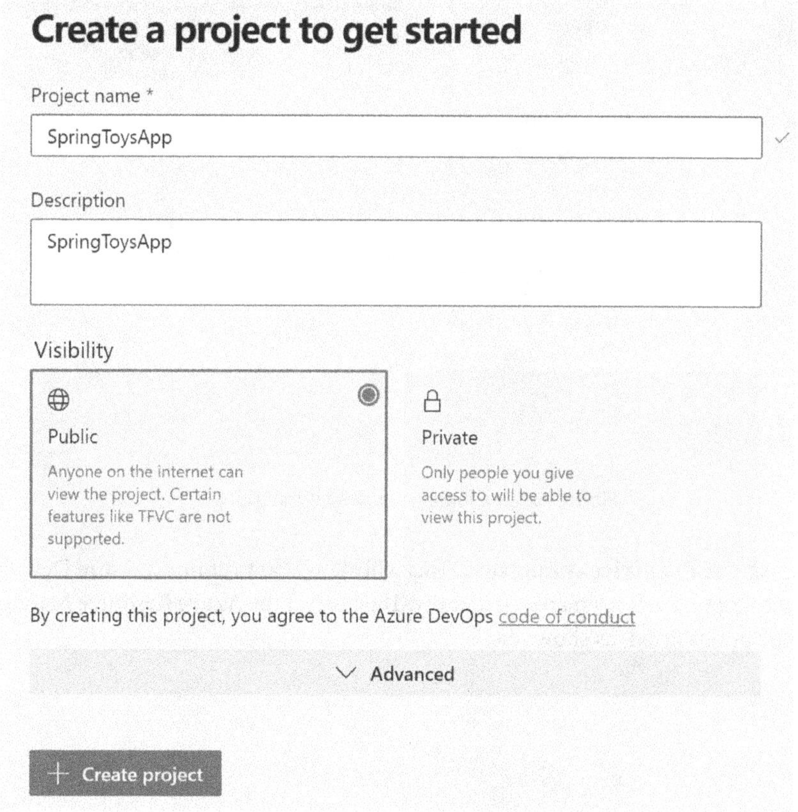
Create a project to get started
Project name *
SpringToysApp
Description
SpringToysApp
Visibility
Public
Anyone on the internet can view the project. Certain features like TFVC are not supported.
Private
Only people you give access to will be able to view this project.
By creating this project, you agree to the Azure DevOps code of conduct
Advanced
Create project

Figure 12.1 – Create a project in Azure DevOps

Now, let's configure the access from Azure DevOps to your Azure subscription. This way, we can deploy resources to the Azure environment. We will use a managed identity and create a *service connection* in Azure DevOps to achieve this.

2. On the **Azure DevOps** initial screen, go to the bottom-left corner and select the **Project settings** option. Once on the settings page of the project, in the left menu, go to the **Pipelines** section and select the **Service connections** option, as shown here:

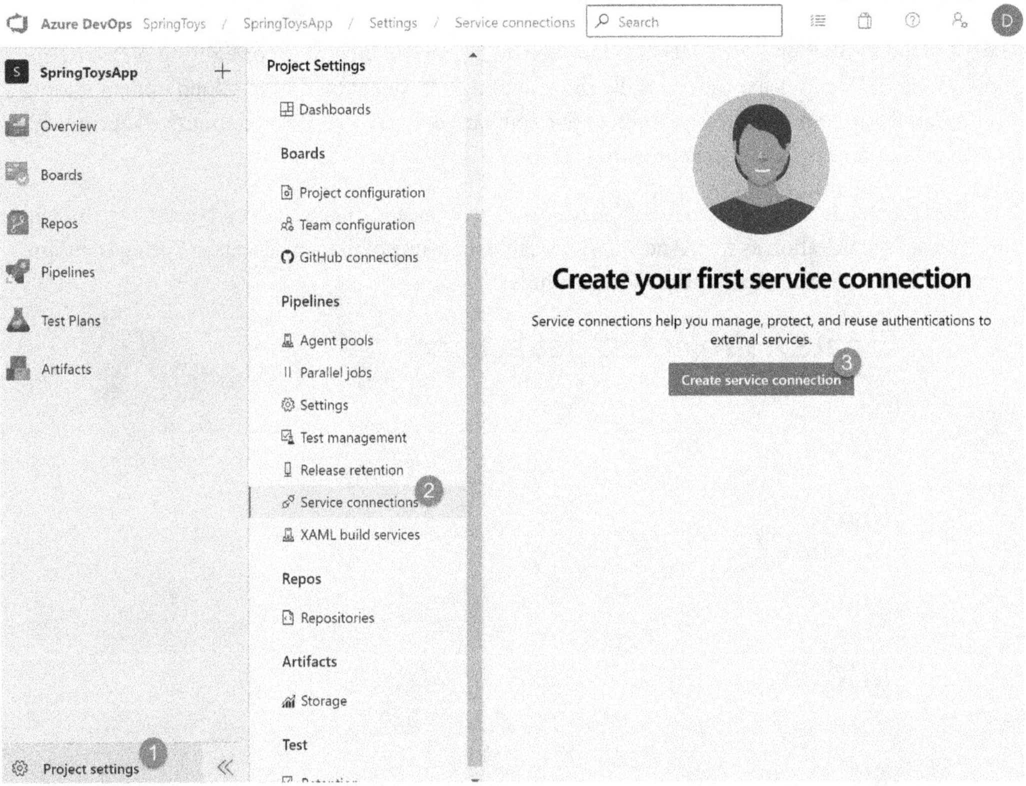
Azure DevOps SpringToys / SpringToysApp / Settings / Service connections
Search
SpringToysApp
Overview
Boards
Repos
Pipelines
Test Plans
Artifacts
Project settings
1
Project Settings
Dashboards
Boards
Project configuration
Team configuration
GitHub connections
Pipelines
Agent pools
Parallel jobs
Settings
Test management
Release retention
Service connections
2
XAML build services
Repos
Repositories
Artifacts
Storage
Test
Create your first service connection
Service connections help you manage, protect, and reuse authentications to external services.
Create service connection
3

Figure 12.2 – Create a service connection

3. Now, select **Create service connection**. This will allow you to connect Azure DevOps to your Azure subscription using a managed identity. Then, select the **Azure Resource Manager** option from the options listed, as shown here:

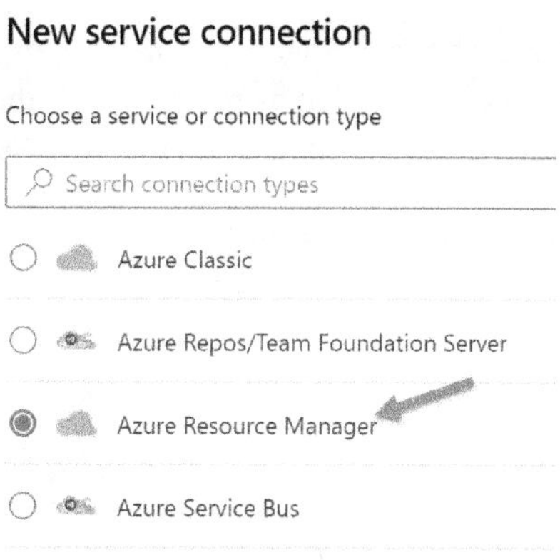
New service connection
Choose a service or connection type
Search connection types
Azure Classic
Azure Repos/Team Foundation Server
Azure Resource Manager
Azure Service Bus

Figure 12.3 – New service connection

Note that Azure Resource manager is just one type of service connection to connect to Azure subscriptions. There are several other types of service connections. You can find them at the following URL: `https://bit.ly/az-pipelines-service-connections`.

4. Then, configure the authentication for the service connection. In this case, select the **Service principal (automatic)** option, as shown here:

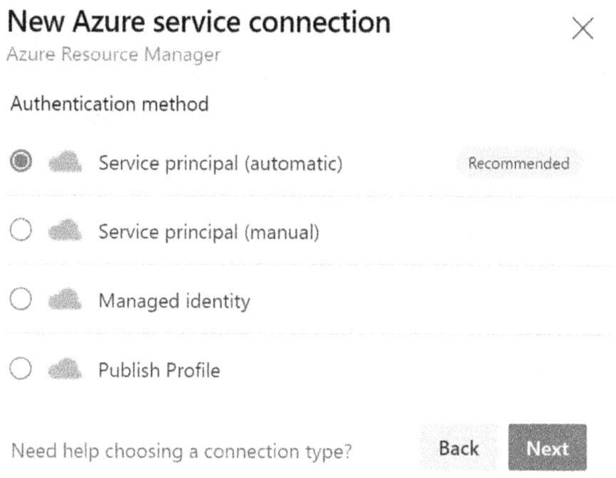
New Azure service connection
Azure Resource Manager
Authentication method
Service principal (automatic)
Recommended
Service principal (manual)
Managed identity
Publish Profile
Need help choosing a connection type?
Back
Next

Figure 12.4 – New service connection – service principal

5. Now provide the parameters for the connection. Ensure you provide the subscription ID, subscription name, tenant ID, and connection name, and enable the checkbox to grant permission to all pipelines. We will name this connection `SpringToysServiceConnection`:

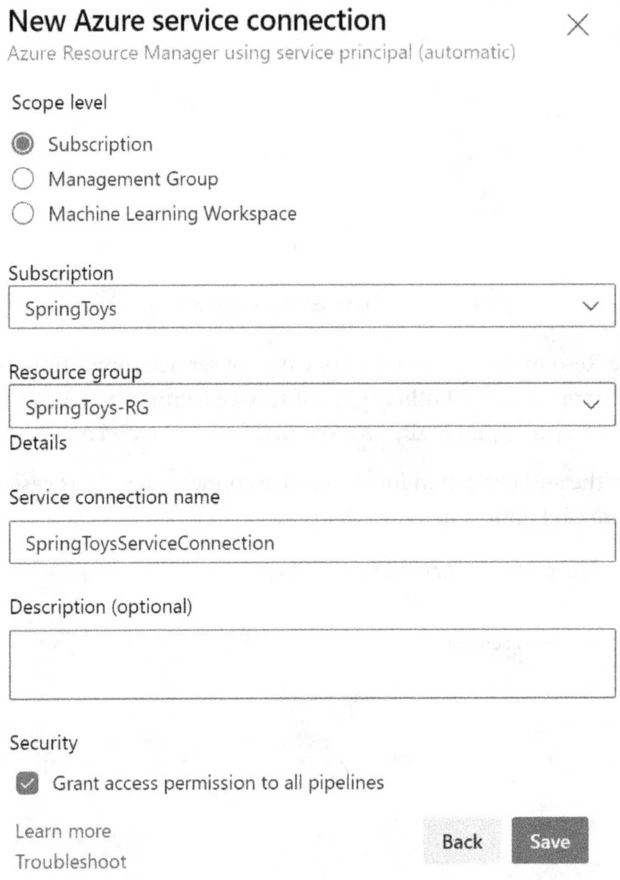
New Azure service connection
Azure Resource Manager using service principal (automatic)
Scope level
Subscription
Management Group
Machine Learning Workspace
Subscription
SpringToys
Resource group
SpringToys-RG
Details
Service connection name
SpringToysServiceConnection
Description (optional)
Security
Grant access permission to all pipelines
Learn more
Troubleshoot
Back
Save

Figure 12.5 – Service connection parameters

The next step is to prepare the code we will use in our pipelines. We can leverage Azure Repos to store Bicep files and ARM templates in Azure DevOps.

Configuring Azure Repos

Azure Repos is a version control service provided by Microsoft as part of its Azure DevOps platform. It helps developers to manage and track changes to their code base, collaborate with team members, and maintain different versions of their code.

Azure Repos provides features such as Git and **Team Foundation Version Control** (**TFVC**) systems, pull requests, code reviews, and more, making it easier for teams to work together on code development and delivery.

To configure Azure Repos, go to the **Azure DevOps** main page, and in the left menu, select the **Repos** option. Then, you will see a page with multiple options to configure your source code. In the left menu, under **Repos**, there's a drop-down menu that includes **Files**, **Commits**, **Pushes**, **Branches**, **Tags**, and **Pull requests**. On the right-hand side, we have options to import a repository:

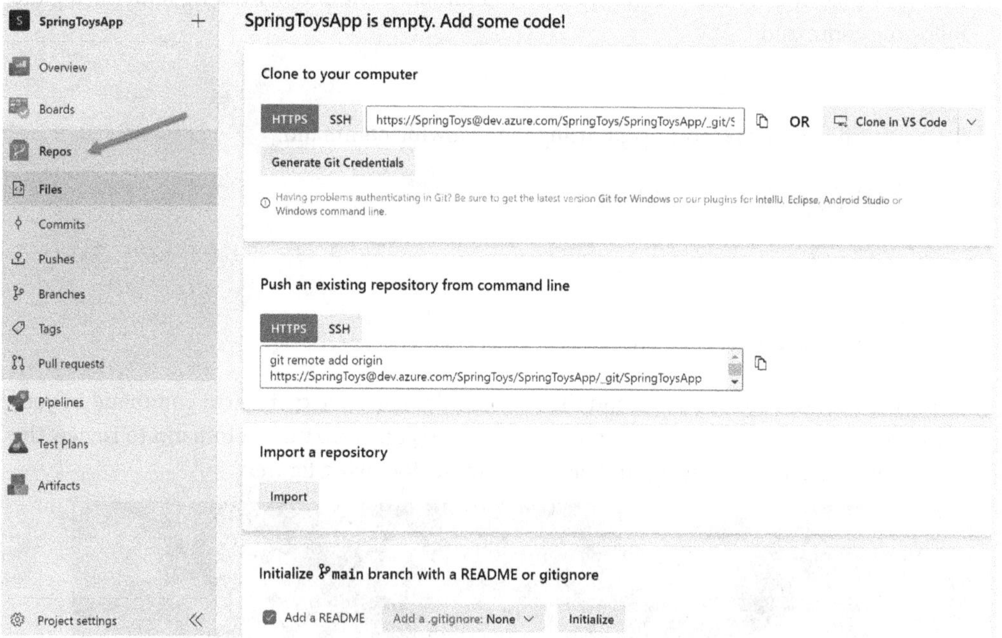
SpringToysApp
Overview
Boards
Repos
Files
Commits
Pushes
Branches
Tags
Pull requests
Pipelines
Test Plans
Artifacts
Project settings
SpringToysApp is empty. Add some code!
Clone to your computer
HTTPS
SSH
https://SpringToys@dev.azure.com/SpringToys/SpringToysApp/_git/S
OR
Clone in VS Code
Generate Git Credentials
Having problems authenticating in Git? Be sure to get the latest version Git for Windows or our plugins for IntelliJ, Eclipse, Android Studio or Windows command line.
Push an existing repository from command line
HTTPS
SSH
git remote add origin
https://SpringToys@dev.azure.com/SpringToys/SpringToysApp/_git/SpringToysApp
Import a repository
Import
Initialize main branch with a README or gitignore
Add a README
Add a .gitignore: None
Initialize

Figure 12.6 – Azure Repos overview

The repository includes three tiers:

- **Tier1**: Contains the platform components such as the virtual network, the SQL Server, storage accounts, and an automation account. You will see Bicep files for each resource here.
- **Tier2**: Contains the application management components. You will find ARM templates for each resource here.
- **Tier3**: Contains the application itself and all the related code.

You can download the code from the following GitHub repository: `https://github.com/PacktPublishing/Azure-Architecture-Explained/tree/main/Chapter12/azure-repos`. Once you have downloaded it to your local machine, let's push it to Azure Repos.

Importing a repository into Azure Repos

To import a repository, follow these steps:

1. First, set the account identity with the following commands:

```
git config --global user.email "you@example.com"
git config --global user.name "Your Name"
```

   And now that we have the source code, we will initialize Git in our working folder using the following command:

```
Git init
```

2. Then, we will add the remote origin using the following command:

```
git remote add origin

https://SpringToys@dev.azure.com/SpringToys/SpringToysApp/_git/
SpringToysApp

git push -u origin --all
```

3. Once you initialize the repo and add the remote origin, use the `git add` command in your Bash to add all the files to the given folder. Use the `git status` command to review the files to be staged to the first commit, as shown in the following figure:

```
$ git status
On branch master

No commits yet

Changes to be committed:
  (use "git rm --cached <file>..." to unstage)
        new file:   Tier1/AutomationAccount/azuredeploy.bicep
        new file:   Tier1/AutomationAccount/azuredeploy.json
        new file:   Tier1/AzureSQLServerwithDatabase/azuredeploy.bicep
        new file:   Tier1/AzureSQLServerwithDatabase/azuredeploy.json
        new file:   Tier1/StorageAccount/azuredeploy.bicep
        new file:   Tier1/StorageAccount/azuredeploy.json
        new file:   Tier1/Vnet/azuredeploy.bicep
        new file:   Tier1/Vnet/azuredeploy.json
        new file:   Tier2/LB/azuredeploy.json
        new file:   Tier2/Vm/azuredeploy.json
        new file:   Tier3/Tier3.zip
```

Figure 12.7 – Files to be committed

4. Now, let's commit the files staged in the local repository: `git commit -m 'your message'`. The following figure shows the output from this command:

```
$ git commit -m "First commit"
[master (root-commit) 18e3cf7] First commit
 11 files changed, 789 insertions(+)
 create mode 100644 Tier1/AutomationAccount/azuredeploy.bicep
 create mode 100644 Tier1/AutomationAccount/azuredeploy.json
 create mode 100644 Tier1/AzureSQLServerwithDatabase/azuredeploy.bicep
 create mode 100644 Tier1/AzureSQLServerwithDatabase/azuredeploy.json
 create mode 100644 Tier1/StorageAccount/azuredeploy.bicep
 create mode 100644 Tier1/StorageAccount/azuredeploy.json
 create mode 100644 Tier1/Vnet/azuredeploy.bicep
 create mode 100644 Tier1/Vnet/azuredeploy.json
 create mode 100644 Tier2/LB/azuredeploy.json
 create mode 100644 Tier2/Vm/azuredeploy.json
 create mode 100644 Tier3/Tier3.zip
```

Figure 12.8 – First commit

5. Finally, push the code to Azure Repos using the following command:

```
git push -u origin --all
```

The following figure shows the output from this command:

```
$ git push -u origin --all
Enumerating objects: 22, done.
Counting objects: 100% (22/22), done.
Delta compression using up to 16 threads
Compressing objects: 100% (19/19), done.
Writing objects: 100% (22/22), 11.19 MiB | 17.22 MiB/s, done.
Total 22 (delta 3), reused 0 (delta 0), pack-reused 0
remote: Analyzing objects... (22/22) (212 ms)
remote: Storing packfile... done (699 ms)
remote: Storing index... done (44 ms)
To https://dev.azure.com/SpringToys/SpringToysApp/_git/SpringToysApp
 * [new branch]      master -> master
branch 'master' set up to track 'origin/master'.
```

Figure 12.9 – Files published to Azure Repos

Now, you should see the files already present in the repository, as shown here:

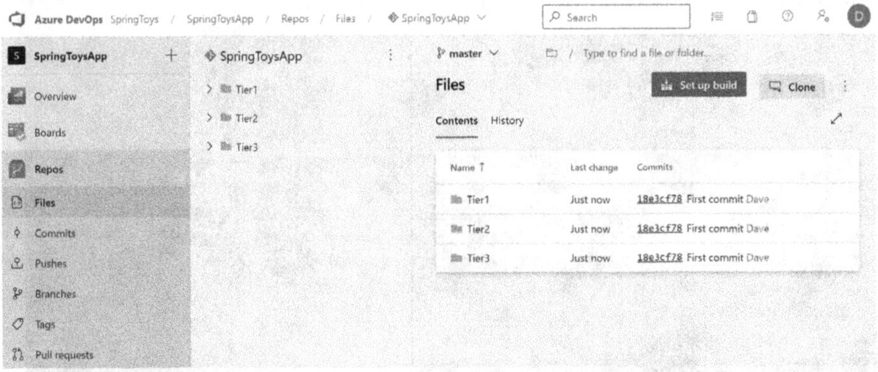
Azure DevOps SpringToys / SpringToysApp / Repos / Files / SpringToysApp
Search
SpringToysApp
Overview
Boards
Repos
Files
Commits
Pushes
Branches
Tags
Pull requests
SpringToysApp
Tier1
Tier2
Tier3
master
Type to find a file or folder...
Files
Set up build
Clone
Contents
History
Name
Last change
Commits
Tier1
Just now
18e3cf78 First commit Dave
Tier2
Just now
18e3cf78 First commit Dave
Tier3
Just now
18e3cf78 First commit Dave

Figure 12.10 – Azure Repos updated

Note all the files are now available in the repository.

As the next step, we will configure a build pipeline in Azure DevOps.

Configuring a build pipeline in Azure DevOps using the Classic Editor

Under Azure Repos, you can configure Azure Pipelines. Follow these steps:

1. Go to the **Pipelines** section and then select **Create Pipeline**:

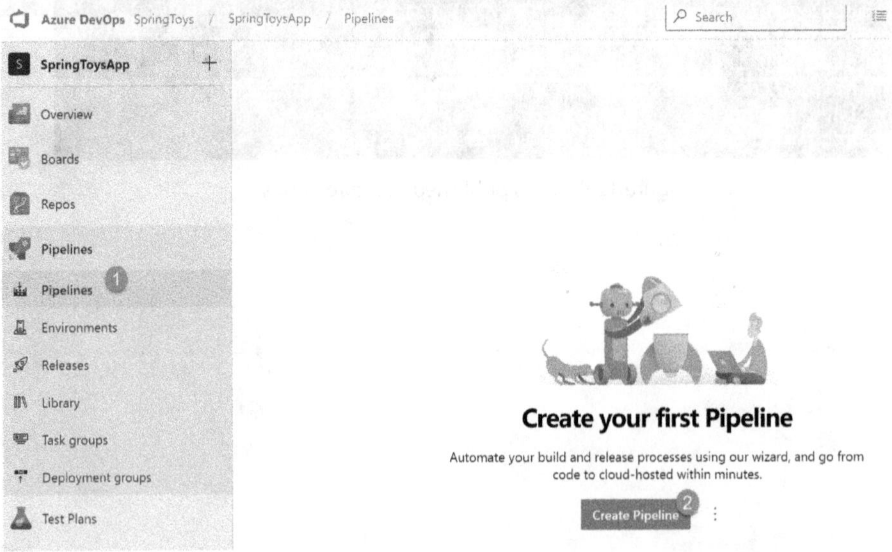
Azure DevOps SpringToys / SpringToysApp / Pipelines
Search
SpringToysApp
Overview
Boards
Repos
Pipelines
Pipelines
1
Environments
Releases
Library
Task groups
Deployment groups
Test Plans
Create your first Pipeline
Automate your build and release processes using our wizard, and go from code to cloud-hosted within minutes.
Create Pipeline
2

Figure 12.11 – Create Pipeline

2. Next, select the **Use the classic editor** option, as shown here:

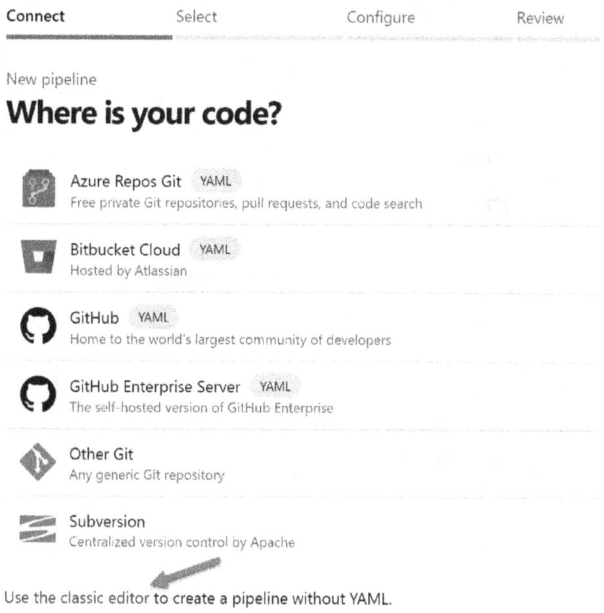

Connect
Select
Configure
Review
New pipeline
Where is your code?
Azure Repos Git YAML
Free private Git repositories, pull requests, and code search
Bitbucket Cloud YAML
Hosted by Atlassian
GitHub YAML
Home to the world's largest community of developers
GitHub Enterprise Server YAML
The self-hosted version of GitHub Enterprise
Other Git
Any generic Git repository
Subversion
Centralized version control by Apache
Use the classic editor to create a pipeline without YAML.

Figure 12.12 – Use the classic editor

3. Then, select the repository and the **master** branch and then select the **Continue** option:

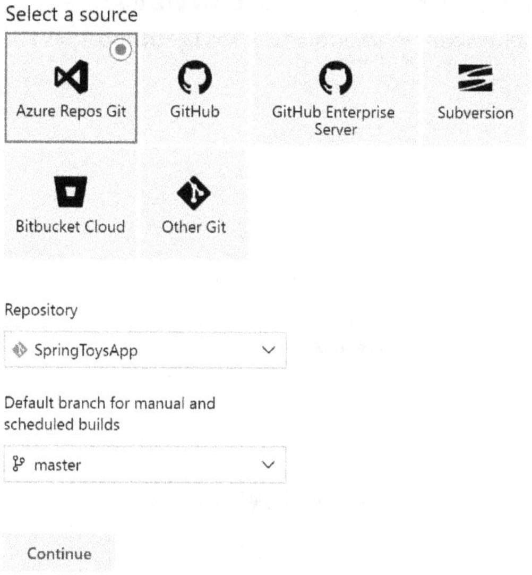

Select a source
Azure Repos Git
GitHub
GitHub Enterprise Server
Subversion
Bitbucket Cloud
Other Git
Repository
SpringToysApp
Default branch for manual and scheduled builds
master
Continue

Figure 12.13 – Azure Pipelines – Select a source

4. Next, select the option to start with an empty job:

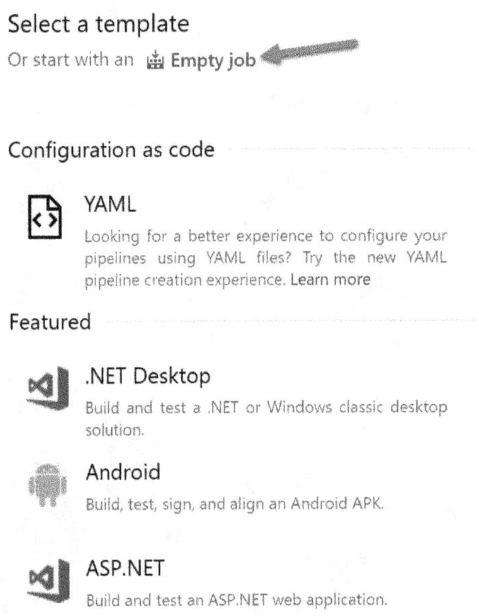
Select a template
Or start with an Empty job
Configuration as code
YAML
Looking for a better experience to configure your pipelines using YAML files? Try the new YAML pipeline creation experience. Learn more
Featured
.NET Desktop
Build and test a .NET or Windows classic desktop solution.
Android
Build, test, sign, and align an Android APK.
ASP.NET
Build and test an ASP.NET web application.

Figure 12.14 – Azure Pipelines – Select a template

5. Now, provide the configuration for the pipeline. Provide a name for this pipeline. We will name it `SpringToysApp-CI`. Then, in the left pane, where it says **Agent job 1**, select this option and then click on the plus icon, +, to add a task to **Agent job 1**, as shown here:

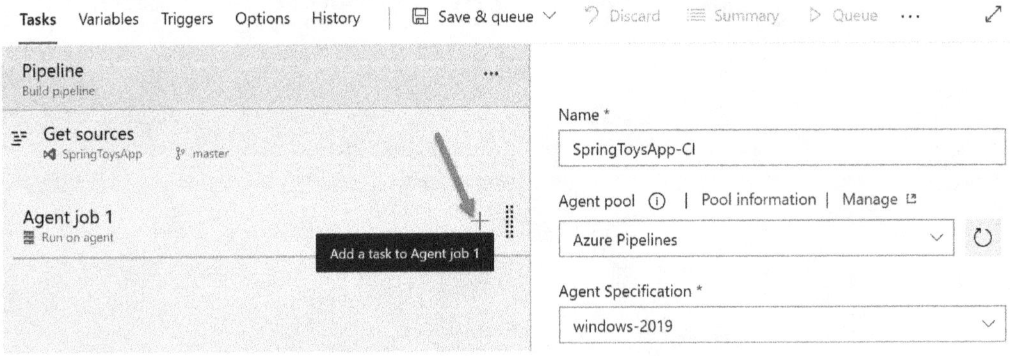
Tasks
Variables
Triggers
Options
History
Save & queue
Discard
Summary
Queue
Pipeline
Build pipeline
Get sources
SpringToysApp
master
Agent job 1
Run on agent
Add a task to Agent job 1
Name *
SpringToysApp-CI
Agent pool
Pool information
Manage
Azure Pipelines
Agent Specification *
windows-2019

Figure 12.15 – Azure Pipelines – Agent job

6. Then, we will add the build artifacts. In the search box, type `publish build` and add it, as shown here:

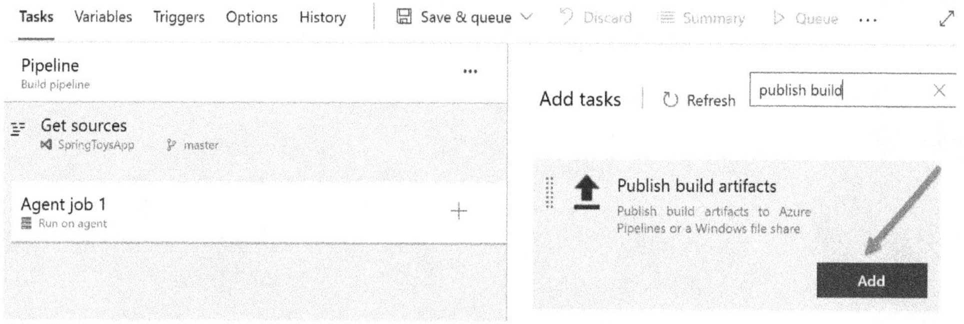

Tasks
Variables
Triggers
Options
History
Save & queue
Discard
Summary
Queue
Pipeline
Build pipeline
Get sources
SpringToysApp
master
Agent job 1
Run on agent
Add tasks
Refresh
publish build
Publish build artifacts
Publish build artifacts to Azure Pipelines or a Windows file share
Add

Figure 12.16 – Azure Pipelines – Publish build artifacts

7. Once added, in the task called **Publish Artifact: drop**, change the display name, for example, `Tier1-CI-AutomationAccount`, as shown here:

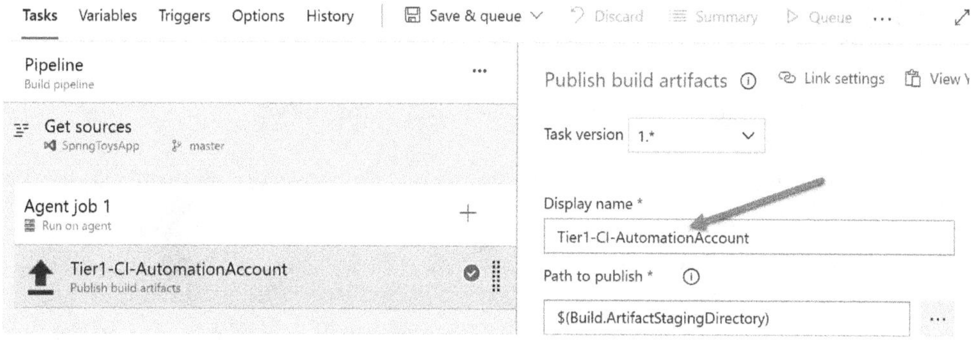

Tasks
Variables
Triggers
Options
History
Save & queue
Discard
Summary
Queue
Pipeline
Build pipeline
Get sources
SpringToysApp
master
Agent job 1
Run on agent
Tier1-CI-AutomationAccount
Publish build artifacts
Publish build artifacts
Link settings
Task version
1.*
Display name *
Tier1-CI-AutomationAccount
Path to publish *
$(Build.ArtifactStagingDirectory)

Figure 12.17 – Azure Pipelines – naming the artifact

8. Next, in the **Path to publish** field, select the options menu (**...**), and look for the folder of the Azure resource where the automation account is located. Expand the repository folder, then look for the **AutomationAccount** folder, and click **OK**:

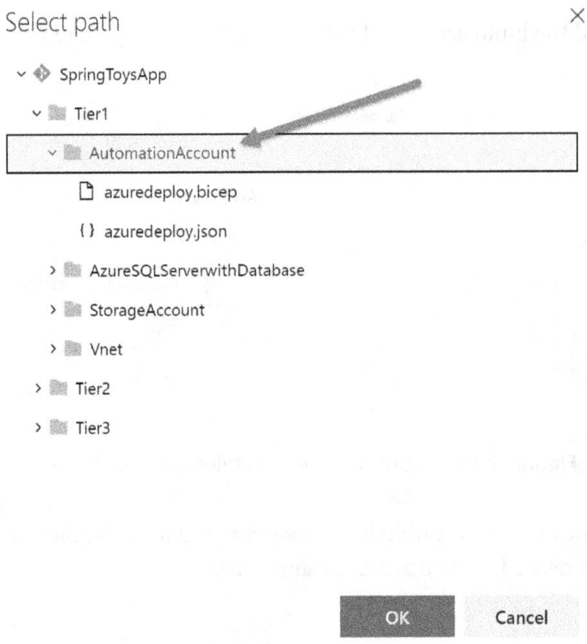
Select path
SpringToysApp
Tier1
AutomationAccount
azuredeploy.bicep
azuredeploy.json
AzureSQLServerwithDatabase
StorageAccount
Vnet
Tier2
Tier3
OK
Cancel

Figure 12.18 – Azure Pipelines – path to the artifact

9. Now save the pipeline. We will manually start a build process to ensure the pipeline works correctly. Go to the **Queue** option:

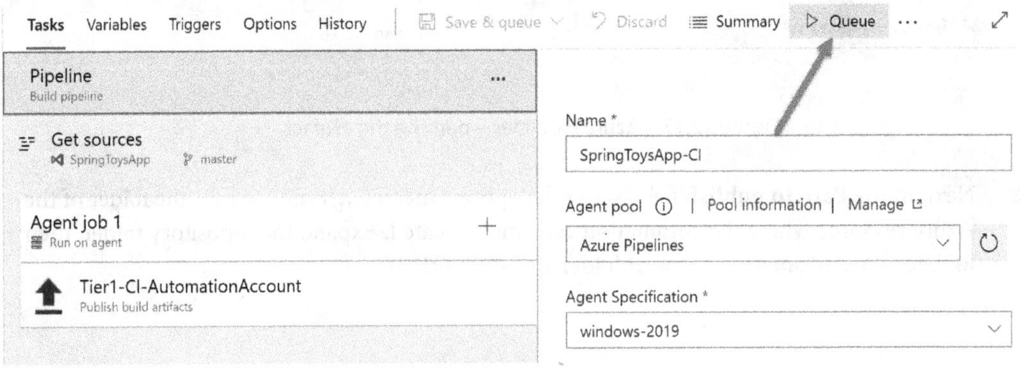
Tasks
Variables
Triggers
Options
History
Save & queue
Discard
Summary
Queue
Pipeline
Build pipeline
Get sources
SpringToysApp
master
Agent job 1
Run on agent
Tier1-CI-AutomationAccount
Publish build artifacts
Name *
SpringToysApp-CI
Agent pool
Pool information
Manage
Azure Pipelines
Agent Specification *
windows-2019

Figure 12.19 – Azure Pipelines – Queue

10. Now run the pipeline:

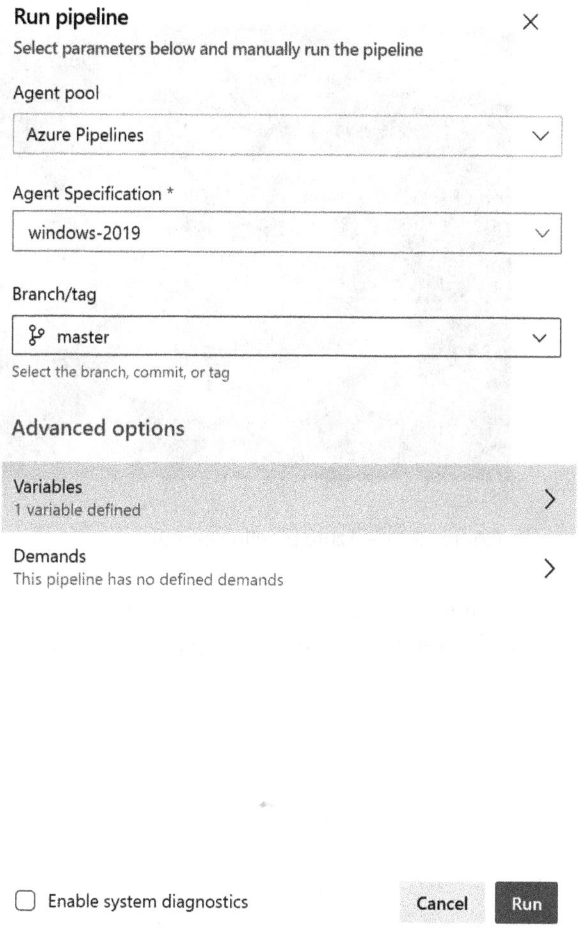
Run pipeline
Select parameters below and manually run the pipeline
Agent pool
Azure Pipelines
Agent Specification *
windows-2019
Branch/tag
master
Select the branch, commit, or tag
Advanced options
Variables
1 variable defined
Demands
This pipeline has no defined demands
Enable system diagnostics
Cancel
Run

Figure 12.20 – Azure Pipelines – Run pipeline

Once you run the pipeline, you should see an output similar to the following figure, where we see the post-job:

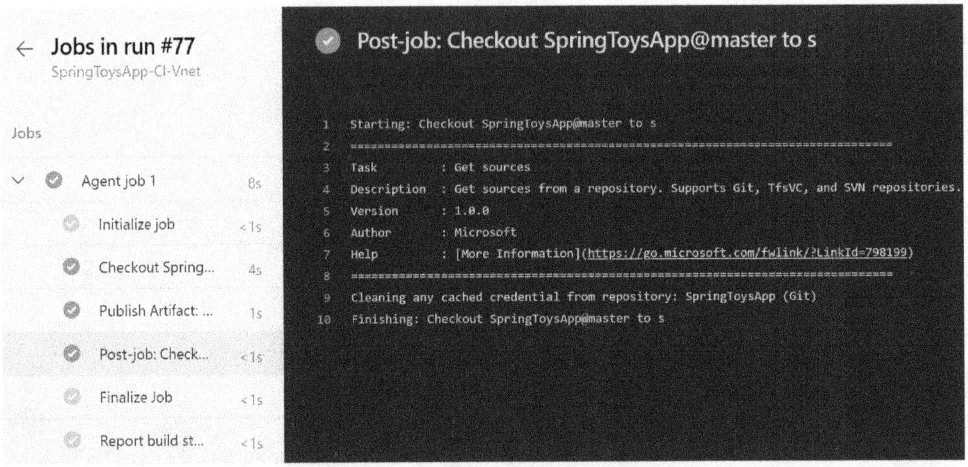
Jobs in run #77
SpringToysApp-CI-Vnet
Jobs
Agent job 1 8s
Initialize job <1s
Checkout Spring... 4s
Publish Artifact: ... 1s
Post-job: Check... <1s
Finalize Job <1s
Report build st... <1s
Post-job: Checkout SpringToysApp@master to s
1 Starting: Checkout SpringToysApp@master to s
2 ==
3 Task : Get sources
4 Description : Get sources from a repository. Supports Git, TfsVC, and SVN repositories.
5 Version : 1.0.0
6 Author : Microsoft
7 Help : [More Information](https://go.microsoft.com/fwlink/?LinkId=798199)
8 ==
9 Cleaning any cached credential from repository: SpringToysApp (Git)
10 Finishing: Checkout SpringToysApp@master to s

Figure 12.21 – Azure pipeline execution

And you can build an Azure pipeline for the rest of the Azure resources. We will perform the same process as before per resource. You should have one build pipeline for each ARM template in the `Tier1` folder, as shown here:

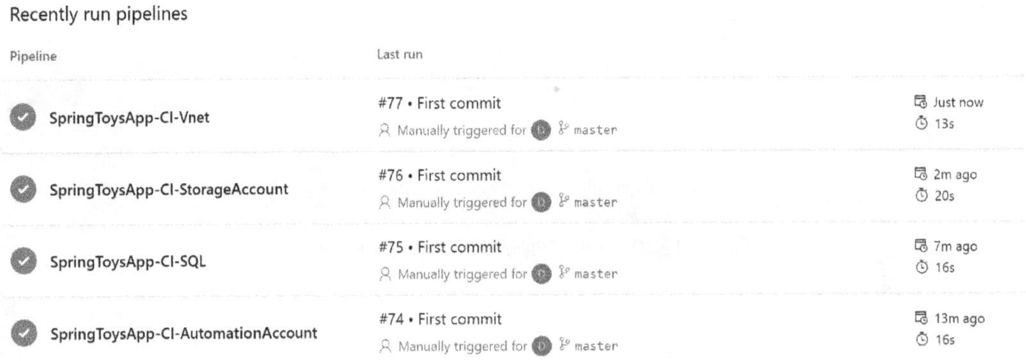
Recently run pipelines
Pipeline
Last run
SpringToysApp-CI-Vnet
#77 • First commit
Manually triggered for master
Just now
13s
SpringToysApp-CI-StorageAccount
#76 • First commit
Manually triggered for master
2m ago
20s
SpringToysApp-CI-SQL
#75 • First commit
Manually triggered for master
7m ago
16s
SpringToysApp-CI-AutomationAccount
#74 • First commit
Manually triggered for master
13m ago
16s

Figure 12.22 – Azure pipelines for various resources

Now, let's configure a release pipeline.

Configuring a release pipeline in Azure DevOps using the Classic Editor

To configure a release pipeline, follow these steps:

1. In Azure DevOps, go to the **Pipelines** section and select the **Releases** option:

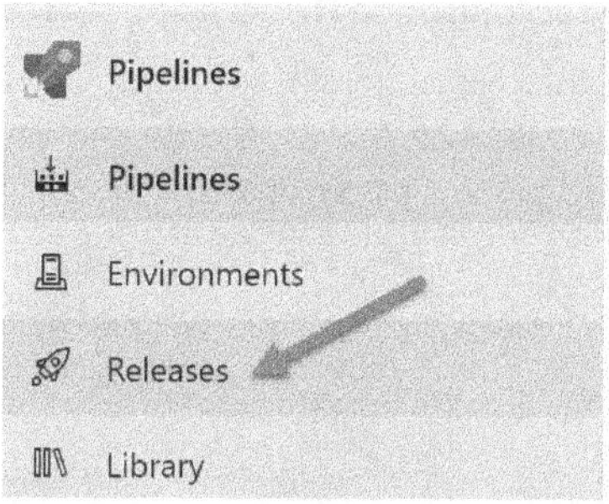
Pipelines
Pipelines
Environments
Releases
Library

Figure 12.23 – Azure Pipelines – Releases

2. Next, create a new pipeline and select the **Empty job** option, as shown here:

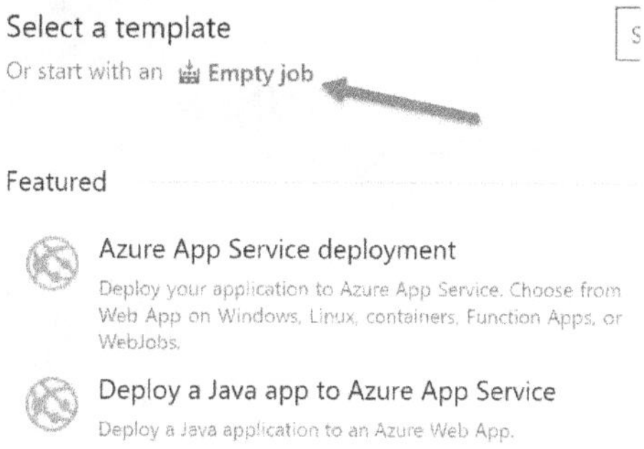
Select a template
Or start with an Empty job
Featured
Azure App Service deployment
Deploy your application to Azure App Service. Choose from Web App on Windows, Linux, containers, Function Apps, or WebJobs.
Deploy a Java app to Azure App Service
Deploy a Java application to an Azure Web App.

Figure 12.24 – Azure Pipelines – Empty job

3. Provide a name for this stage, `Dev`, and save it:

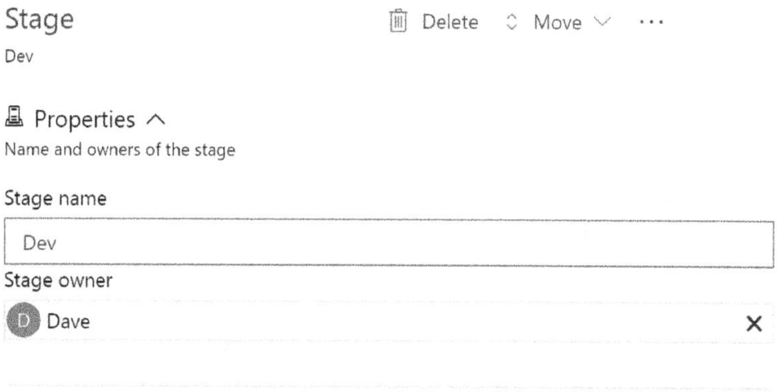
Stage
Dev
Delete
Move
Properties
Name and owners of the stage
Stage name
Dev
Stage owner
Dave

Figure 12.25 – Azure Pipelines – Stage

4. Then, we will add the artifacts. These artifacts will be the Bicep files: the automation account, the SQL database, the storage account, and the virtual network. On the same page, select the + **Add an artifact** option:

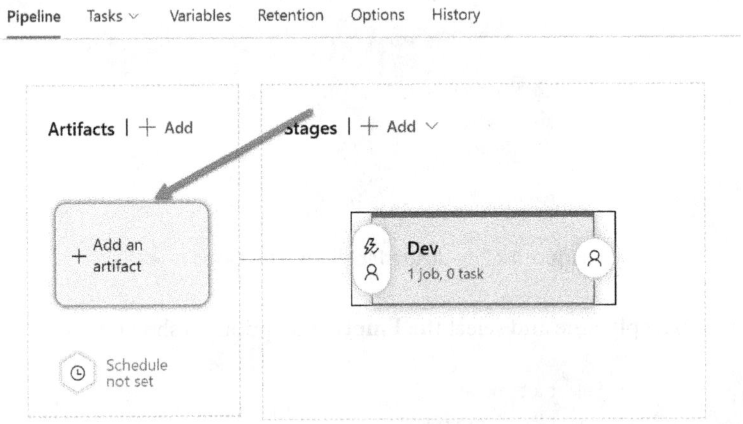
Pipeline
Tasks
Variables
Retention
Options
History
Artifacts | + Add
Stages | + Add
Add an artifact
Dev
1 job, 0 task
Schedule not set

Figure 12.26 – Azure Pipelines – Artifacts

5. Select the build pipeline related to the resources in `Tier1`, and add four artifacts, one artifact per build pipeline:

Add an artifact
Source type
Build
Azure Repos ...
GitHub
TFVC
5 more artifact types
Project *
SpringToysApp
Source (build pipeline) *
SpringToysApp-CI-AutomationAccount
Default version *
Latest
Source alias *
_SpringToysApp-CI-AutomationAccount
The artifacts published by each version will be available for deployment in release pipelines. The latest successful build of SpringToysApp-CI-AutomationAccount published the following artifacts: AutomationAccount.
Add

Figure 12.27 – Azure Pipelines – configuring artifacts

6. Repeat the preceding process until you have added the four build pipelines as artifacts. Once you add the four pipelines as artifacts, you should see them in the **Artifacts** section, as shown here:

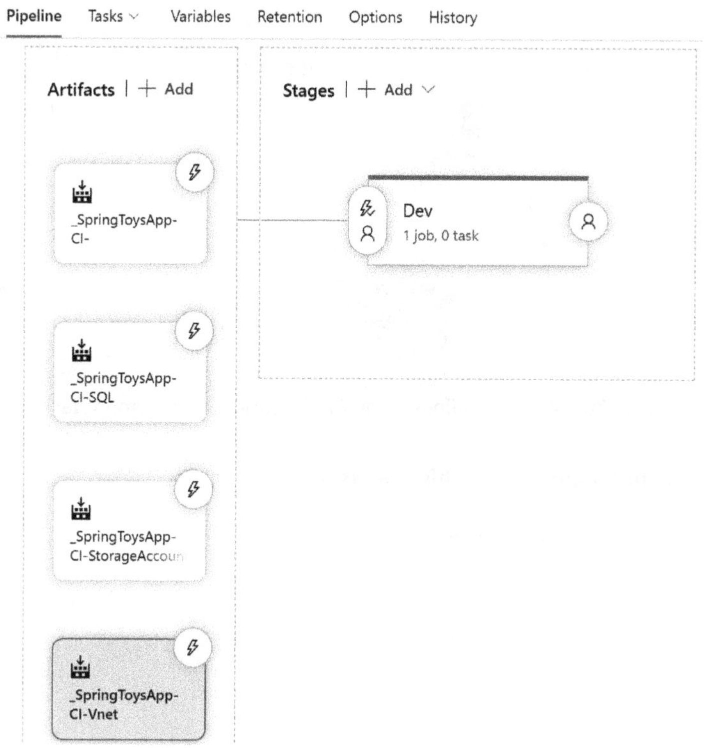
Pipeline
Tasks
Variables
Retention
Options
History
Artifacts | + Add
Stages | + Add
_SpringToysApp-
CI-
_SpringToysApp-
CI-SQL
_SpringToysApp-
CI-StorageAccou
_SpringToysApp-
CI-Vnet
Dev
1 job, 0 task

Figure 12.28 – Azure Pipelines – release and artifacts

7. Now, add a new task in the **Dev** stage. You can select the shortcut in the **Dev** stage, as shown here:

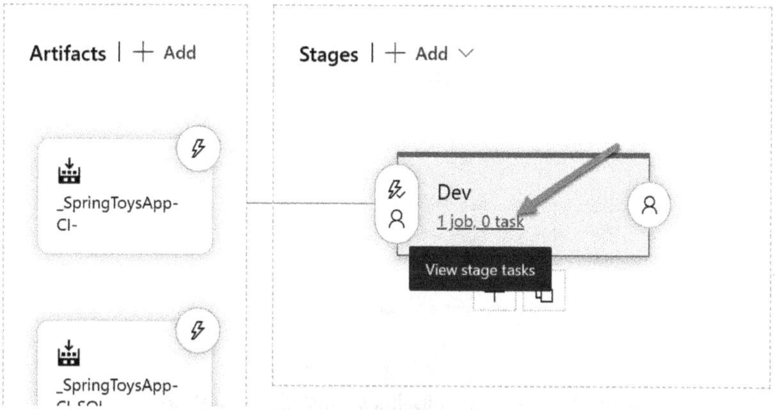
Artifacts | + Add
Stages | + Add
_SpringToysApp-
CI-
_SpringToysApp-
Dev
1 job, 0 task
View stage tasks

Figure 12.29 – Azure Pipelines – add tasks

8. Then, select the + option next to **Agent job** and look for **ARM template deployment**. Then add it:

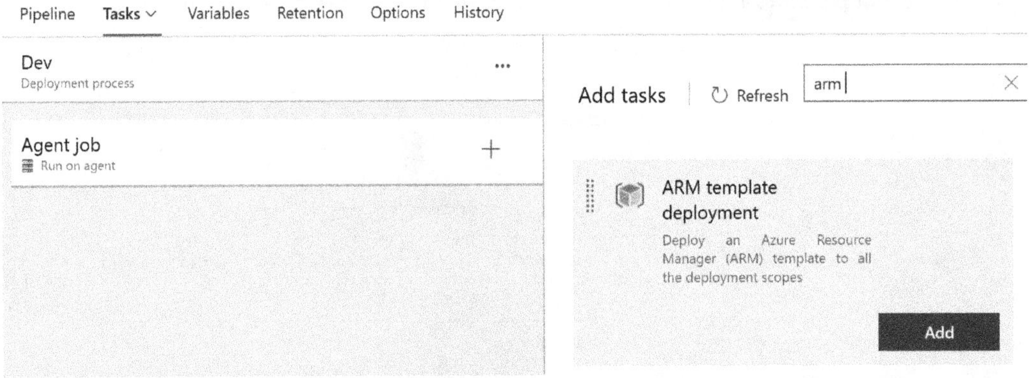
Pipeline
Tasks
Variables
Retention
Options
History
Dev
Deployment process
Agent job
Run on agent
Add tasks
Refresh
arm
ARM template deployment
Deploy an Azure Resource Manager (ARM) template to all the deployment scopes
Add

Figure 12.30 – Azure Pipelines – add ARM template deployment task

9. Now configure the parameters for this task, as shown here:

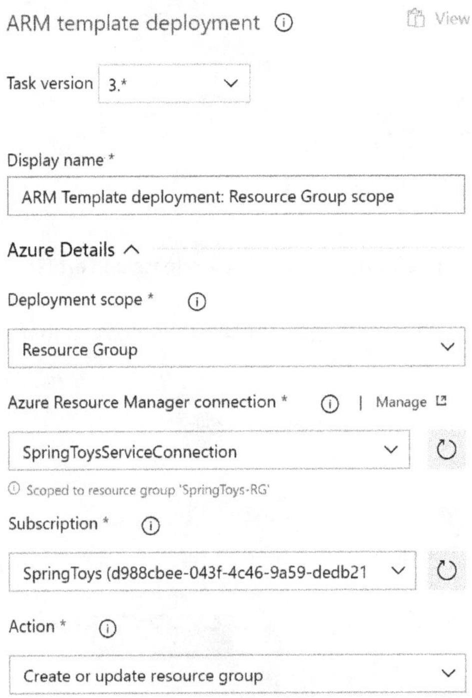
ARM template deployment
Task version 3.*
Display name *
ARM Template deployment: Resource Group scope
Azure Details
Deployment scope *
Resource Group
Azure Resource Manager connection *
Manage
SpringToysServiceConnection
Scoped to resource group 'SpringToys-RG'
Subscription *
SpringToys (d988cbee-043f-4c46-9a59-dedb21
Action *
Create or update resource group

Figure 12.31 – Azure Pipelines – task configuration

For the template, we will reference the Bicep files:

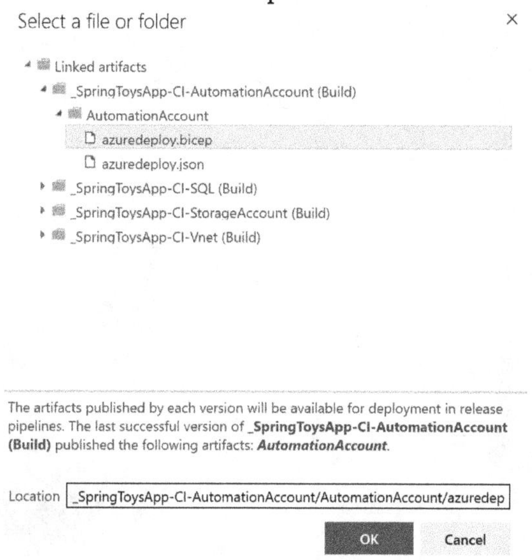
Select a file or folder
Linked artifacts
_SpringToysApp-CI-AutomationAccount (Build)
AutomationAccount
azuredeploy.bicep
azuredeploy.json
_SpringToysApp-CI-SQL (Build)
_SpringToysApp-CI-StorageAccount (Build)
_SpringToysApp-CI-Vnet (Build)
The artifacts published by each version will be available for deployment in release pipelines. The last successful version of _SpringToysApp-CI-AutomationAccount (Build) published the following artifacts: AutomationAccount.
Location
_SpringToysApp-CI-AutomationAccount/AutomationAccount/azuredep
OK
Cancel

Figure 12.32 – Azure Pipelines – configuring the path to the artifact

10. Now, save it and create a release:

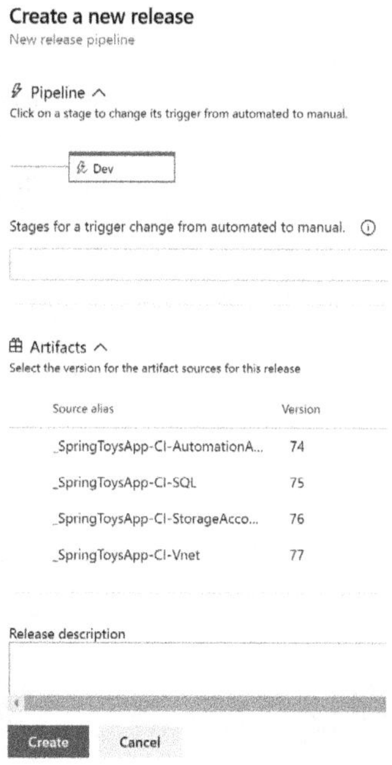
Create a new release
New release pipeline
Pipeline
Click on a stage to change its trigger from automated to manual.
Dev
Stages for a trigger change from automated to manual.
Artifacts
Select the version for the artifact sources for this release
Source alias
Version
_SpringToysApp-CI-AutomationA...
74
_SpringToysApp-CI-SQL
75
_SpringToysApp-CI-StorageAcco...
76
_SpringToysApp-CI-Vnet
77
Release description
Create
Cancel

Figure 12.33 – Azure Pipelines – Create a new release

After a few seconds, you should see the results, as shown here:

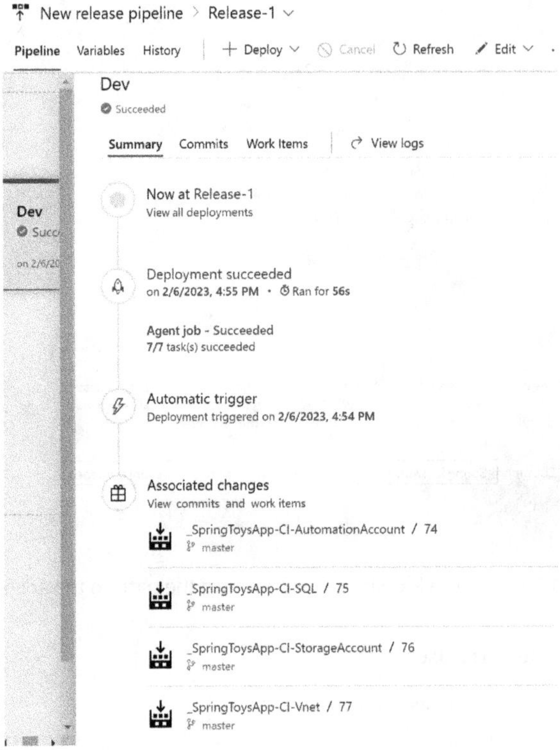
New release pipeline > Release-1
Pipeline Variables History + Deploy Cancel Refresh Edit
Dev
Succeeded
Summary Commits Work Items View logs
Now at Release-1
View all deployments
Deployment succeeded
on 2/6/2023, 4:55 PM • Ran for 56s
Agent job - Succeeded
7/7 task(s) succeeded
Automatic trigger
Deployment triggered on 2/6/2023, 4:54 PM
Associated changes
View commits and work items
_SpringToysApp-CI-AutomationAccount / 74
master
_SpringToysApp-CI-SQL / 75
master
_SpringToysApp-CI-StorageAccount / 76
master
_SpringToysApp-CI-Vnet / 77
master

Figure 12.34 – Azure Pipelines – release result

11. You can also verify the result using the Azure portal. Go to the resource group and validate that the automation account has been created:

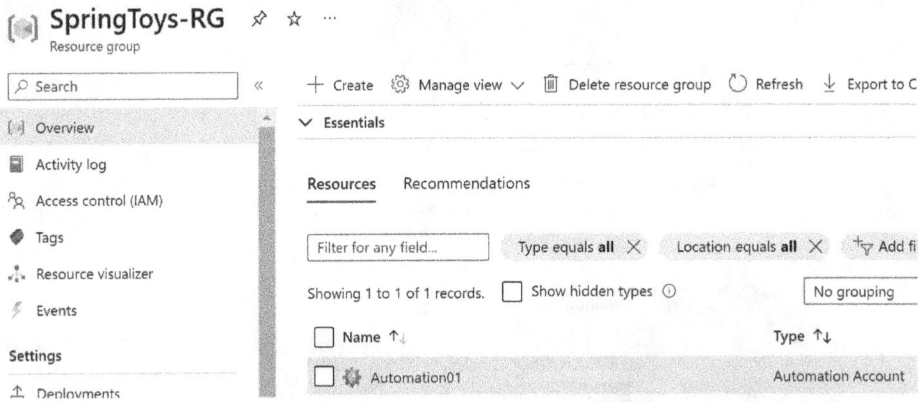
SpringToys-RG
Resource group
Search
+ Create Manage view Delete resource group Refresh Export to C
Overview
Activity log
Access control (IAM)
Tags
Resource visualizer
Events
Settings
Essentials
Resources Recommendations
Filter for any field...
Type equals all
Location equals all
Showing 1 to 1 of 1 records.
Show hidden types
No grouping
Name
Type
Automation01
Automation Account

Figure 12.35 – Azure portal – resource group overview

Now that you have validated all the processes, you can select **Enable continuous integration**. This option is enabled in the build pipeline, as shown here:

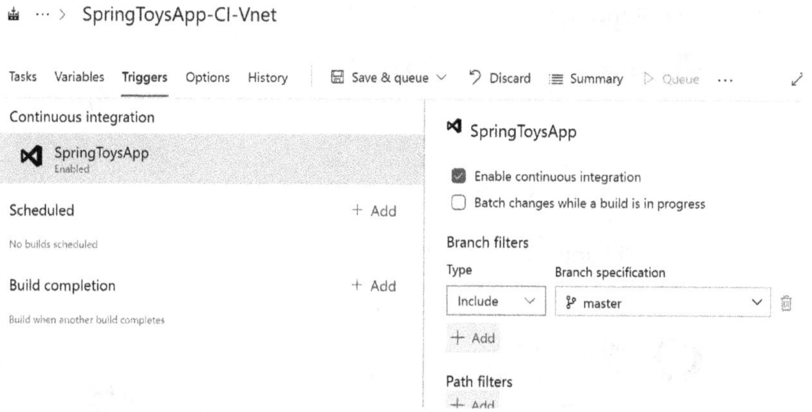
SpringToysApp-CI-Vnet
Tasks
Variables
Triggers
Options
History
Save & queue
Discard
Summary
Queue
Continuous integration
SpringToysApp
Enabled
Scheduled
+ Add
No builds scheduled
Build completion
+ Add
Build when another build completes
SpringToysApp
Enable continuous integration
Batch changes while a build is in progress
Branch filters
Type
Branch specification
Include
master
+ Add
Path filters

Figure 12.36 – Azure Pipelines – Enable continuous integration

Hopefully, this gives you an idea of how you can configure Azure Pipelines and leverage Bicep files in a modularized approach with Azure DevOps.

In the next section, we will review how you can create an Azure pipeline using YAML.

Configuring Azure Pipelines with YAML

In this section, we will perform the steps to configure an Azure pipeline with YAML:

1. Sign in to your Azure DevOps account and navigate to the project where you want to create the pipeline.
2. Then, click on the **Pipelines** tab on the left, and click on the **New pipeline** button to start creating a new pipeline, as shown here:

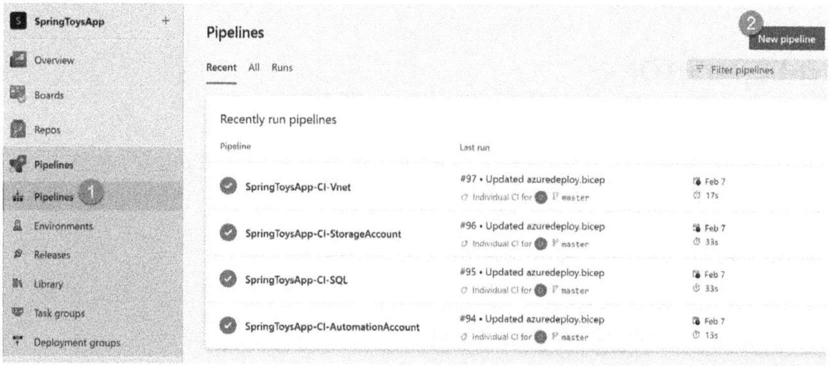
SpringToysApp
Overview
Boards
Repos
Pipelines
Pipelines
Environments
Releases
Library
Task groups
Deployment groups
Pipelines
New pipeline
Recent
All
Runs
Filter pipelines
Recently run pipelines
Pipeline
Last run
SpringToysApp-CI-Vnet
#97 • Updated azuredeploy.bicep
SpringToysApp-CI-StorageAccount
#96 • Updated azuredeploy.bicep
SpringToysApp-CI-SQL
#95 • Updated azuredeploy.bicep
SpringToysApp-CI-AutomationAccount
#94 • Updated azuredeploy.bicep

Figure 12.37 – Azure Pipelines – New pipeline

3. In the **Where is your code?** section, select the repository where your code is located. Azure DevOps supports various version control systems such as Azure Repos, GitHub, and Bitbucket.

   We will select **Azure Repos Git YAML**, as shown here:

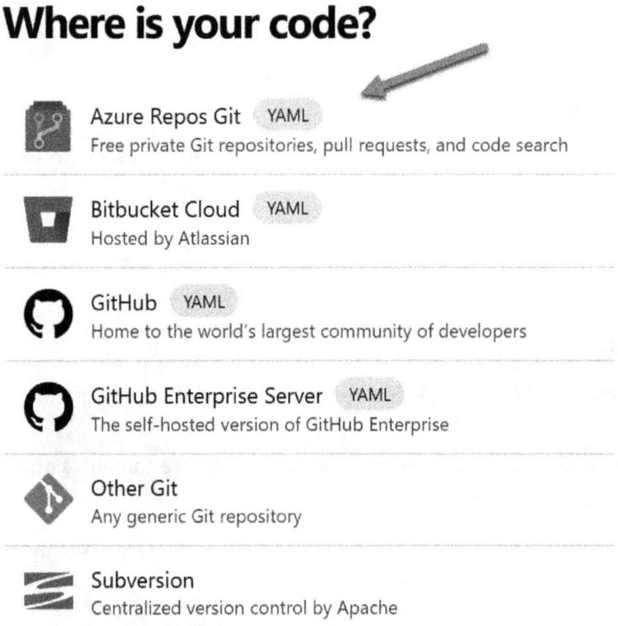
Where is your code?
Azure Repos Git
YAML
Free private Git repositories, pull requests, and code search
Bitbucket Cloud
YAML
Hosted by Atlassian
GitHub
YAML
Home to the world's largest community of developers
GitHub Enterprise Server
YAML
The self-hosted version of GitHub Enterprise
Other Git
Any generic Git repository
Subversion
Centralized version control by Apache

Figure 12.38 – Azure Pipelines – Where is your code?

   Azure DevOps will analyze your repository and suggest a YAML template based on the project type and existing configuration. If the suggested template matches your requirements, you can choose it.

4. Next, we will select the repository named **SpringToysApp** from Azure Repos:

New pipeline
Select a repository
Filter by keywords
SpringToysApp
SpringToysApp

Figure 12.39 – Azure Pipelines – Select a repository

5. Then, in the pipeline configuration wizard, select the **Starter pipeline** option, as shown here:

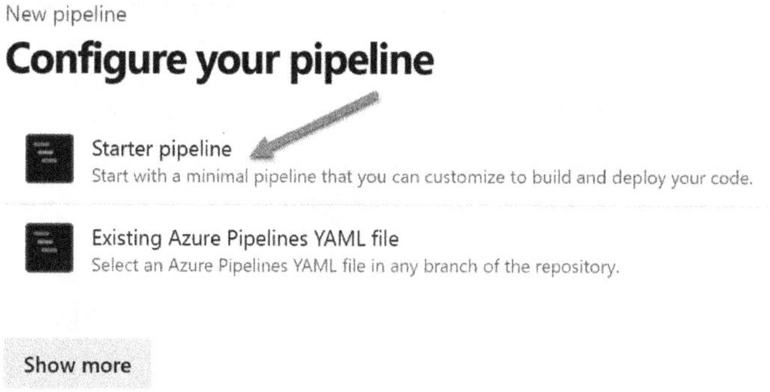
New pipeline
Configure your pipeline
Starter pipeline
Start with a minimal pipeline that you can customize to build and deploy your code.
Existing Azure Pipelines YAML file
Select an Azure Pipelines YAML file in any branch of the repository.
Show more

Figure 12.40 – Azure Pipelines – Starter pipeline

6. Azure DevOps will generate a basic YAML pipeline file for you. Then, you can replace the content of the generated file with your own pipeline configuration. You can manually edit the YAML file or copy and paste your existing YAML configuration into the editor.

   The following figure shows the example of the starter pipeline:

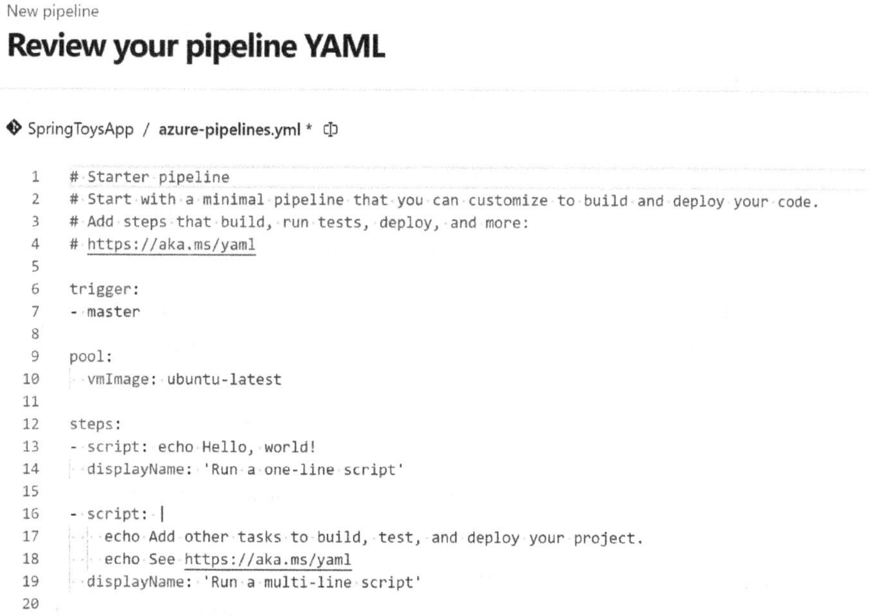
New pipeline
Review your pipeline YAML
SpringToysApp / azure-pipelines.yml *
Starter pipeline
Start with a minimal pipeline that you can customize to build and deploy your code.
Add steps that build, run tests, deploy, and more:
https://aka.ms/yaml
trigger:
- master
pool:
vmImage: ubuntu-latest
steps:
- script: echo Hello, world!
displayName: 'Run a one-line script'
- script: |
echo Add other tasks to build, test, and deploy your project.
echo See https://aka.ms/yaml
displayName: 'Run a multi-line script'

Figure 12.41 – Azure Pipelines – starter pipeline

From here, you can define the stages, jobs, and steps of your pipeline.

Each stage represents a phase in your pipeline, such as build, test, or deployment. Each job represents a unit of work within a stage, and steps define the tasks to be performed in each job.

7. Next, we will replace the code with our code, which creates the automation account, as shown here:

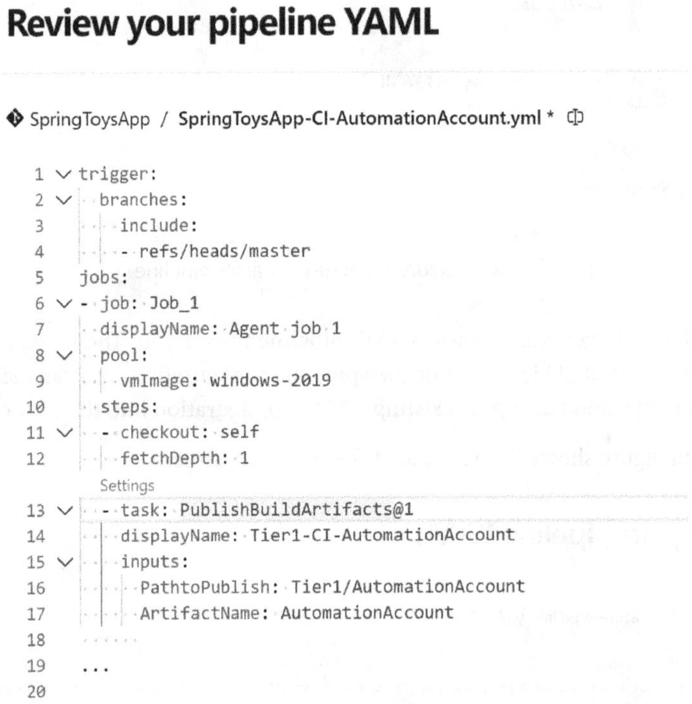
Review your pipeline YAML
SpringToysApp / SpringToysApp-CI-AutomationAccount.yml *
trigger:
  branches:
    include:
    - refs/heads/master
jobs:
- job: Job_1
  displayName: Agent job 1
  pool:
    vmImage: windows-2019
  steps:
  - checkout: self
    fetchDepth: 1
Settings
  - task: PublishBuildArtifacts@1
    displayName: Tier1-CI-AutomationAccount
    inputs:
      PathtoPublish: Tier1/AutomationAccount
      ArtifactName: AutomationAccount
...

Figure 12.42 – Azure Pipelines – update pipeline

You can specify the tasks and configuration for each step in your pipeline. Tasks can include build tasks, test runners, or deployment actions. Azure DevOps provides a wide range of pre-defined tasks that you can use, or you can create custom tasks if needed.

You can configure any necessary variables, triggers, or conditions for your pipeline. Variables allow you to define values that can be used across different stages and jobs. Triggers determine when the pipeline should run, such as on every commit or on a schedule. Conditions control the flow of your pipeline based on specific criteria.

8. Once you have defined your pipeline configuration, click on the **Save and run** button to save your pipeline and trigger the first run. You can also save the pipeline without triggering a run by clicking on the **Save** button:

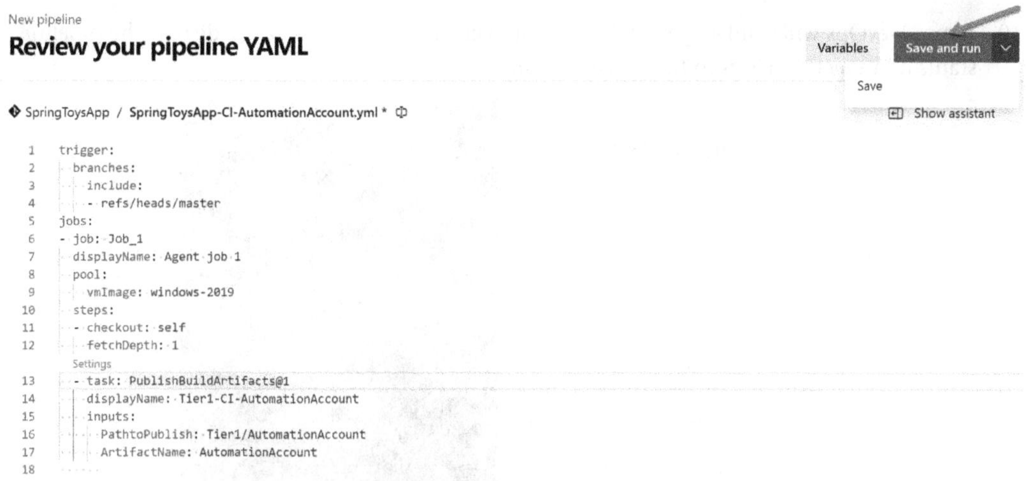
New pipeline
Review your pipeline YAML
Variables
Save and run
Save
Show assistant
SpringToysApp / SpringToysApp-CI-AutomationAccount.yml *
trigger:
branches:
include:
- refs/heads/master
jobs:
- job: Job_1
displayName: Agent job 1
pool:
vmImage: windows-2019
steps:
- checkout: self
fetchDepth: 1
Settings
- task: PublishBuildArtifacts@1
displayName: Tier1-CI-AutomationAccount
inputs:
PathtoPublish: Tier1/AutomationAccount
ArtifactName: AutomationAccount

Figure 12.43 – Azure Pipelines – Save and run

9. Once you select the **Save and run** button, you will be able to specify the commit message and run your pipeline, as shown here:

Save and run
Saving will commit SpringToysApp-CI-AutomationAccount.yml to the repository.
Commit message
Set up CI with Azure Pipelines YAML
Optional extended description
Add an optional description...
Commit directly to the master branch
Create a new branch for this commit
Save and run

Figure 12.44 – Azure Pipelines – Save and run

10. Azure DevOps will validate your YAML file and create the pipeline. It will display the pipeline status and execution logs in the pipeline dashboard:

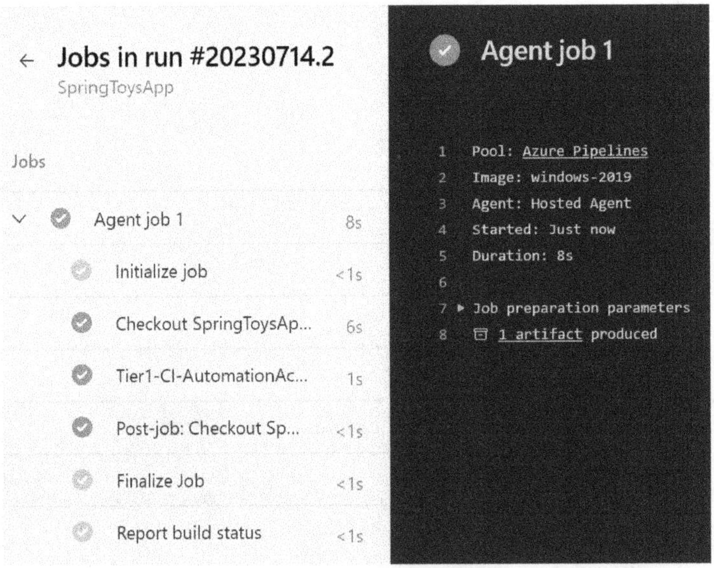
Jobs in run #20230714.2
SpringToysApp
Jobs
Agent job 1
8s
Initialize job
<1s
Checkout SpringToysAp...
6s
Tier1-CI-AutomationAc...
1s
Post-job: Checkout Sp...
<1s
Finalize Job
<1s
Report build status
<1s
Agent job 1
Pool: Azure Pipelines
Image: windows-2019
Agent: Hosted Agent
Started: Just now
Duration: 8s
Job preparation parameters
1 artifact produced

Figure 12.45 – Azure Pipelines – Jobs in run

Now we have successfully created an Azure Pipeline using YAML. You can further customize and enhance your pipeline by adding more stages, jobs, and steps, or by incorporating advanced features such as environment deployments, approvals, or parallelism as per your project requirements.

Summary

With Azure Pipelines, organizations can automate their software development pipeline, reducing the risk of errors and freeing up development teams to focus on creating high-quality software.

This chapter reviewed how you can get started with Azure DevOps and configure your repository using Azure Repos. Then, we created a few build pipelines and release pipelines. Finally, we created a release pipeline and verified the creation of a resource in our Azure environment.

The next chapter will review how you can enable CI and continuous deployment with Azure Pipelines.

13

Continuous Integration and Deployment in Azure DevOps

In today's fast-paced world, businesses need to deliver software faster and with better quality to stay ahead of the competition. This is where the practice of DevOps comes in, which combines development and operations to streamline the software delivery process. Azure DevOps is a popular platform that provides end-to-end tools and services to support DevOps practices.

In this chapter, we will delve into the two crucial aspects of DevOps – **continuous integration** (**CI**) and **continuous delivery** (**CD**) – and see how they can be implemented using Azure DevOps. We will explore the benefits of CI and CD and how they can help teams to automate the building, testing, and deployment process, leading to faster delivery of high-quality software. Whether you are a developer, a DevOps engineer, or a project manager, this chapter will provide you with valuable insights into how Azure DevOps can help you achieve your DevOps goals.

In this chapter, we'll cover the following main topics:

- DevOps transformation – achieving reliable and efficient software development through CI and CD practices
- CI in Azure DevOps
- CD in Azure DevOps

Let's get started!

DevOps transformation – achieving reliable and efficient software development through CI and CD practices

Organizations need to stay ahead of the curve by adopting new practices and technologies. One such approach is DevOps, which emphasizes collaboration between development and operations teams to deliver high-quality software faster. SpringToys, a hypothetical company, has recently embarked on its DevOps transformation journey and wants to ensure the reliability and quality of its software development process to deliver new features to customers quickly and efficiently.

Let's say SpringToys wants to ensure its systems are stable and allow the release of new features to reach customers more quickly and efficiently.

In this context, SpringToys can consider CI and CD as foundational practices for ensuring the reliability and quality of its software development.

CI and CD are related but different software development practices. CI is a programming approach that mandates developers to merge code into a shared repository multiple times throughout the day. Each integration is then verified by an automated building and testing process to catch and fix any issues as soon as possible. It includes the adoption of tools for the following:

- **Source control**: Establish a source control system, such as Git, to store the code base and manage code changes.
- **Automated builds**: Set up an automated build system, such as Jenkins or Azure DevOps, that integrates with the source control system and builds the code whenever changes are pushed to the repository.
- **Automated tests**: Implement automated tests, such as unit tests, integration tests, and functional tests, to validate the code changes and catch any issues as soon as possible.
- **Integration**: Ensure developers integrate their code changes into the shared repository several times daily. The build system should automatically run the tests and build the code whenever changes are made.

On the other hand, CD refers to the practice of automating the process of delivering code changes to production. It builds on top of CI by adding additional testing and deployment stages, such as integration testing, acceptance testing, and performance testing. Once these tests are passed, the code is automatically deployed to production.

With continuous delivery, the software does not have to be (automatically) deployed to production, whereas with continuous deployment, it is automatically deployed all the way to production if all tests pass. CD involves the following additional steps:

- **Staging environment**: Set up a staging environment, a replica of the production environment, where code changes can be deployed and tested before they are released to the public

- **Automated deployment**: Automate the deployment process, using tools such as Ansible or Terraform, to deploy code changes from the build system to the staging environment
- **Continuous testing**: Continuously run tests, such as integration tests, acceptance tests, and performance tests, to validate that the code changes are working as expected
- **Release approval**: Set up a release approval process where a designated person or group must review and approve the code changes before they are deployed to production
- **Continuous monitoring**: Continuously monitor the production environment to detect and resolve any issues as soon as they arise

Put simply, CI focuses on integrating code changes and ensuring that the code remains in a working state, while CD focuses on automating the entire deployment pipeline, from integration to production.

When discussing CI, it is essential to understand the concept of triggers. Triggers in CI play a critical role in ensuring the reliability and quality of software development in Azure DevOps.

Triggers automate the CI process, making it easier to build, test, and deploy code changes frequently and consistently. This helps to reduce manual errors and increase efficiency.

By automatically building, testing, and deploying code changes, triggers help to increase confidence in the code and ensure that it meets quality standards.

Think of Azure DevOps triggers as events that prompt an automated build and deployment process. When a CI trigger is fired, the code is automatically built, tested, and deployed to a target environment.

And there are various ways to trigger a CI pipeline in Azure DevOps, for example, if a developer pushes changes to the source control repository. You can also schedule builds so that a pipeline can be scheduled to run at specific times. External events such as a message in a queue or a webhook can also trigger CI pipelines.

Now that we clearly understand CI and CD and how they relate to one another, let's look at how we can configure CI and CD in Azure DevOps.

CI in Azure DevOps using the Classic Editor

In the previous chapter, we configured Azure Repos as the source control for the project in Azure DevOps and also configured a few build pipelines. However, each build pipeline has to be run manually.

In this example, we have four build pipelines, one for each resource: the automation account, SQL database, storage account, and virtual network. Each pipeline has its own build definition, including the *Agent job* and an artifact, as shown here:

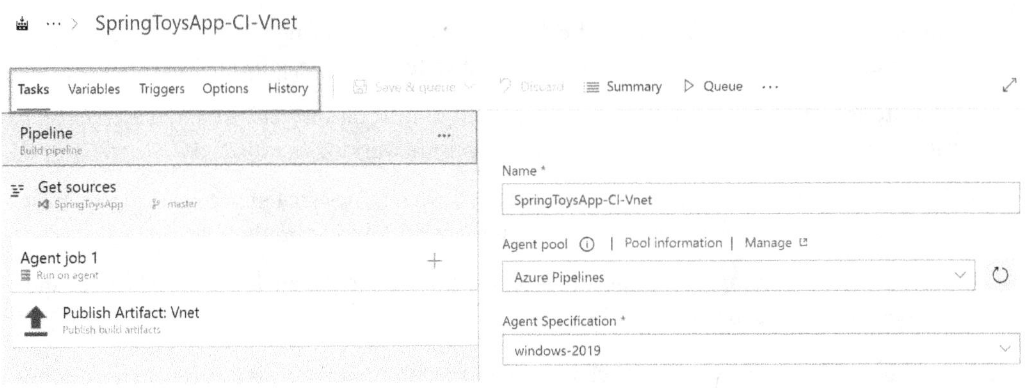
SpringToysApp-CI-Vnet
Tasks Variables Triggers Options History
Save & queue
Discard
Summary
Queue
Pipeline
Build pipeline
Get sources
SpringToysApp
master
Agent job 1
Run on agent
Publish Artifact: Vnet
Publish build artifacts
Name *
SpringToysApp-CI-Vnet
Agent pool
Pool information | Manage
Azure Pipelines
Agent Specification *
windows-2019

Figure 13.1 – Example Azure pipeline – build pipeline

Note that in the top menu, there are multiple options such as **Tasks**, **Variables**, **Triggers**, **Options**, and **History**. The **Triggers** section includes the controls to enable CI, as shown here:

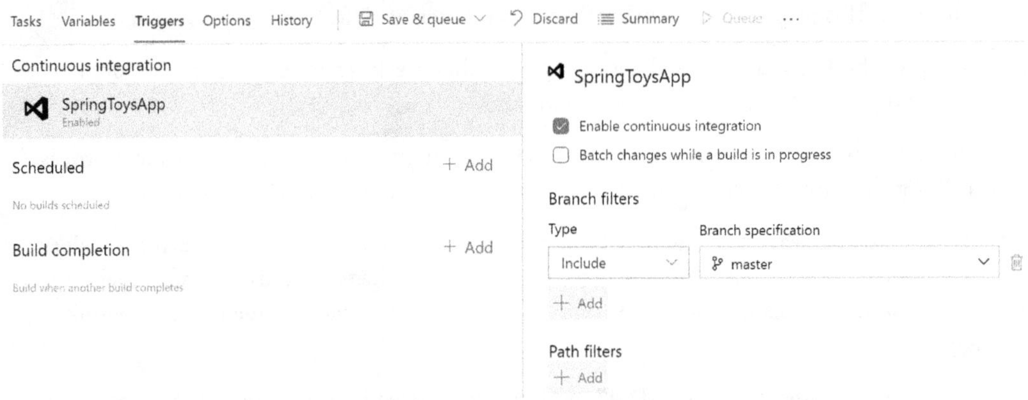
Tasks Variables Triggers Options History
Save & queue
Discard
Summary
Queue
Continuous integration
SpringToysApp
Enabled
Scheduled
+ Add
No builds scheduled
Build completion
+ Add
Build when another build completes
SpringToysApp
Enable continuous integration
Batch changes while a build is in progress
Branch filters
Type
Branch specification
Include
master
+ Add
Path filters
+ Add

Figure 13.2 – Azure pipeline – Triggers section

Triggers can be configured in several ways, including the following:

- **CI trigger**: Triggered automatically when a code change is pushed to the source control repository.
- **Scheduled trigger**: Enables scheduling the build pipeline to run at a specified time and date.
- **Pull request trigger**: Runs the build pipeline whenever a pull request is raised in the source control repository.
- **Path filters trigger**: Specifies the path or paths in the source control repository that will trigger the build pipeline.

- **YAML pipeline trigger**: Configures the build pipeline using a YAML file in the source control repository and runs the pipeline automatically when changes are made to the YAML file.
- **Artifacts trigger**: Runs the build pipeline whenever an artifact is published from another pipeline.
- **API trigger**: Triggers the build pipeline through a REST API call.
- **Gated check-in trigger**: Similar to the CI trigger, but requires manual approval before a build can be triggered. This is useful for code changes that must be reviewed before being committed to the source control repository.

In the pipeline configuration, go to the **Triggers** section, select the **Enable continuous integration** option, and set the **Branch specification** option to **master**, as shown here:

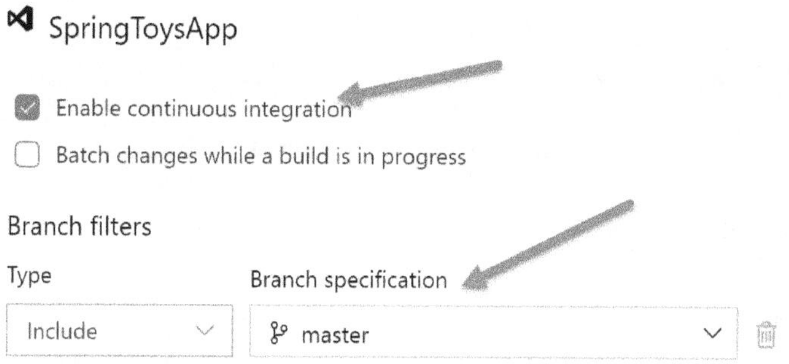
SpringToysApp
Enable continuous integration
Batch changes while a build is in progress
Branch filters
Type
Branch specification
Include
master

Figure 13.3 – Azure pipeline – Enable continuous integration

Then, save the changes. Now, perform the same configuration for all the build pipelines.

Once you enable CI in the four build pipelines, you can make a code change in the repo and check the result.

In this example, we will modify the Bicep file of the virtual network. And we will update the API version as shown here:

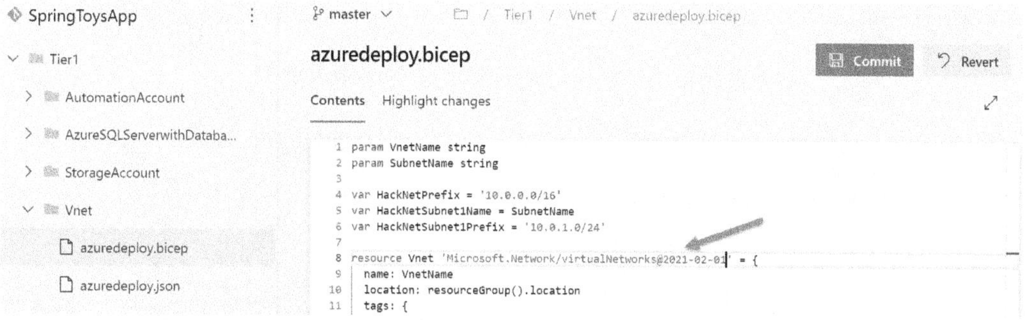
SpringToysApp
Tier1
AutomationAccount
AzureSQLServerwithDataba...
StorageAccount
Vnet
azuredeploy.bicep
azuredeploy.json
master
Tier1 / Vnet / azuredeploy.bicep
azuredeploy.bicep
Commit
Revert
Contents
Highlight changes
param VnetName string
param SubnetName string
var HackNetPrefix = '10.0.0.0/16'
var HackNetSubnet1Name = SubnetName
var HackNetSubnet1Prefix = '10.0.1.0/24'
resource Vnet 'Microsoft.Network/virtualNetworks@2021-02-01' = {
name: VnetName
location: resourceGroup().location
tags: {

Figure 13.4 – Azure Repos – modifying the Bicep file

Then, commit the change, and you will see that the jobs for each pipeline have been triggered, as shown here:

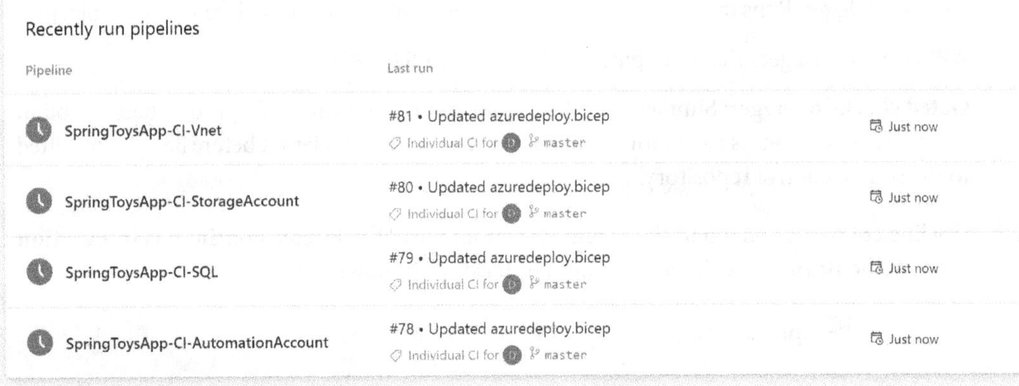
Recently run pipelines
Pipeline
Last run
SpringToysApp-CI-Vnet
#81 • Updated azuredeploy.bicep
Individual CI for
master
Just now
SpringToysApp-CI-StorageAccount
#80 • Updated azuredeploy.bicep
Individual CI for
master
Just now
SpringToysApp-CI-SQL
#79 • Updated azuredeploy.bicep
Individual CI for
master
Just now
SpringToysApp-CI-AutomationAccount
#78 • Updated azuredeploy.bicep
Individual CI for
master
Just now

Figure 13.5 – Azure pipelines

After a few seconds, you should see the jobs complete, as shown here:

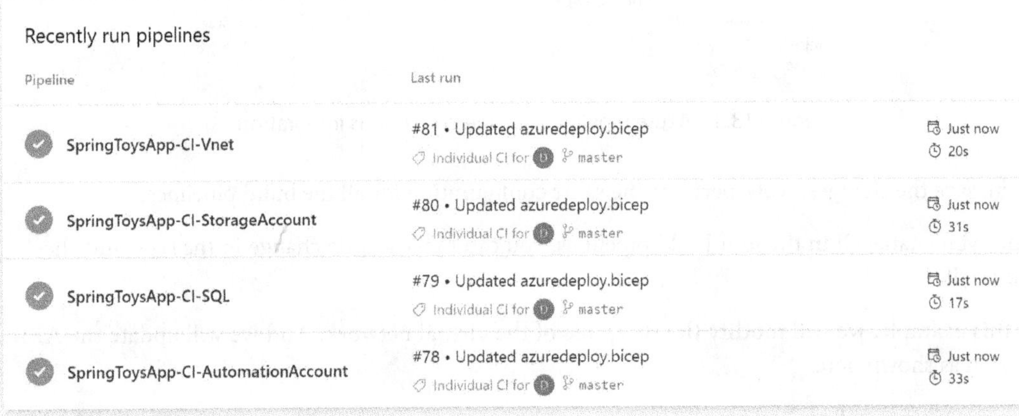
Recently run pipelines
Pipeline
Last run
SpringToysApp-CI-Vnet
#81 • Updated azuredeploy.bicep
Individual CI for
master
Just now
20s
SpringToysApp-CI-StorageAccount
#80 • Updated azuredeploy.bicep
Individual CI for
master
Just now
31s
SpringToysApp-CI-SQL
#79 • Updated azuredeploy.bicep
Individual CI for
master
Just now
17s
SpringToysApp-CI-AutomationAccount
#78 • Updated azuredeploy.bicep
Individual CI for
master
Just now
33s

Figure 13.6 – Azure Pipelines – Recently run pipelines

Once we verify that the trigger is properly working, it is time to start working on the CD configuration.

The following section will review how to set up CD in Azure DevOps.

CD in Azure DevOps

Achieving consistent and efficient value delivery has become an essential requirement for organizations. To continuously enhance the experience for SpringToys customers, it is imperative to establish a process of regular and flawless software releases.

CD automates the process of building, testing, configuring, and deploying code from a build environment to a production environment. Multiple testing or staging environments can be established through a release pipeline to automate the infrastructure setup and deploy new builds.

These environments provide opportunities for longer-running integration, load, and user acceptance testing, leading to a more comprehensive evaluation of the code changes.

We created a release pipeline in the previous chapter to deploy the code to the Azure environment. The following figure shows the release pipeline:

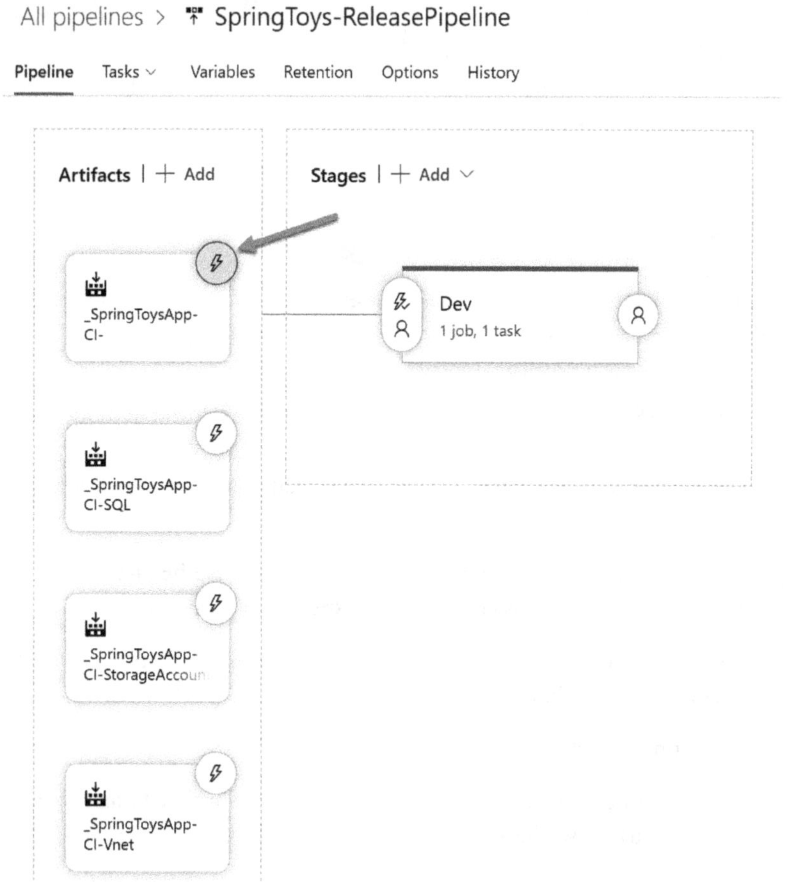
All pipelines > SpringToys-ReleasePipeline
Pipeline
Tasks
Variables
Retention
Options
History
Artifacts | + Add
Stages | + Add
_SpringToysApp-CI-
Dev
1 job, 1 task
_SpringToysApp-CI-SQL
_SpringToysApp-CI-StorageAccoun
_SpringToysApp-CI-Vnet

Figure 13.7 – SpringToys release pipeline

Note the icon in the top-right corner of each of the artifacts. If you select this option, you will see the controls to enable continuous deployment.

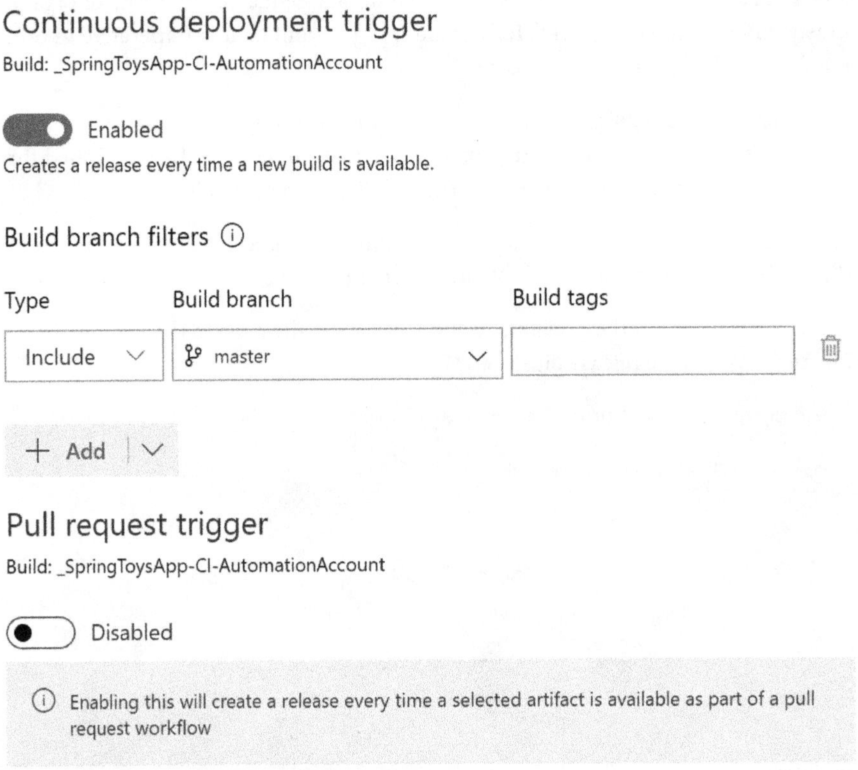
Continuous deployment trigger
Build: _SpringToysApp-CI-AutomationAccount
Enabled
Creates a release every time a new build is available.
Build branch filters
Type
Build branch
Build tags
Include
master
Add
Pull request trigger
Build: _SpringToysApp-CI-AutomationAccount
Disabled
Enabling this will create a release every time a selected artifact is available as part of a pull request workflow

Figure 13.8 – SpringToys release pipeline – continuous deployment trigger

Continuous deployment triggers allow you to automate the creation of releases every time a new build artifact becomes available.

You can set up deployment for specific target branches by using **Build branch** filters. Releases will only be triggered if the Git push contains changes in the specified branch.

For example, selecting the **master** or **main** branch will initiate a release for any Git push that includes one or more commits to the `main` branch.

You can enable this option for each artifact.

Go back to the release pipeline and clone the existing **Dev** stage using the **Clone** option under the stage. Then, name it `Prod`, as shown here:

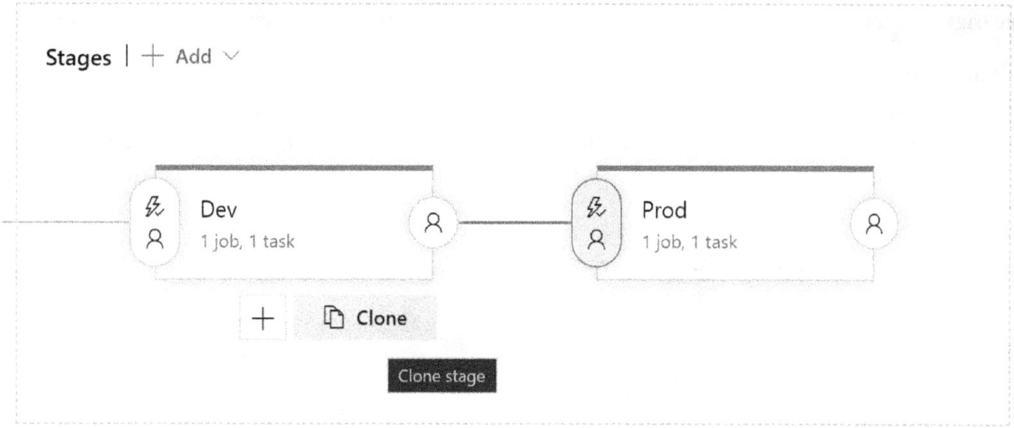
Stages | + Add
Dev
1 job, 1 task
Prod
1 job, 1 task
Clone
Clone stage

Figure 13.9 – Dev and Prod stages

Then, select the **Prod** stage, and you will see **Pre-deployment conditions**, as shown here:

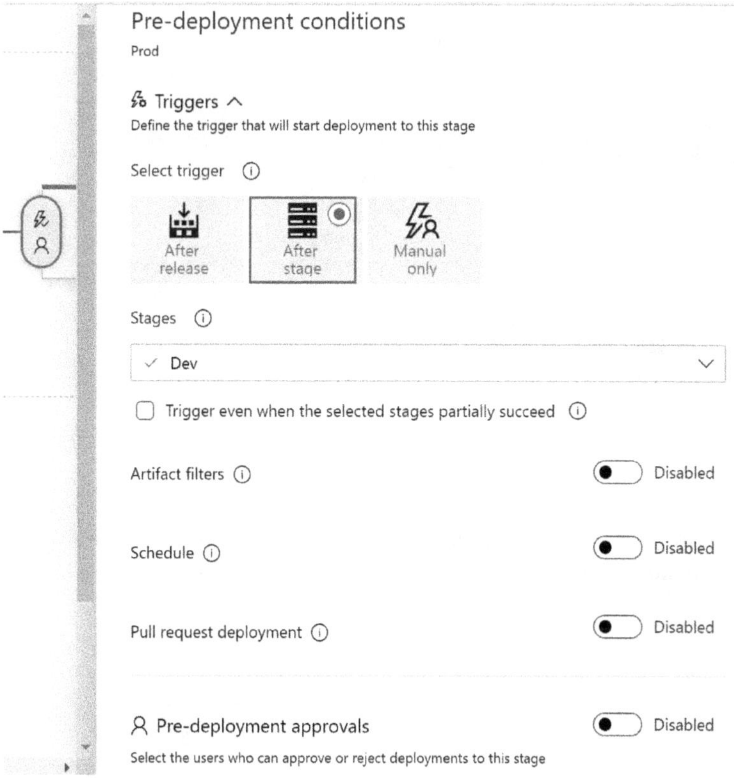
Pre-deployment conditions
Prod
Triggers
Define the trigger that will start deployment to this stage
Select trigger
After release
After stage
Manual only
Stages
Dev
Trigger even when the selected stages partially succeed
Artifact filters
Disabled
Schedule
Disabled
Pull request deployment
Disabled
Pre-deployment approvals
Disabled
Select the users who can approve or reject deployments to this stage

Figure 13.10 – Pre-deployment conditions

Stage triggers allow you to establish specific conditions that trigger a deployment to a particular stage.

You can set up the trigger for your stage by selecting the **Select trigger** option. This will determine how the deployment to the stage is initiated automatically. You can use the **Stages** dropdown to trigger a release following a successful deployment to the selected stage. If you only want manual triggering, select **Manual only**.

In this example, we will select the **After stage** trigger and select the **Dev** stage. Then, we will create a new release, as shown in the following figure:

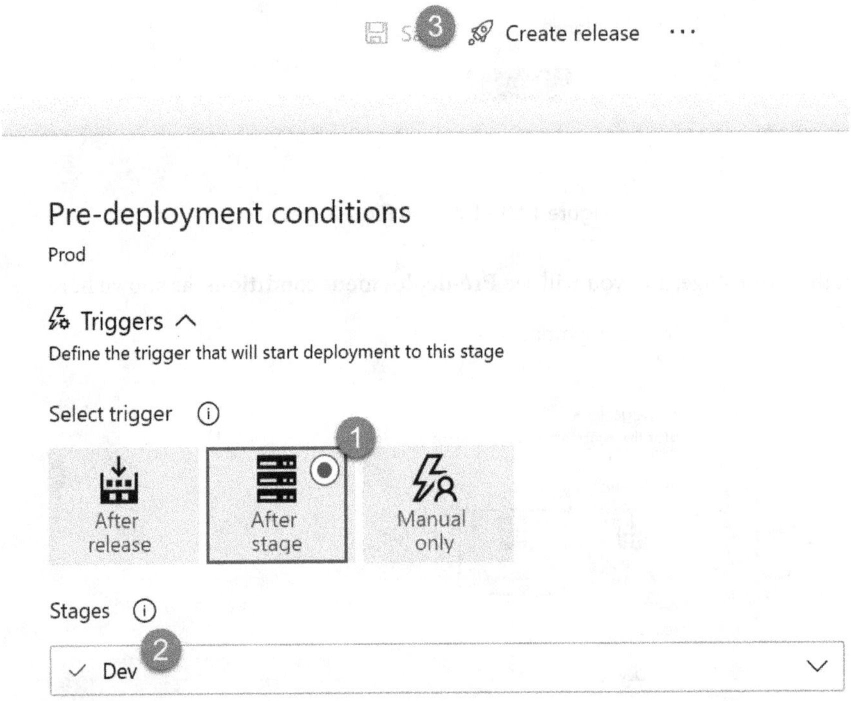
3
Create release
Pre-deployment conditions
Prod
Triggers
Define the trigger that will start deployment to this stage
Select trigger
1
After release
After stage
Manual only
Stages
2
Dev

Figure 13.11 – Pre-deployment conditions – Triggers

When you create a new release, you can define additional configurations such as changing the trigger from automated to manual, choose specific versions of the artifacts, and provide a release description, as shown here:

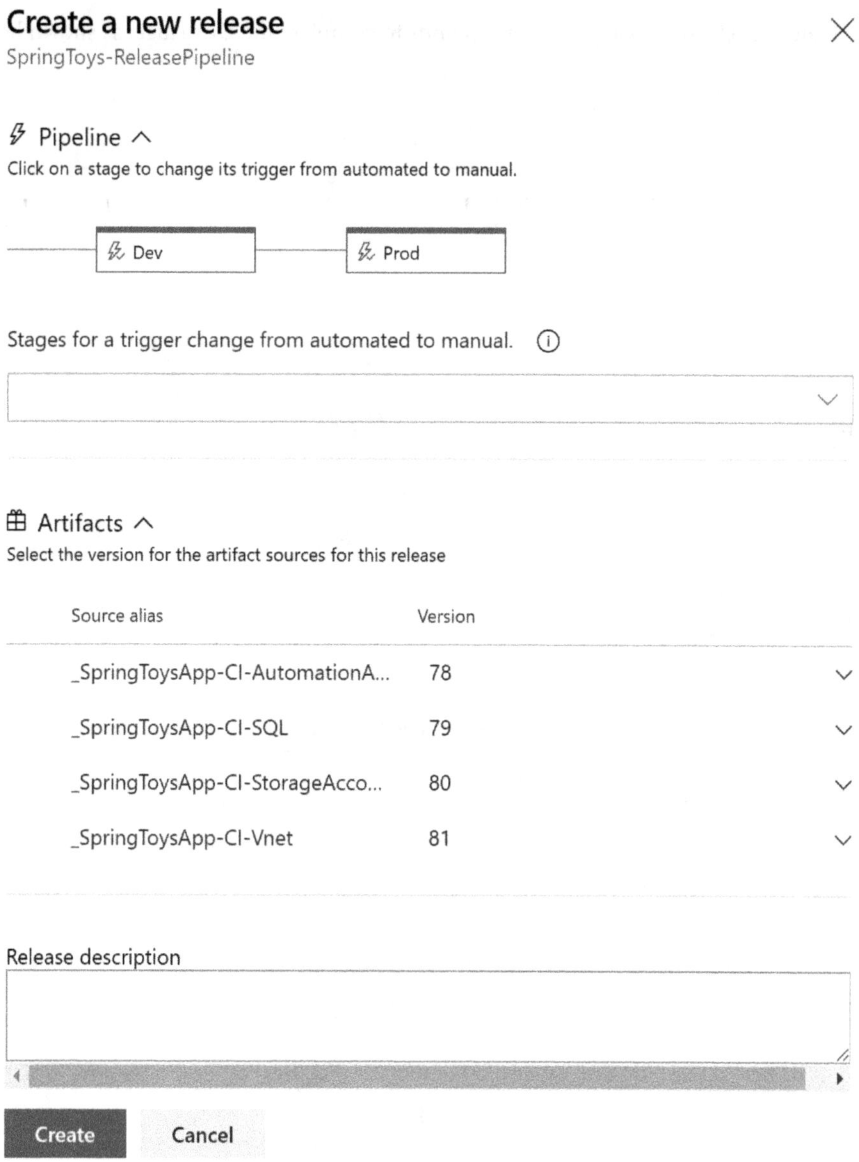
Create a new release
SpringToys-ReleasePipeline
Pipeline
Click on a stage to change its trigger from automated to manual.
Dev
Prod
Stages for a trigger change from automated to manual.
Artifacts
Select the version for the artifact sources for this release
Source alias
Version
_SpringToysApp-CI-AutomationA...
78
_SpringToysApp-CI-SQL
79
_SpringToysApp-CI-StorageAcco...
80
_SpringToysApp-CI-Vnet
81
Release description
Create
Cancel

Figure 13.12 – New release

Once you create the release, it will take a few seconds to complete the **Dev** stage, as shown here:

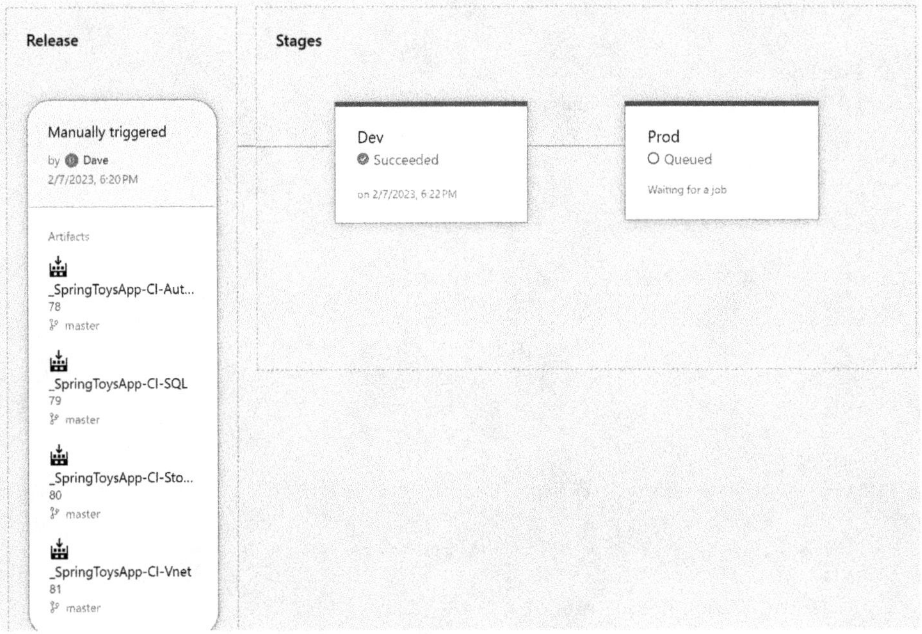
Release
Stages
Manually triggered
by Dave
2/7/2023, 6:20 PM
Artifacts
_SpringToysApp-CI-Aut...
78
master
_SpringToysApp-CI-SQL
79
master
_SpringToysApp-CI-Sto...
80
master
_SpringToysApp-CI-Vnet
81
master
Dev
Succeeded
on 2/7/2023, 6:22 PM
Prod
Queued
Waiting for a job

Figure 13.13 – Dev stage succeeded

Then, the **Prod** stage will automatically start:

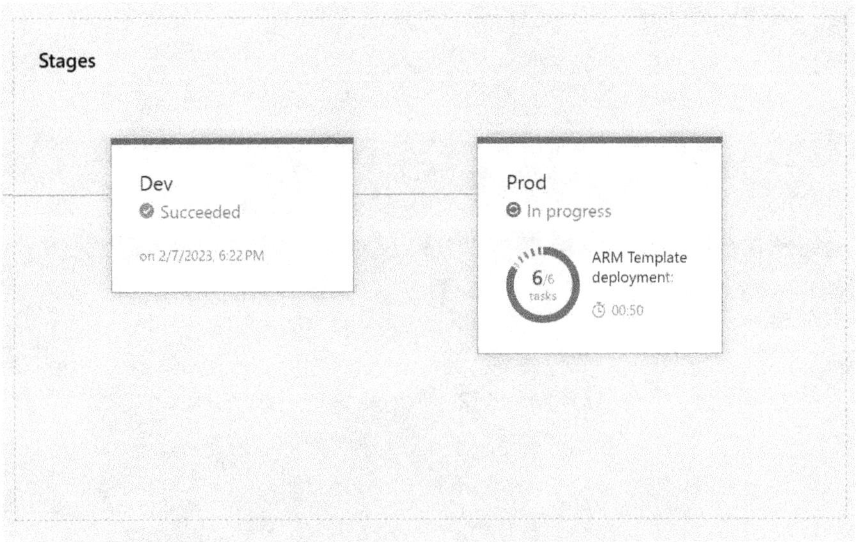
Stages
Dev
Succeeded
on 2/7/2023, 6:22 PM
Prod
In progress
6/6
tasks
ARM Template deployment:
00:50

Figure 13.14 – Prod stage in progress

After a minute, you should see that the **Prod** stage has been completed:

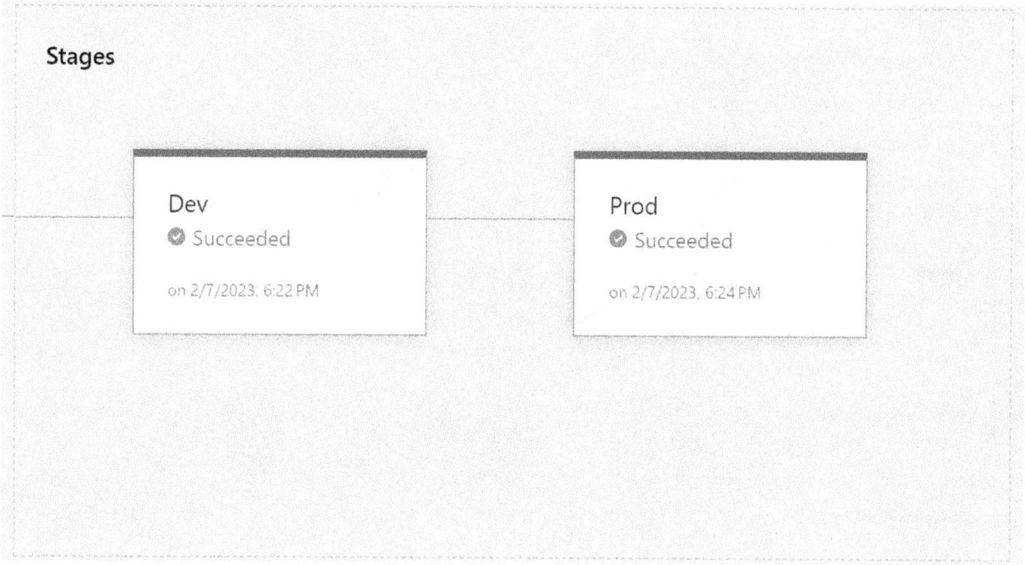
Stages
Dev
Succeeded
on 2/7/2023, 6:22 PM
Prod
Succeeded
on 2/7/2023, 6:24 PM

Figure 13.15 – Prod stage succeeded

Next, we will make a change in the Azure repo and update the API version of the virtual network again, as shown here:

```
resource Vnet 'Microsoft.Network/virtualNetworks@2022-07-01' = {
  name: VnetName
  location: resourceGroup().location
```

Figure 13.16 – Updating the Bicep file

Then, commit the change and go to the **Release** section. You should see the new release in progress, as shown here:

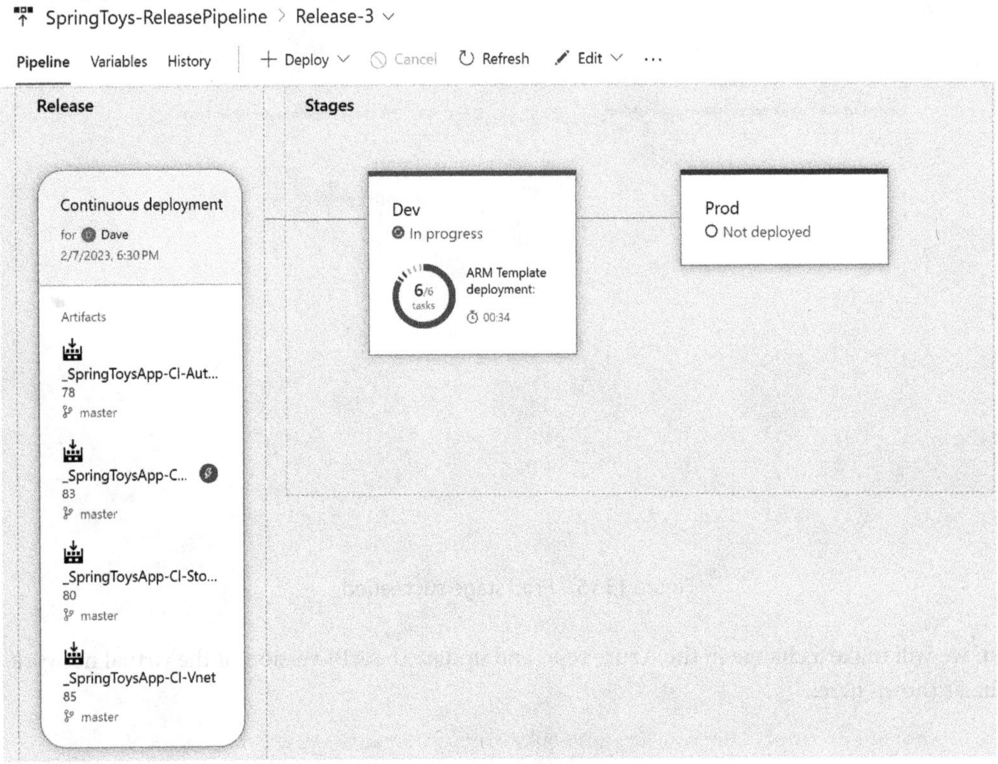
SpringToys-ReleasePipeline > Release-3
Pipeline
Variables
History
+ Deploy
Cancel
Refresh
Edit
Release
Stages
Continuous deployment
for Dave
2/7/2023, 6:30 PM
Artifacts
_SpringToysApp-CI-Aut...
78
master
_SpringToysApp-C...
83
master
_SpringToysApp-CI-Sto...
80
master
_SpringToysApp-CI-Vnet
85
master
Dev
In progress
6/6
tasks
ARM Template deployment:
00:34
Prod
Not deployed

Figure 13.17 – New release pipeline in progress

After a minute, you will see both stages in the **Succeeded** state:

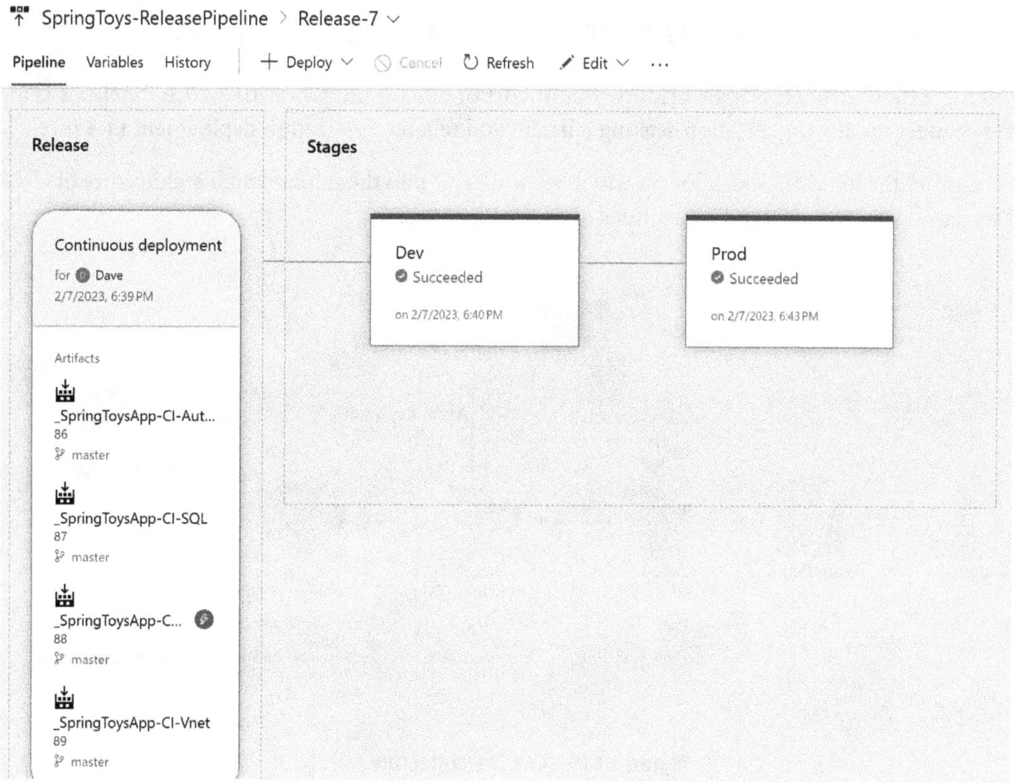
SpringToys-ReleasePipeline > Release-7
Pipeline Variables History + Deploy Cancel Refresh Edit ...
Release
Stages
Continuous deployment
for Dave
2/7/2023, 6:39 PM
Artifacts
_SpringToysApp-CI-Aut...
86
master
_SpringToysApp-CI-SQL
87
master
_SpringToysApp-C...
88
master
_SpringToysApp-CI-Vnet
89
master
Dev
Succeeded
on 2/7/2023, 6:40 PM
Prod
Succeeded
on 2/7/2023, 6:43 PM

Figure 13.18 – Stages – succeeded state

As you can see, one of the key benefits of using Azure Pipelines is ensuring that code changes are validated and integrated regularly, reducing the risk of defects, and enabling faster feedback to developers.

Azure Pipelines provides a flexible and configurable release pipeline, allowing teams to deploy to multiple testing or staging environments and automate infrastructure creation, testing, and deployment.

Note that the classic editor approach to configuring build completion triggers in the UI was previously used for YAML pipelines. However, it is no longer recommended. Instead, it is now recommended to specify pipeline triggers directly within the YAML file.

In the next section, we will look at a CI/CD baseline architecture and leverage YAML-based pipelines.

CI/CD baseline architecture using Azure Pipelines

Adopting a high-level DevOps workflow that incorporates CI and CD with Azure Pipelines is a game-changer for any organization seeking efficient and reliable application deployment in Azure.

Let's consider the following workflow in which we will delve into the fundamental architecture of CI/CD by harnessing the powerful capabilities of Azure Pipelines.

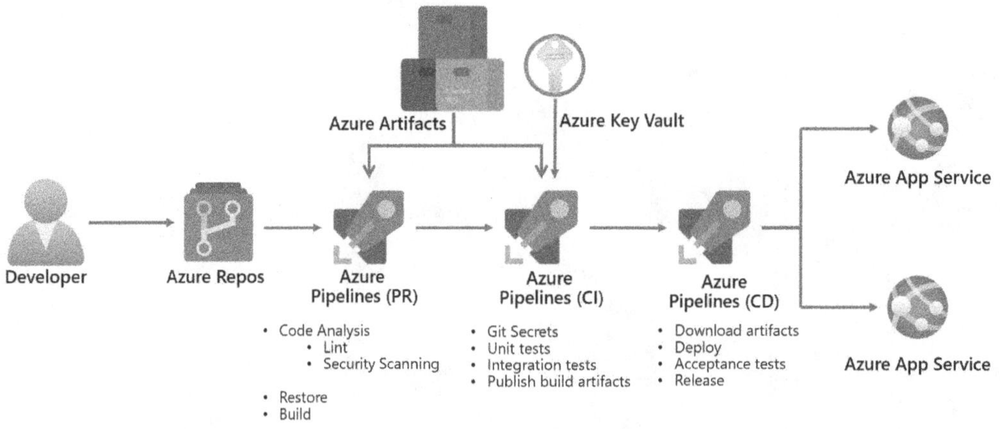

Azure Artifacts
Azure Key Vault
Developer
Azure Repos
Azure Pipelines (PR)
Azure Pipelines (CI)
Azure Pipelines (CD)
Azure App Service
Azure App Service
• Code Analysis
• Lint
• Security Scanning
• Restore
• Build
• Git Secrets
• Unit tests
• Integration tests
• Publish build artifacts
• Download artifacts
• Deploy
• Acceptance tests
• Release

Figure 13.19 – CI/CD architecture

First, let's understand the components of the preceding architecture:

- **Azure Repos**: Provides a definitive suite of version control tools essential for efficient code management. Embracing version control from the outset, regardless of project scale, is imperative for optimal software development. Version control systems are indispensable software tools that precisely monitor code modifications throughout the development process.
- **Azure Pipelines**: Empowers building, testing, packaging, and releasing application and infrastructure code. It comprises three pipelines:
  - **Pull request build validation pipelines**: Validate code pre-merge by performing linting, building, and unit testing.
  - **CI pipelines**: Execute post-merge validation, similar to PR Pipelines, with the addition of integration testing and artifact publishing upon success. This includes code analysis, unit tests, integration tests, and publish build artifacts.
  - **CD pipelines**: Handle artifact deployment, acceptance testing, and production release.

- **Azure Artifacts Feeds**: Allow efficient management and sharing of software packages (e.g., Maven, npm, NuGet). Facilitate versioning, promoting, and retiring packages to ensure teams use the latest, secure versions.
- **Key Vault**: Securely manages sensitive data for the solution, including secrets, encryption keys, and certificates. Used here to store application secrets, accessed through the pipeline with Key Vault tasks or linked secrets.

Now, let's analyze the dataflow in the preceding architecture:

- **PR pipeline**: A **pull request** (**PR**) to Azure Repos Git triggers a PR pipeline. This pipeline conducts quality checks, including building the code with dependencies, analyzing the code using tools such as static code analysis, linting, and security scanning, and performing unit tests. If any checks fail, the pipeline ends, and the developer must address the issues. If all checks pass, a PR review is required. If the review fails, the pipeline ends, and the developer must make the necessary changes. Successful checks and reviews lead to a successful PR merge.
- **CI pipeline**: A merge to Azure Repos Git triggers a CI pipeline. It includes the same checks as the PR pipeline, with the addition of integration tests. These tests do not require deploying the solution since build artifacts are not yet created. If integration tests need secrets, the pipeline fetches them from Azure Key Vault. If any checks fail, the pipeline ends, and the developer must make the required changes. A successful run results in the creation and publishing of build artifacts.
- **CD pipeline trigger**: The publishing of artifacts triggers the CD pipeline.
- **CD release to staging**: The CD pipeline downloads the build artifacts from the CI pipeline and deploys the solution to a staging environment. Acceptance tests are then run to validate the deployment. If any acceptance test fails, the pipeline ends and the developer must address the issues. Successful tests may include a manual validation step to ensure a person or group validates the deployment before continuing the pipeline.
- **CD release to production**: If manual intervention occurs or is not implemented, the pipeline releases the solution to production. Smoke tests are performed in production to verify the release's functionality. If a manual intervention results in a cancellaion, the release fails, or smoke tests fail, the pipeline initiates a rollback, and the developer must make the necessary changes.

Additionally, organizations could leverage Azure Monitor to collect observability data, such as logs and metrics, for analysis of health, performance, and usage data. On top of that, Application Insights can be utilized as it collects application-specific monitoring data such as traces, and Azure Log Analytics stores all the collected data.

Consider using an Azure DevOps multistage pipeline to break down your CI/CD process into distinct stages representing various development cycle phases. This approach enhances deployment visibility and facilitates seamless integration of approvals and checks.

In the upcoming section, we will show you precisely how to build a YAML pipeline with multiple stages, effectively dividing your CI/CD process.

Building a multistage YAML pipeline

Consider a scenario in which SpringToys wants to deploy its core application leveraging build and release pipelines using YAML. This enables users to access identical pipeline features as those utilizing the visual designer but with the added benefit of a markup file that can be managed like any other source file.

To add YAML build definitions to a project, include the corresponding files in the repository's root. Azure DevOps further offers default templates for popular project types and a YAML designer, streamlining the task of defining build and release tasks. Organizations can use these robust capabilities to expedite and optimize their development process.

The following practical demonstration will review how to configure a multistage CI/CD pipeline as code with YAML in Azure DevOps.

We will follow these steps:

- Configure a new project in Azure DevOps
- Configure a CI/CD pipeline with YAML

Let's start with the configuration of a new project in Azure DevOps.

Configuring a new project in Azure DevOps

To configure a new project, use the following steps:

1. First, in Azure DevOps, we will create a new project named `SpringToysEShop`, as shown here:

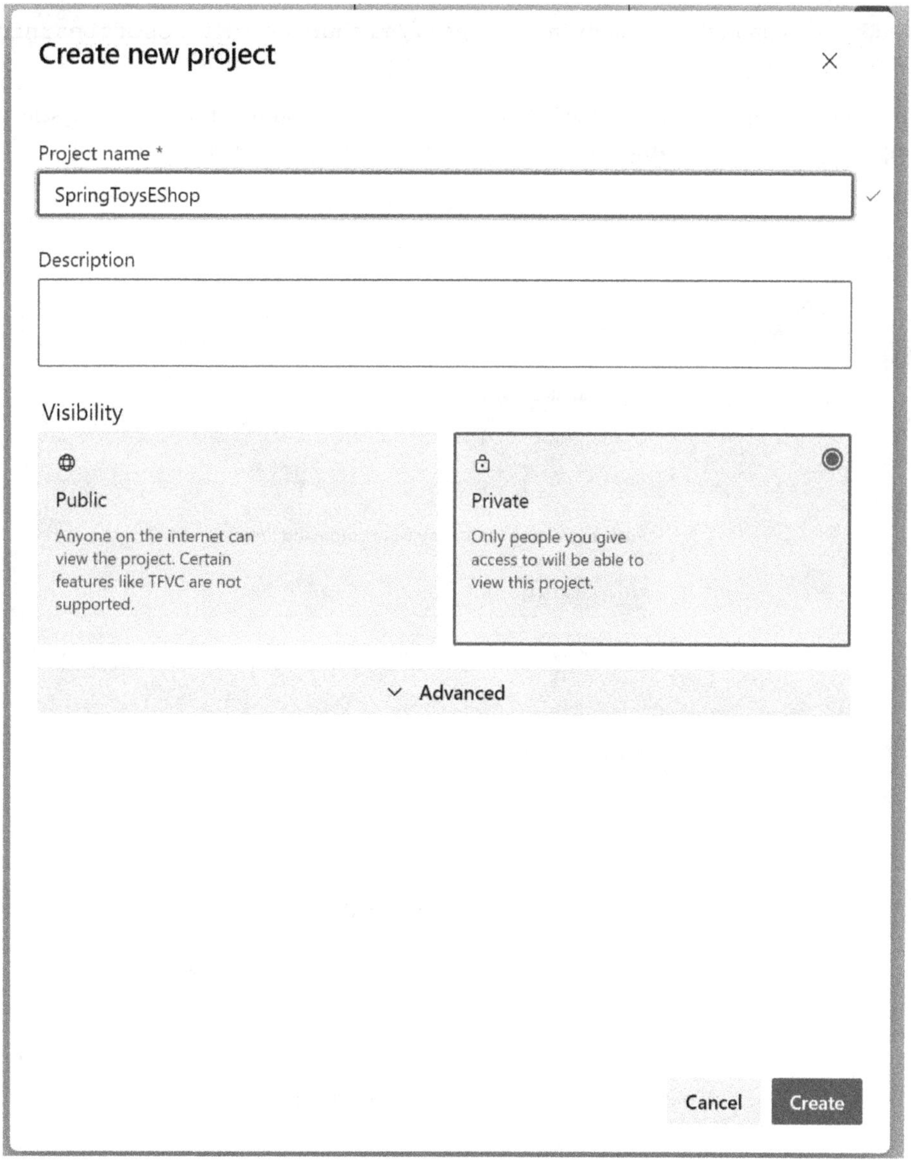
Create new project
Project name *
SpringToysEShop
Description
Visibility
Public
Anyone on the internet can view the project. Certain features like TFVC are not supported.
Private
Only people you give access to will be able to view this project.
Advanced
Cancel
Create

Figure 13.20 – Create new project – Azure DevOps

2. Next, we will import the repository from `https://github.com/MicrosoftLearning/eShopOnWeb.git`.

   To import the repository, go to the left menu in the project and select **Repos**. Then, select the **Import** option from the **Import a repository** section, as shown here:

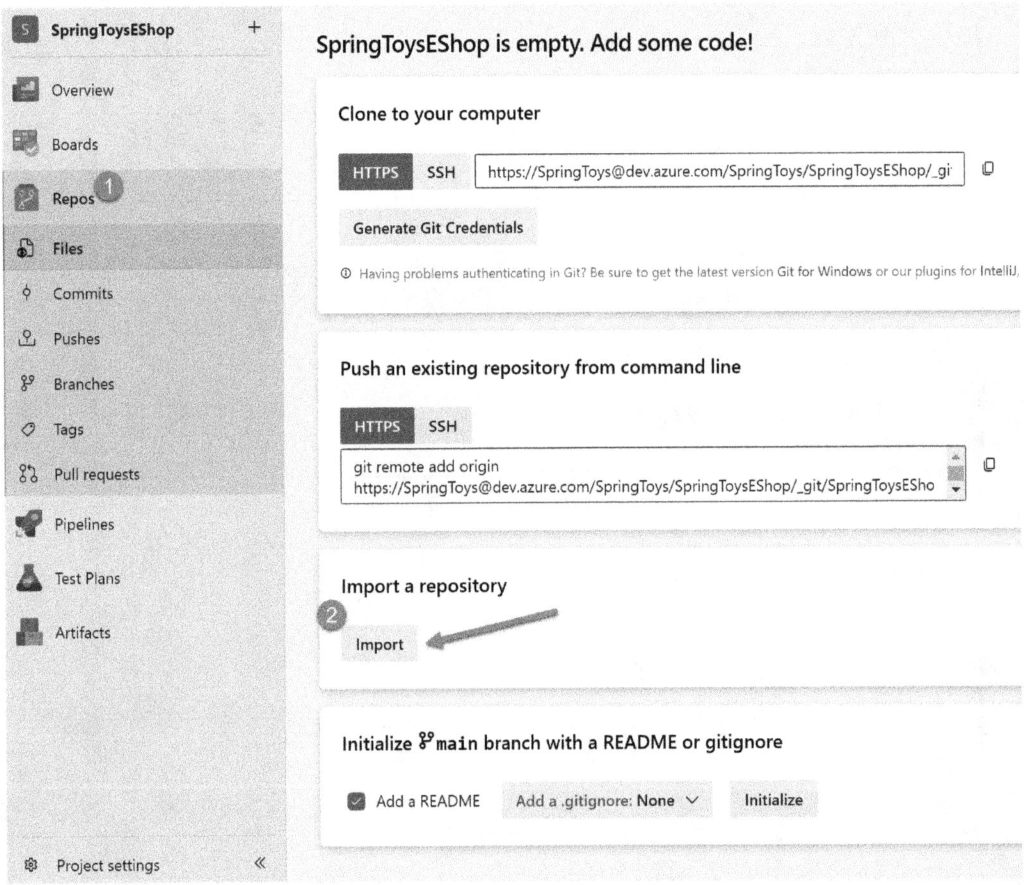
SpringToysEShop
Overview
Boards
Repos
Files
Commits
Pushes
Branches
Tags
Pull requests
Pipelines
Test Plans
Artifacts
Project settings
SpringToysEShop is empty. Add some code!
Clone to your computer
HTTPS
SSH
https://SpringToys@dev.azure.com/SpringToys/SpringToysEShop/_gi
Generate Git Credentials
Having problems authenticating in Git? Be sure to get the latest version Git for Windows or our plugins for IntelliJ,
Push an existing repository from command line
HTTPS
SSH
git remote add origin
https://SpringToys@dev.azure.com/SpringToys/SpringToysEShop/_git/SpringToysESho
Import a repository
Import
Initialize main branch with a README or gitignore
Add a README
Add a .gitignore: None
Initialize

Figure 13.21 – Import a repository – Azure Repos

3. Next, provide the URL of the repository and click **Import**:

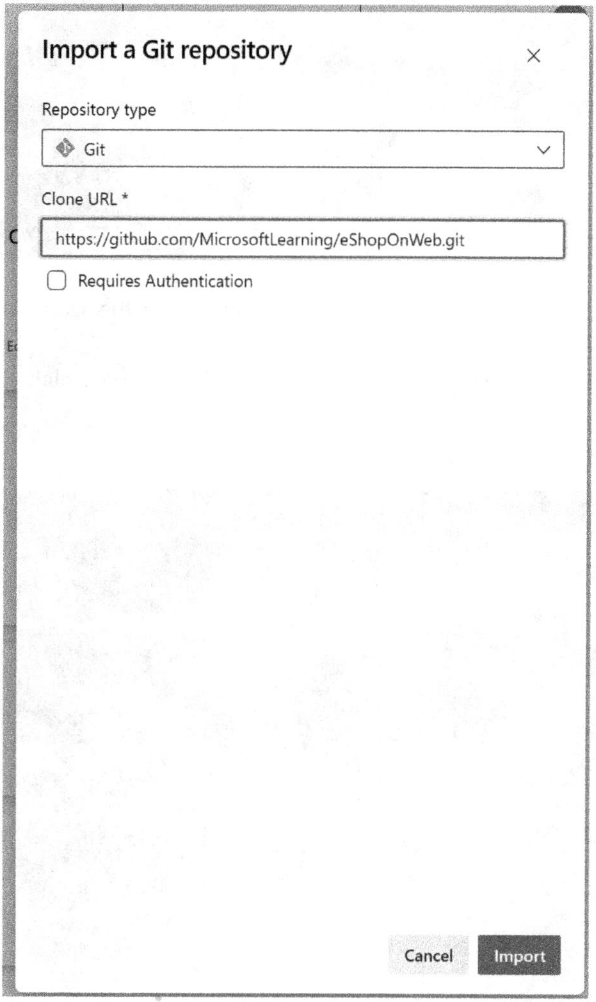
Import a Git repository
Repository type
Git
Clone URL *
https://github.com/MicrosoftLearning/eShopOnWeb.git
Requires Authentication
Cancel
Import

Figure 13.22 – Import a Git repository

After a few seconds, you should see that the import process was successful.

4. Now, we will create an App Service. First, go to `https://shell.azure.com` and ensure you select the **Bash** option. We will create a new resource group using the following command:

```
az group create --name 'SpringToys-RG' --location 'eastus'
```

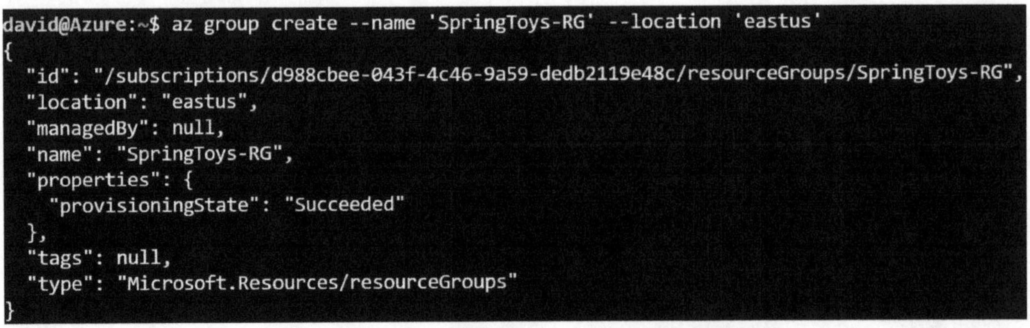
david@Azure:~$ az group create --name 'SpringToys-RG' --location 'eastus'
{
"id": "/subscriptions/d988cbee-043f-4c46-9a59-dedb2119e48c/resourceGroups/SpringToys-RG",
"location": "eastus",
"managedBy": null,
"name": "SpringToys-RG",
"properties": {
"provisioningState": "Succeeded"
},
"tags": null,
"type": "Microsoft.Resources/resourceGroups"
}

Figure 13.23 – Creating a new resource group using the Azure CLI

5. Next, we will use the following command to create an App Service plan:

```
az appservice plan create --resource-group 'SpringToys-RG'
--name SpringToysAppPlan --sku B3
```

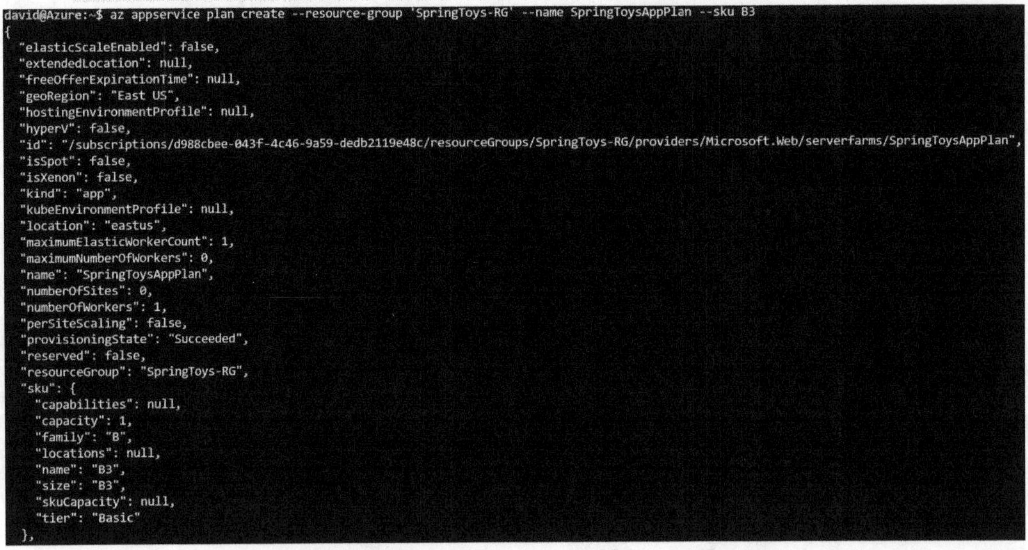
david@Azure:~$ az appservice plan create --resource-group 'SpringToys-RG' --name SpringToysAppPlan --sku B3
{
"elasticScaleEnabled": false,
"extendedLocation": null,
"freeOfferExpirationTime": null,
"geoRegion": "East US",
"hostingEnvironmentProfile": null,
"hyperV": false,
"id": "/subscriptions/d988cbee-043f-4c46-9a59-dedb2119e48c/resourceGroups/SpringToys-RG/providers/Microsoft.Web/serverfarms/SpringToysAppPlan",
"isSpot": false,
"isXenon": false,
"kind": "app",
"kubeEnvironmentProfile": null,
"location": "eastus",
"maximumElasticWorkerCount": 1,
"maximumNumberOfWorkers": 0,
"name": "SpringToysAppPlan",
"numberOfSites": 0,
"numberOfWorkers": 1,
"perSiteScaling": false,
"provisioningState": "Succeeded",
"reserved": false,
"resourceGroup": "SpringToys-RG",
"sku": {
"capabilities": null,
"capacity": 1,
"family": "B",
"locations": null,
"name": "B3",
"size": "B3",
"skuCapacity": null,
"tier": "Basic"
},

Figure 13.24 – Creating an App Service plan using the Azure CLI

6. Once the App Service plan has been created, we will create the App Service instance using the following command:

```
az webapp create --resource-group 'SpringToys-RG' --plan
'SpringToysAppPlan' --name 'SpringToysEShop'
```

david@Azure:~$ az webapp create --resource-group 'SpringToys-RG' --plan 'SpringToysAppPlan' --name 'SpringToysEShop'
{
"availabilityState": "Normal",
"clientAffinityEnabled": true,
"clientCertEnabled": false,
"clientCertExclusionPaths": null,
"clientCertMode": "Required",
"cloningInfo": null,
"containerSize": 0,
"customDomainVerificationId": "769D6483E5E13DB3893D240F8B34AF137FBD218655D0215A51AC6D8489E5EA2E",
"dailyMemoryTimeQuota": 0,
"defaultHostName": "springtoyseshop.azurewebsites.net",
"enabled": true,
"enabledHostNames": [
"springtoyseshop.azurewebsites.net",
"springtoyseshop.scm.azurewebsites.net"
],
"extendedLocation": null,
"ftpPublishingUrl": "ftps://waws-prod-blu-369.ftp.azurewebsites.windows.net/site/wwwroot",
"hostNameSslStates": [
{
"certificateResourceId": null,
"hostType": "Standard",
"ipBasedSslResult": null,
"ipBasedSslState": "NotConfigured",
"name": "springtoyseshop.azurewebsites.net",
"sslState": "Disabled",
"thumbprint": null,
"toUpdate": null,
"toUpdateIpBasedSsl": null,
"virtualIp": null
},

Figure 13.25 – Creating an Azure App Service using the Azure CLI

We have created our Azure DevOps project, imported the repository, and configured the Azure resources needed. Next, we will configure CI/CD pipelines with YAML.

Configuring CI/CD pipelines with YAML

In this section, we will configure the CI/CD pipelines using YAML. The first step is to add a YAML build definition to our project.

1. In the left menu, go to **Pipelines** and create a new pipeline, as shown here:

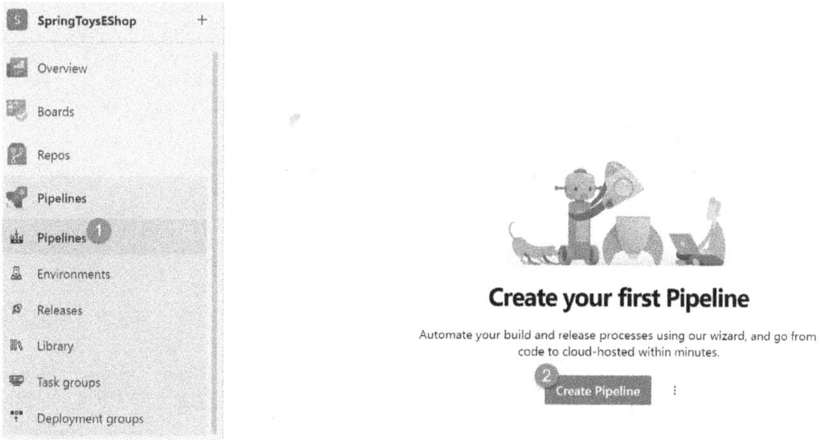
SpringToysEShop
Overview
Boards
Repos
Pipelines
Pipelines
1
Environments
Releases
Library
Task groups
Deployment groups
Create your first Pipeline
Automate your build and release processes using our wizard, and go from code to cloud-hosted within minutes.
2
Create Pipeline

Figure 13.26 – Creating a new pipeline

2. In the **Where is your code?** tab, select **Azure Repos Git**:

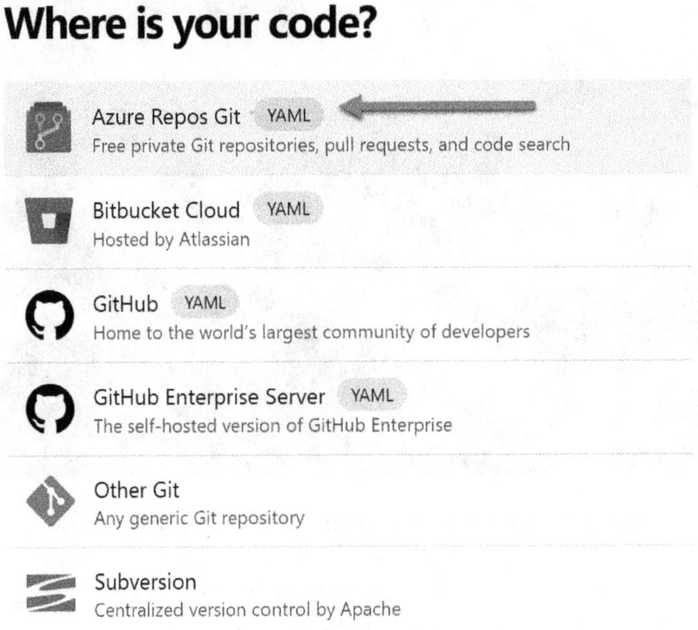

Where is your code?
Azure Repos Git YAML
Free private Git repositories, pull requests, and code search
Bitbucket Cloud YAML
Hosted by Atlassian
GitHub YAML
Home to the world's largest community of developers
GitHub Enterprise Server YAML
The self-hosted version of GitHub Enterprise
Other Git
Any generic Git repository
Subversion
Centralized version control by Apache

Figure 13.27 – Selecting the source code for the pipeline

3. Then, select the repository that was imported previously:

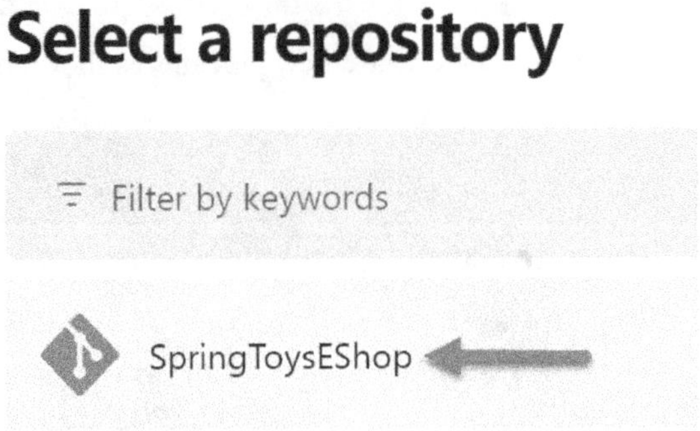

Select a repository
Filter by keywords
SpringToysEShop

Figure 13.28 – Select a repository

4. Next, in the **Configure your pipeline** tab, select the option named **Existing Azure Pipelines YAML file**:

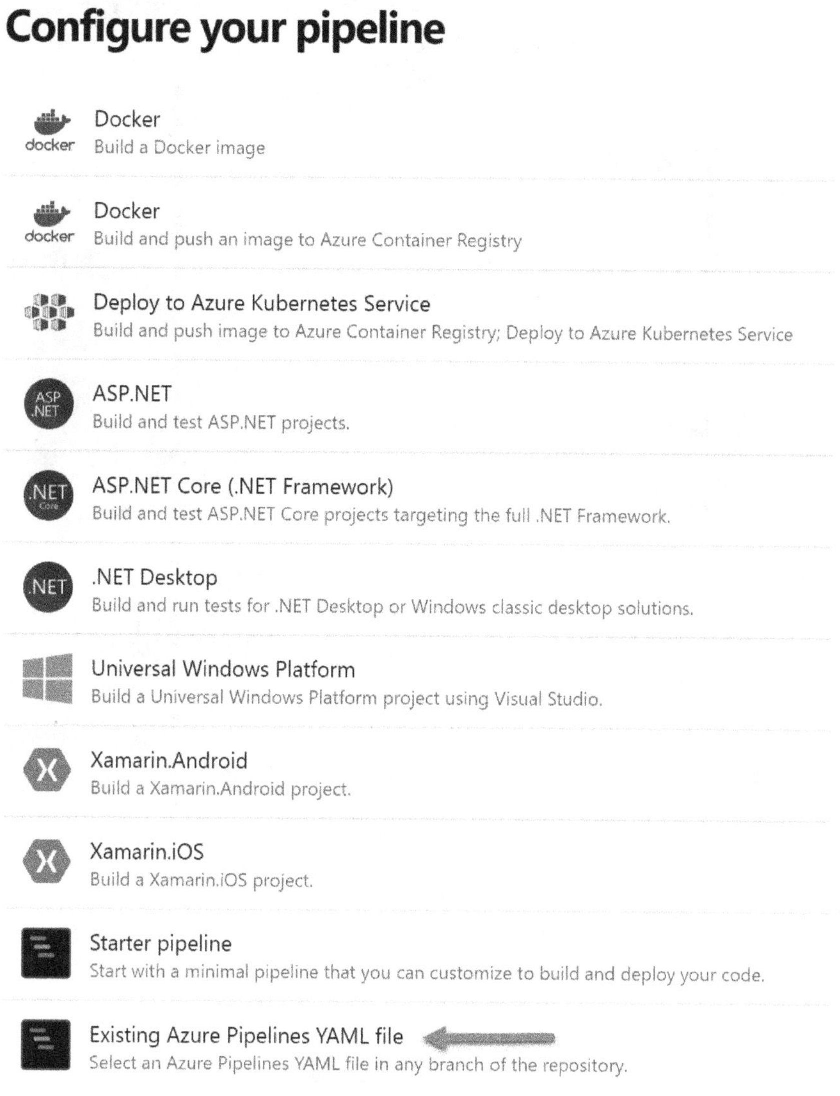
Configure your pipeline
docker
Docker
Build a Docker image
docker
Docker
Build and push an image to Azure Container Registry
Deploy to Azure Kubernetes Service
Build and push image to Azure Container Registry; Deploy to Azure Kubernetes Service
ASP .NET
ASP.NET
Build and test ASP.NET projects.
.NET Core
ASP.NET Core (.NET Framework)
Build and test ASP.NET Core projects targeting the full .NET Framework.
.NET
.NET Desktop
Build and run tests for .NET Desktop or Windows classic desktop solutions.
Universal Windows Platform
Build a Universal Windows Platform project using Visual Studio.
X
Xamarin.Android
Build a Xamarin.Android project.
X
Xamarin.iOS
Build a Xamarin.iOS project.
Starter pipeline
Start with a minimal pipeline that you can customize to build and deploy your code.
Existing Azure Pipelines YAML file
Select an Azure Pipelines YAML file in any branch of the repository.

Figure 13.29 – Configure your pipeline

5. Now, under **Select an existing YAML file**, provide the following parameters – for **Branch**, choose **main**, and for **Path**, provide the location of the YAML file, then select **Continue**:

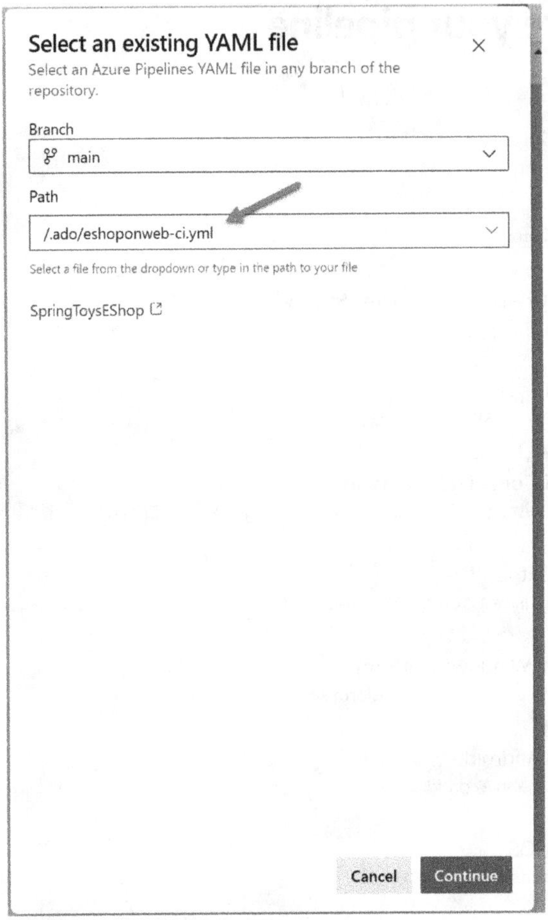
Select an existing YAML file
Select an Azure Pipelines YAML file in any branch of the repository.
Branch
main
Path
/.ado/eshoponweb-ci.yml
Select a file from the dropdown or type in the path to your file
SpringToysEShop
Cancel
Continue

Figure 13.30 – Select an existing YAML file

The following code shows the YAML definition:

```
#NAME THE PIPELINE SAME AS FILE (WITHOUT ".yml")
# trigger:
# - main

resources:
  repositories:
    - repository: self
      trigger: none
```

```
stages:
- stage: Build
  displayName: Build .Net Core Solution
  jobs:
  - job: Build
    pool:
      vmImage: ubuntu-latest
    steps:
    - task: DotNetCoreCLI@2
      displayName: Restore
      inputs:
        command: 'restore'
        projects: '**/*.sln'
        feedsToUse: 'select'

    - task: DotNetCoreCLI@2
      displayName: Build
      inputs:
        command: 'build'
        projects: '**/*.sln'

    - task: DotNetCoreCLI@2
      displayName: Test
      inputs:
        command: 'test'
        projects: 'tests/UnitTests/*.csproj'

    - task: DotNetCoreCLI@2
      displayName: Publish
      inputs:
        command: 'publish'
        publishWebProjects: true
        arguments: '-o $(Build.ArtifactStagingDirectory)'

    - task: PublishBuildArtifacts@1
      displayName: Publish Artifacts ADO - Website
      inputs:
        pathToPublish: '$(Build.ArtifactStagingDirectory)'
        artifactName: Website

    - task: PublishBuildArtifacts@1
      displayName: Publish Artifacts ADO - Bicep
      inputs:
```

```
        PathtoPublish: '$(Build.SourcesDirectory)/.azure/bicep/
webapp.bicep'
        ArtifactName: 'Bicep'
        publishLocation: 'Container'
```

The preceding YAML definition is designed to build and deploy a .NET Core solution. It follows a clear and concise structure, ensuring effective execution:

- The pipeline is named the same as the YAML file
- The pipeline is not triggered automatically based on repository events as it has `trigger: none` specified

The pipeline has two stages, starting with the `Build` stage:

- In the `Build` stage, there is a single job named `Build` that runs on an Ubuntu agent (`vmImage: ubuntu-latest`)
- The `Build` job begins with a `Restore` task using `DotNetCoreCLI` to restore dependencies specified in the `.sln` file
- The next task is the `Build` task, which uses `DotNetCoreCLI` to build the .NET Core solution
- Following that, a `Test` task using `DotNetCoreCLI` to run unit tests is located in the `tests/UnitTests` directory
- The `Publish` task using `DotNetCoreCLI` is then executed to publish the built artifacts, specifically web projects, to the specified location (`$(Build.ArtifactStagingDirectory)`)

Two `PublishBuildArtifacts` tasks are used to publish the generated artifacts:

- The first task publishes the website artifacts located in the `$(Build.ArtifactStagingDirectory)" to the "Website` artifact.
- The second task publishes the `webapp.bicep` file, located in the `.azure/bicep` directory of the repository, to the `Bicep` artifact. The artifact is published within a container.

This pipeline follows the best practices for CI/CD pipelines as code in Azure DevOps, ensuring that the .NET Core solution is built, tested, and artifacts are published for further deployment or distribution.

6. Next, select the **Run** option to start the Build Pipeline process:

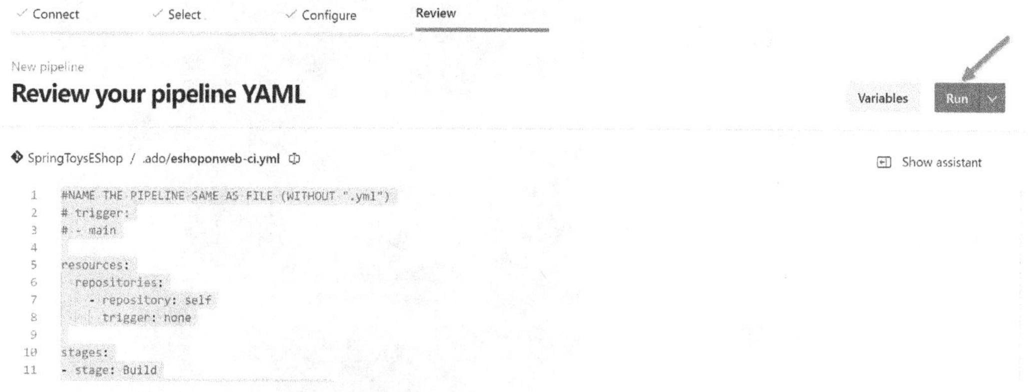
Connect
Select
Configure
Review
New pipeline
Review your pipeline YAML
Variables
Run
SpringToysEShop / .ado/eshoponweb-ci.yml
Show assistant
#NAME THE PIPELINE SAME AS FILE (WITHOUT ".yml")
trigger:
- main
resources:
repositories:
- repository: self
trigger: none
stages:
- stage: Build

Figure 13.31 – Review your pipeline YAML

You should see the job as shown here:

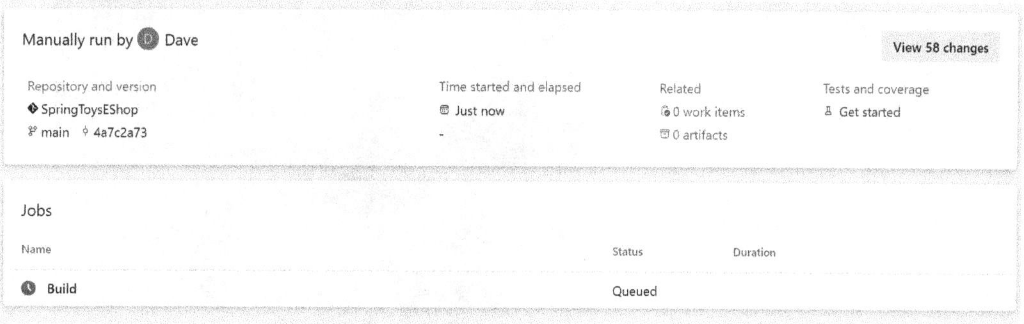
Manually run by Dave
View 58 changes
Repository and version
SpringToysEShop
main 4a7c2a73
Time started and elapsed
Just now
Related
0 work items
0 artifacts
Tests and coverage
Get started
Jobs
Name
Status
Duration
Build
Queued

Figure 13.32 – Job summary

The job will build the .Net Core solution. The build process might take a few minutes to complete.

← Jobs in run #20230723.1
SpringToysEShop
Build .Net Core Solution
Build 3m 41s
Initialize job 1s
Checkout SpringToysES... 1s
Restore 1m 57s
Build 20s
Test 15s
Publish 56s
Publish Artifacts ADO - ... 8s
Publish Artifacts ADO ... <1s
Post-job: Checkout Sp... <1s
Finalize Job <1s
Report build status <1s
Finalize Job
1 Starting: Finalize Job
2 Cleaning up task key
3 Start cleaning up orphan processes.
4 Terminate orphan process: pid (1760) (dotnet)
5 Terminate orphan process: pid (1884) (dotnet)
6 Terminate orphan process: pid (1909) (dotnet)
7 Finishing: Finalize Job

Figure 13.33 – Jobs in run

Once the job is complete, we can add CD to our pipeline.

Enabling CD to the YAML pipeline

We will now implement CD into the YAML-based definition of the pipeline that was previously created in the preceding task. This will ensure the pipeline automates the seamless and efficient delivery of changes to the deployment environment, providing a streamlined and reliable process for your development and deployment workflow.

Follow these steps:

1. On the **Pipeline run** pane, select the *options* menu and select the **Edit pipeline** option, as shown here:

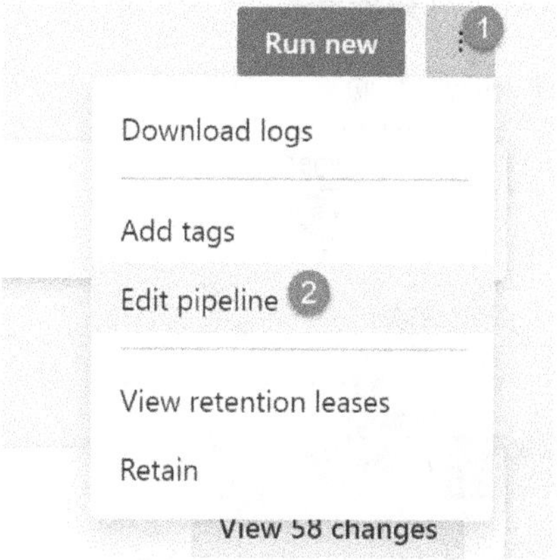
Run new
1
Download logs
Add tags
Edit pipeline
2
View retention leases
Retain
View 58 changes

Figure 13.34 – Options to edit the pipeline

2. Next, we will edit the YAML file. Go to the end of the pipeline definition, and we will add a few lines to define the `Release` stage. We will add the following:

```
- stage: Deploy
  displayName: Deploy to an Azure Web App
  jobs:
  - job: Deploy
    pool:
      vmImage: 'windows-2019'
    steps:
```

3. Now, in the right section of the pipeline, you will see a list of tasks. Look for the **Azure App Service deploy** task and select it.

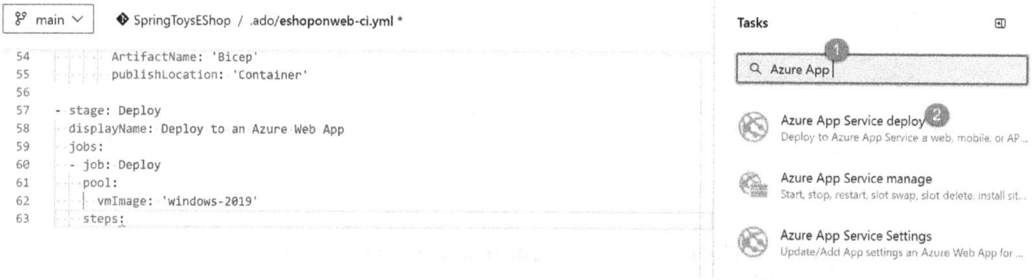
main
SpringToysEShop / .ado/eshoponweb-ci.yml *
ArtifactName: 'Bicep'
publishLocation: 'Container'
- stage: Deploy
displayName: Deploy to an Azure Web App
jobs:
- job: Deploy
pool:
vmImage: 'windows-2019'
steps:
Tasks
Azure App
1
Azure App Service deploy
2
Deploy to Azure App Service a web, mobile, or AP...
Azure App Service manage
Start, stop, restart, slot swap, slot delete, install sit...
Azure App Service Settings
Update/Add App settings an Azure Web App for ...

Figure 13.35 – Adding a task to the pipeline

4. Then, in **Connection type**, select **Azure Resource Manager**, provide the **Azure subscription** details, and then click **Authorize**:

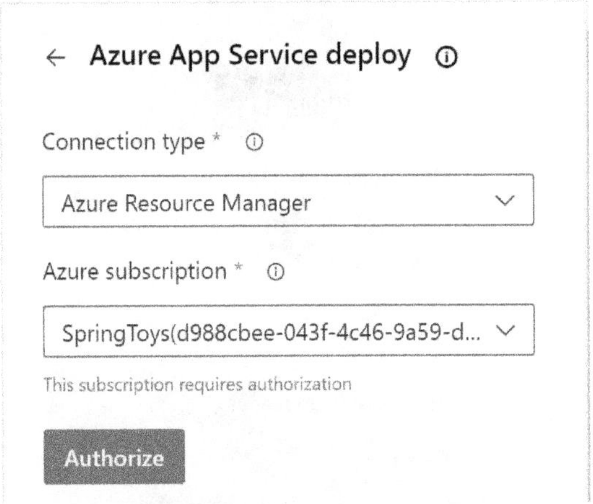
Azure App Service deploy
Connection type *
Azure Resource Manager
Azure subscription *
SpringToys(d988cbee-043f-4c46-9a59-d...
This subscription requires authorization
Authorize

Figure 13.36 – Configuring Azure App Service deploy

Utilize the identical user account used during the deployment of Azure resources for authentication.

5. Next, in **App Service name**, select the App Service previously deployed:

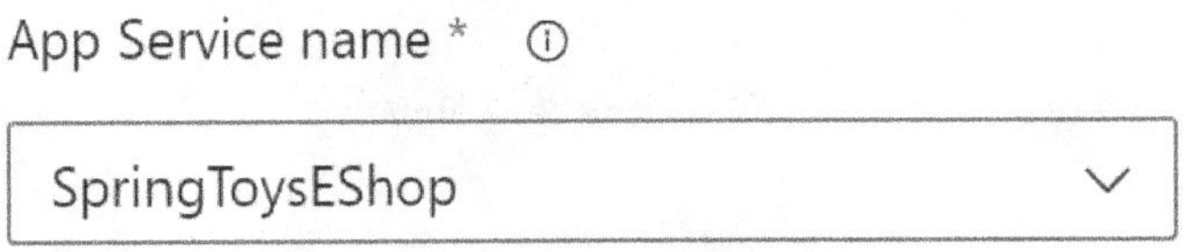
App Service name *
SpringToysEShop

Figure 13.37 – App Service name

6. Then, in the **Package or folder** option, provide the following value:

```
$(Build.ArtifactStagingDirectory)/**/Web.zip
```

Package or folder *
$(Build.ArtifactStagingDirectory)/**/Web.zip

Figure 13.387 – Package or folder option

7. Then, click **Add**:

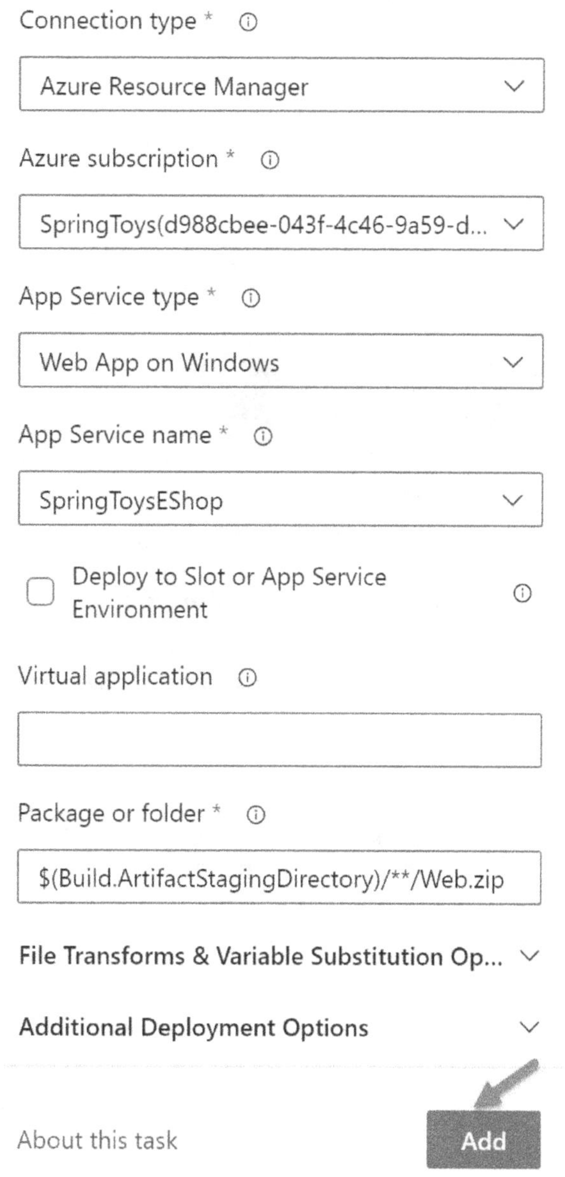
Connection type *
Azure Resource Manager
Azure subscription *
SpringToys(d988cbee-043f-4c46-9a59-d...
App Service type *
Web App on Windows
App Service name *
SpringToysEShop
Deploy to Slot or App Service Environment
Virtual application
Package or folder *
$(Build.ArtifactStagingDirectory)/**/Web.zip
File Transforms & Variable Substitution Op...
Additional Deployment Options
About this task
Add

Figure 13.39 – Task configuration

Note that the task has been added to the YAML pipeline. Ensure you add the corresponding indentation. The recently added code should look as follows:

```
- stage: Deploy
  displayName: Deploy to an Azure Web App
  jobs:
  - job: Deploy
    pool:
      vmImage: 'windows-2019'
    steps:
    Settings
    - task: AzureRmWebAppDeployment@4
      inputs:
        ConnectionType: 'AzureRM'
        azureSubscription: 'SpringToys(d988cbee-043f-4c46-9a59-dedb2119e48c)'
        appType: 'webApp'
        WebAppName: 'SpringToysEShop'
        packageForLinux: '$(Build.ArtifactStagingDirectory)/**/Web.zip'
```

Figure 13.40 – YAML pipeline

8. Next, we will add a new task called **Download build artifacts**. This task will be added before the **Azure App Service deploy** task.
9. In the **Tasks** pane, look for the **Download build artifacts** task and select it:

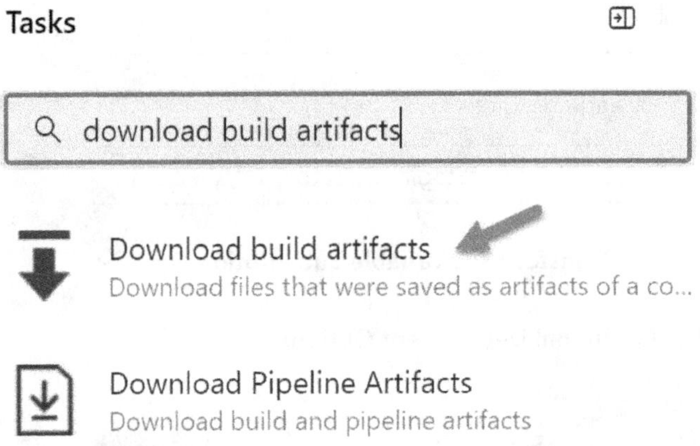
Tasks
download build artifacts
Download build artifacts
Download files that were saved as artifacts of a co...
Download Pipeline Artifacts
Download build and pipeline artifacts

Figure 13.41 – Adding the "Download build artifacts" pipeline task

10. Then, provide the configuration as shown in the following figure and click **Add**:

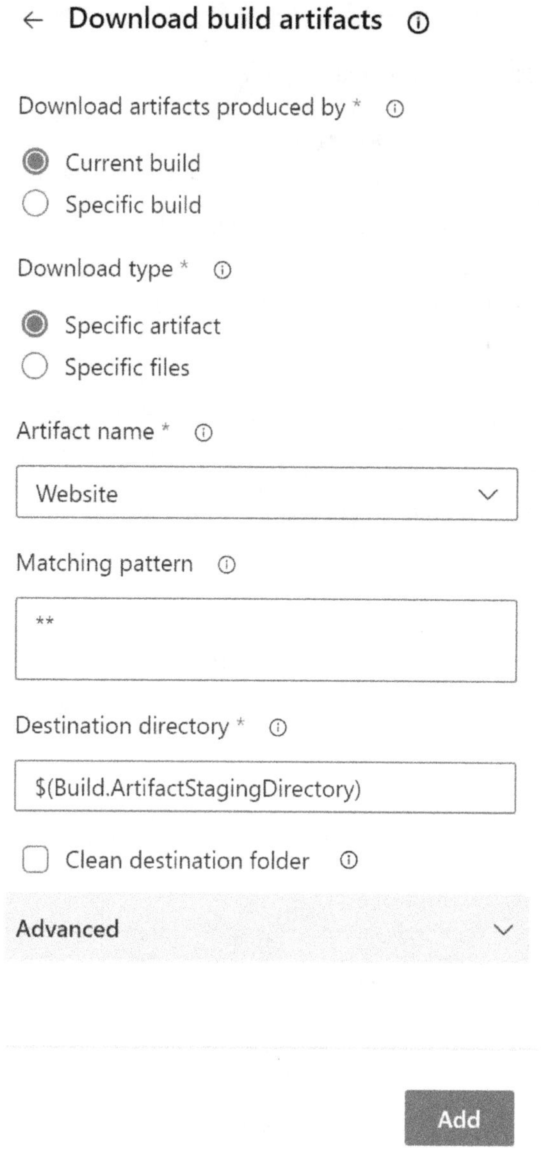
Download build artifacts
Download artifacts produced by *
Current build
Specific build
Download type *
Specific artifact
Specific files
Artifact name *
Website
Matching pattern
**
Destination directory *
$(Build.ArtifactStagingDirectory)
Clean destination folder
Advanced
Add

Figure 13.42 – Configuring the task

Here's the code added from this task in line 64:

```
- task: DownloadBuildArtifacts@0
  inputs:
    buildType: 'current'
    downloadType: 'single'
    artifactName: 'Website'
    downloadPath: '$(Build.ArtifactStagingDirectory)'
```

Here's the complete section added from the previous tasks:

```
57 - stage: Deploy
58   displayName: Deploy to an Azure Web App
59   jobs:
60   - job: Deploy
61     pool:
62       vmImage: 'windows-2019'
63     steps:
       Settings
64     - task: DownloadBuildArtifacts@1
65       inputs:
66         buildType: 'current'
67         downloadType: 'single'
68         artifactName: 'Website'
69         downloadPath: '$(Build.ArtifactStagingDirectory)'
70
       Settings
71     - task: AzureRmWebAppDeployment@4
72       inputs:
73         ConnectionType: 'AzureRM'
74         azureSubscription: 'SpringToys(d988cbee-043f-4c46-9a59-dedb2119e48c)'
75         appType: 'webApp'
76         WebAppName: 'SpringToysEShop'
77         packageForLinux: '$(Build.ArtifactStagingDirectory)/**/Web.zip'
```

Figure 13.43 – YAML pipeline updated

11. Now, save the pipeline and run it. Note we now have two stages, the **Build .Net Core Solution** stage and the **Deploy to an Azure Web App** stage:

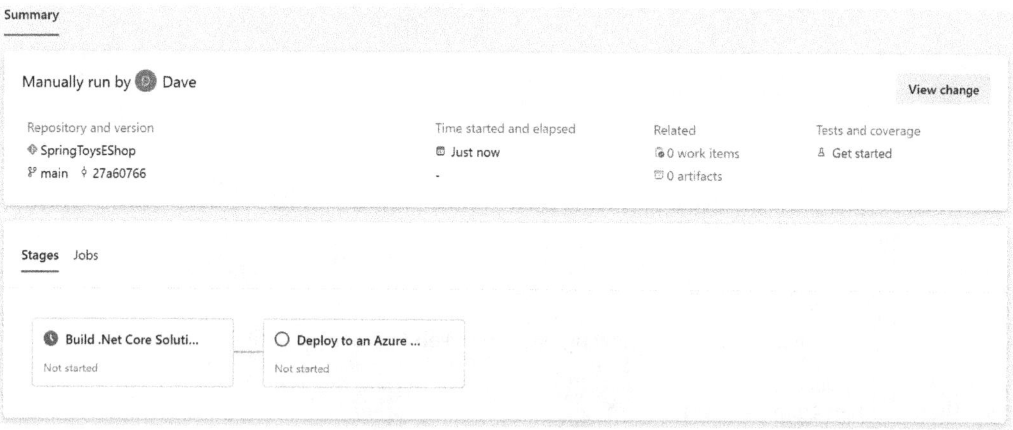
Summary
Manually run by Dave
View change
Repository and version
SpringToysEShop
main 27a60766
Time started and elapsed
Just now
Related
0 work items
0 artifacts
Tests and coverage
Get started
Stages Jobs
Build .Net Core Soluti...
Not started
Deploy to an Azure ...
Not started

Figure 13.44 – Multistage pipeline

After the build pipeline completes, you will see a message that the deploy pipeline needs permission to access a resource before this run can continue to deploy to an Azure Web App, as shown here:

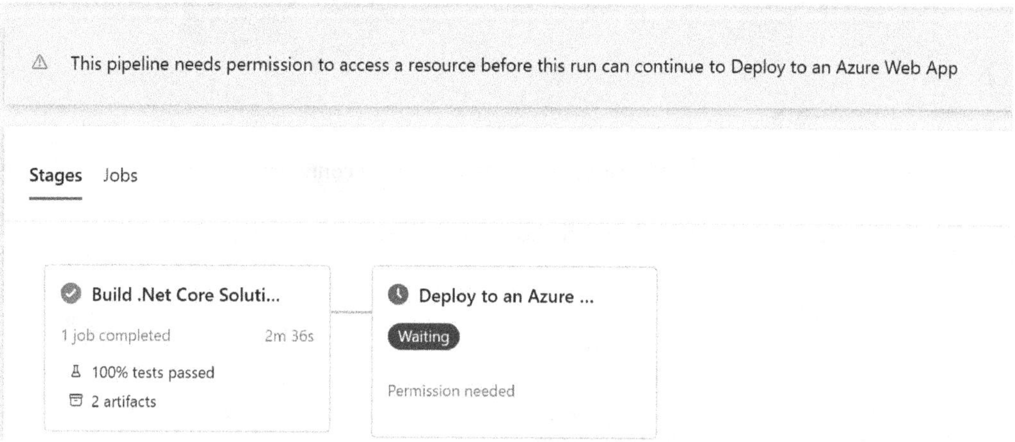
This pipeline needs permission to access a resource before this run can continue to Deploy to an Azure Web App
Stages Jobs
Build .Net Core Soluti...
1 job completed
2m 36s
100% tests passed
2 artifacts
Deploy to an Azure ...
Waiting
Permission needed

Figure 13.45 – Execution of YAML pipeline

12. Select the **View** option and then **Permit**:

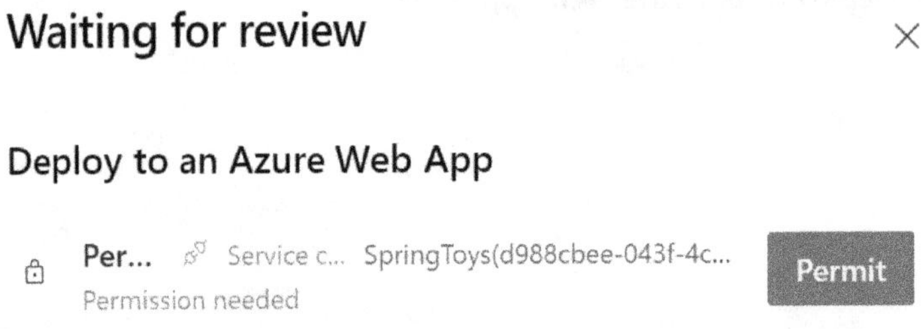
Waiting for review
Deploy to an Azure Web App
Per...
Service c...
SpringToys(d988cbee-043f-4c...
Permission needed
Permit

Figure 13.46 – Reviewing the Azure Web App deployment

13. Then confirm again, as shown here:

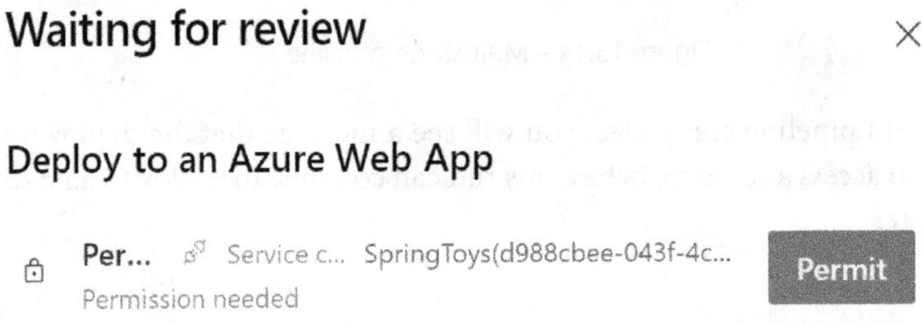
Waiting for review
Deploy to an Azure Web App
Per...
Service c...
SpringToys(d988cbee-043f-4c...
Permission needed
Permit

Figure 13.47 – Permit access to use the Service connection

After a few minutes, the deployment should be completed:

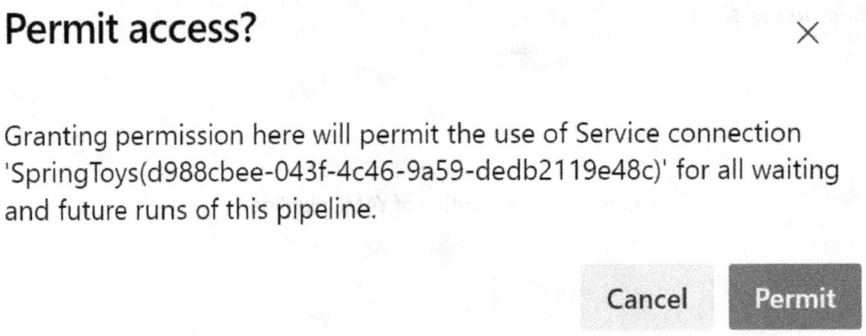
Permit access?
Granting permission here will permit the use of Service connection 'SpringToys(d988cbee-043f-4c46-9a59-dedb2119e48c)' for all waiting and future runs of this pipeline.
Cancel
Permit

Figure 13.48 – Stages Azure pipeline

14. Go to the **Azure Portal to the App Service** instance, and you should see the application running; in this demonstration, this is the URL of the website, `springtoyseshop.azurewebsites.net`, as shown here:

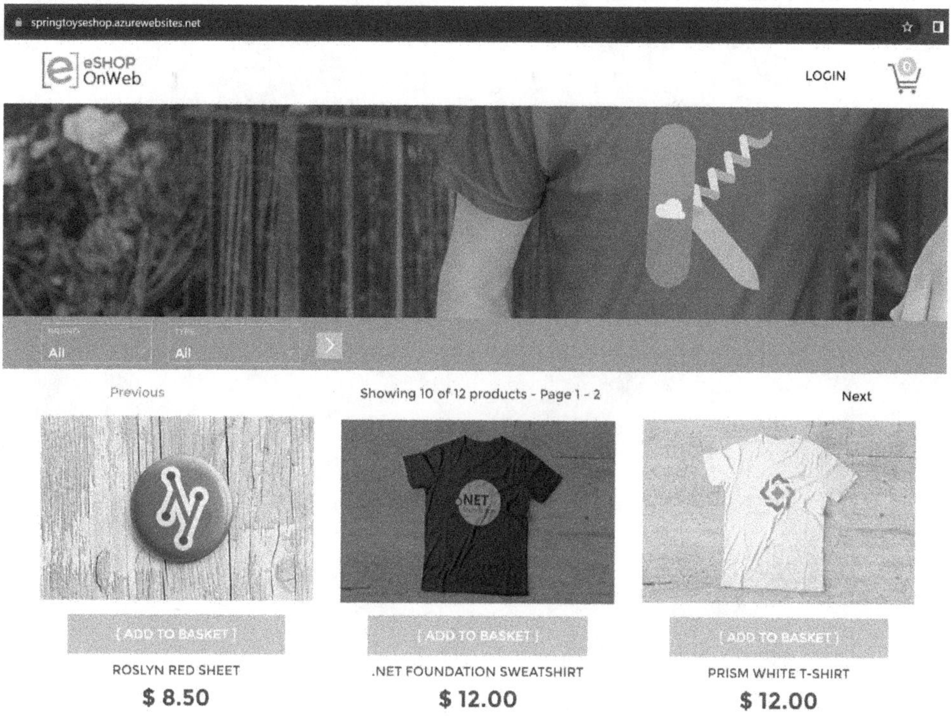
springtoyseshop.azurewebsites.net
eSHOP
OnWeb
LOGIN
All
All
Previous
Showing 10 of 12 products - Page 1 - 2
Next
[ADD TO BASKET]
[ADD TO BASKET]
[ADD TO BASKET]
ROSLYN RED SHEET
$ 8.50
.NET FOUNDATION SWEATSHIRT
$ 12.00
PRISM WHITE T-SHIRT
$ 12.00

Figure 13.49 – Deployment of application

Note that the application is successfully running, and we have completed the configuration of the CI/CD pipelines with YAML.

Summary

Incorporating CI and CD into your software development process can significantly improve your software delivery pipeline's quality, speed, and efficiency. With Azure Pipelines, organizations can take advantage of a comprehensive and scalable platform that offers a range of tools and features to automate and streamline the software delivery process.

By implementing CI/CD, teams can detect issues early in the development process and quickly resolve them, resulting in fewer bugs and more stable releases.

In contrast, CD automates the release of software modifications to production, freeing up teams' time and resources to concentrate on creating and delivering new functionality.

In the next chapter, we will share some of the best practices and tips from the field that we have learned along our journey to help you optimize your Azure cloud environment.

14

Tips from the Field

If you are responsible for managing and maintaining your company's cloud computing environment, such as Microsoft Azure, ensuring that the cloud environment is secure and available and performing optimally to make the most of its cloud computing resources is vital. Therefore, adopting best practices for implementing and managing an Azure cloud environment is essential.

In the following sections, we will share a few of the best practices and tips from the field we've learned along our journey to ensure your environment is secure, compliant, and scalable, while also reducing costs and minimizing the risk of outages and security breaches. Furthermore, the following best practices will help you stay up-to-date with the latest trends and technologies in the Azure cloud, ensuring that your environment remains efficient and effective over time.

Adopting Azure cloud best practices is vital for the health and security of your organization's cloud environment; it also helps you stay ahead of the competition and meet the evolving needs of your organization.

The first step is to establish a framework for managing access, permissions, and roles for different users of cloud services, monitoring usage and performance, and ensuring compliance with regulatory and security requirements through Azure governance.

Azure governance

Imagine that the SpringToys IT team was tasked with implementing Azure governance. They were assigned to ensure that all the resources and data stored in Azure were protected and managed efficiently. The IT team was excited but intimidated by the task's magnitude.

We reviewed Azure governance as a crucial aspect of working in a cloud environment as it helps ensure that the environment is secure and compliant. Governance will also enforce standardization so that resources and processes are consistent across the company. This can improve efficiency, reduce costs, and minimize the risk of errors and inconsistencies.

In companies with limited resources, forming a cloud core team that includes at least one expert in Azure governance is highly recommended. This dedicated expert can actively listen to the IT team's concerns and provide invaluable guidance on implementing Azure governance best practices.

Determining where to begin establishing a governance baseline foundation for your organization is crucial for success. A strong governance foundation sets the tone for a secure, compliant, and efficient cloud environment and lays the groundwork for future growth and success.

It's vital to assess your current governance needs and understand the specific regulations, standards, and requirements your organization must comply with. From there, you can develop a governance plan outlining the policies, processes, and technologies used to manage and maintain your cloud environment.

The plan should be reviewed regularly to ensure that it stays up-to-date with the latest trends and best practices and remains aligned with your organization's changing needs. By taking the time to establish a governance baseline foundation, you can ensure that your cloud environment is well managed and supports the success of your organization.

You can leverage the **Cloud Adoption Framework Governance Benchmark Tool** to assess where you are in terms of governance: `https://cafbaseline.com/`.

There are a few other tools you can utilize to define your governance strategy: `https://bit.ly/az-governance-tools`.

The first step is to define a clear organizational hierarchy. This means defining the different roles and responsibilities within your organization and ensuring that each role has the least access to those Azure resources that are required to do their job. This will help you to ensure that your Azure initiatives are aligned with your overall organizational goals. Initiatives can be tied to regulations such as NIST-SP 800-53, PCI DSS, HIPAA, HITRUST, and so on.

> **Remember**
>
> Azure initiatives are a set of Azure policies and include actions that you can use to manage and govern your Azure resources.

Conversely, Azure Policy can be utilized to enforce compliance and consistency across your Azure environment. With Azure Policy, the SpringToys IT team could set and enforce rules for the resources deployed in Azure, such as naming conventions, resource types, and locations.

Another best practice is to consider **Infrastructure as Code** (**IaC**) for Azure (for example, ARM templates or the Bicep language) to automate the deployment of resources and ensure that they are deployed consistently and in compliance with Azure governance policies. The SpringToys IT team would be thrilled to learn they could save time and reduce human error using ARM templates or Bicep files.

You can check the latest updates on Azure Bicep in the following GitHub repository: `https://github.com/azure/bicep`.

Also, in the following repositories, you will find practical examples to deploy workloads:

- `https://github.com/daveRendon/azinsider`
- `https://github.com/Azure/azure-quickstart-templates`

Lastly, it is recommended to leverage Azure landing zones with the Bicep language. You can follow the latest updates on the Azure landing zones in the following repository: `https://github.com/Azure/ALZ-Bicep`

An essential component of Azure governance is Azure Blueprints, in preview as of the time of writing. Consider using them to create templates that define the configuration of your Azure resources, and then use those templates to quickly and easily deploy new resources.

And remember to monitor and audit your Azure environment continuously. I can't stress enough the importance of monitoring what is happening in your Azure environment and tracking changes. By monitoring and auditing, the SpringToys IT team could quickly identify and resolve any issues before they become more significant problems.

Azure monitoring

When hosting workloads in the cloud, establishing clear ownership and accountability for your monitoring environment is of utmost importance. Consider creating a dedicated team in your organization responsible for managing and maintaining your monitoring solution and ensuring they have the necessary skills and resources to perform their duties effectively.

Establishing clear communication channels between the monitoring team and other stakeholders in your organization, such as developers and operations teams, is essential to ensure everyone knows the monitoring goals and objectives and can work together to achieve them.

Planning and designing your Azure Monitor environment is crucial for your cloud strategy. The monitoring goals and objectives should be clearly defined, and you should identify the data types you need to collect to achieve those goals. It's also important to consider your monitoring solution's performance and availability requirements and ensure it meets your organization's needs.

Cost plays a critical role in this scenario. Many organizations have configured Azure Monitor/Log Analytics to receive billions of log messages every day, often hitting the free 5 GB limit within a few days. As a result, customers are unpleasantly surprised by the significantly high Azure Monitor costs reflected in their bills.

Setting up effective alerts in Azure Monitor is one of the core recommendations to ensure that you can detect and respond to issues promptly. Define clear alert criteria, such as thresholds for specific metrics or events that can trigger an alert. And ensure that alerts are configured to provide relevant information and context, such as the severity of the issue and potential impact on users or applications.

Bear in mind that Azure Monitor provides a *Dynamic thresholds* feature, which enables you to set thresholds based on historical data and trends rather than static values. This can help to reduce the number of false alerts and ensure that alerts are triggered only when there is a genuine issue. Additionally, consider the Azure Monitor Action Group feature to define a set of actions that should be taken when an alert is triggered, such as sending an email or SMS message or running a specific script.

Consider the case in which SpringToys implements a monitoring strategy for its online store. Let's say the online store is experiencing a sudden surge in traffic due to a popular sale. As a result, the monitoring system generates an alert indicating that the website's response time has increased beyond a certain threshold. However, this alert may not be as handy as expected during such an event.

You can leverage alert suppression mechanisms to prevent the generation of unnecessary or duplicate alerts in a monitoring system. These mechanisms are available in Azure Monitor. You can use an alert suppression mechanism to temporarily suppress or delay the alert until the surge in traffic has subsided. This can be done by setting up a suppression rule based on the time of day or other conditions.

Consider using Azure Monitor suppression rules to prevent duplicate or unnecessary alerts from being triggered, such as when multiple alerts are triggered for the same issue. Prioritizing alerts based on their severity and potential impact is vital to address the critical problems first. By following these best practices, you can ensure that your Azure Monitor alerts are configured to enable you to detect and respond to issues quickly and effectively, while minimizing false positives and unnecessary alerts.

You can find the most relevant Azure security monitoring tools at the following URL: `https://bit.ly/az-security-monitoring-tools`.

Now, we will review how you can adopt best practices for managing access to resources using **Azure Active Directory** (**Azure AD**). This is crucial for ensuring the security and compliance of your cloud environment.

Identity management and protection

Role-based access control (**RBAC**) is a powerful feature in Azure AD that allows administrators to assign different levels of access to other users or groups based on their roles within the organization. Using features such as RBAC and **multi-factor authentication** (**MFA**), SpringToys can ensure that access to sensitive resources is granted only to authorized users, helping reduce the risk of security incidents, and complying with relevant regulations and standards.

Regularly reviewing and revoking access to resources is another critical best practice for managing access using Azure AD. This way, SpringToys can ensure that users have only the access they need and that security policies are being enforced effectively.

Remember that Azure AD and Microsoft Sentinel work together to provide a comprehensive security solution for organizations using the Microsoft Azure cloud platform.

By integrating Azure AD with Microsoft Sentinel, SpringToys can leverage the strengths of both solutions to enhance their overall security posture. Microsoft Sentinel can use the information about access control and authentication from Azure AD to provide a complete picture of security events and identify potential threats.

Azure AD and Microsoft Sentinel provide organizations with a powerful security solution that helps protect against security threats and meet compliance requirements.

Adopting best practices for using Microsoft Sentinel is essential for maintaining security and protecting the assets of SpringToys and ensuring that they are leveraging their full potential to protect against security threats and meet their compliance requirements. These best practices can include, for example, configuring alerts, creating playbooks, and conducting regular security assessments to identify any potential security gaps.

Regularly reviewing and updating the configurations and policies in Microsoft Sentinel is also essential to ensure that they are aligned with the changing needs of the organization and the evolving security landscape. The previously mentioned tips will help SpringToys improve its security posture and protect the organization from security threats.

Microsoft Sentinel and Azure networking services complement each other by helping organizations maintain the security and reliability of their cloud environment. By integrating Microsoft Sentinel with Azure networking services, SpringToys can leverage the strengths of both solutions to enhance its overall security posture.

Microsoft Sentinel can use information about network traffic and communication patterns from Azure networking services to detect potential security threats and respond accordingly. At the same time, Azure networking services can be configured and managed to support the security policies and requirements defined by Microsoft Sentinel.

Conversely, you can leverage Microsoft Sentinel and Azure networking services to provide a comprehensive security solution that helps protect against security threats and maintain a reliable and efficient cloud environment.

Azure networking services provide the backbone for communication between resources in the cloud, on-premises resources, or a hybrid or multi-cloud environment. Therefore, leveraging components such as service endpoints to secure access to resources or network security groups to control inbound and outbound traffic is crucial, as well as using ExpressRoute to provide a dedicated and secure connection to the Azure network.

Azure networking

Ensure that Azure networking services are configured to meet the organization's security, performance, and compliance requirements. Regularly reviewing and updating the configurations and policies of Azure networking services is also essential to ensure that they are aligned with the changing needs of the organization and the evolving security landscape.

It is crucial to emphasize that shared networking resources such as ExpressRoute circuits and VPN gateways should only be accessible to a restricted group of privileged users – specifically, network administrators.

Azure networking services are one of the core aspects when working with containerized applications. Both are essential components of a cloud environment, and they work together to ensure the reliability, scalability, and security of an organization's cloud environment.

Network security is crucial to securing any cloud deployment, including those in Azure. Azure provides many network security features to help organizations secure workloads, such as virtual networks, network security groups, and Azure Firewall. However, it's essential to implement best practices when designing and configuring your Azure network to ensure your workloads are secure and protected from potential threats.

One common best practice for securing your Azure network is implementing network segmentation. Network segmentation involves dividing your network into smaller segments or subnets, each with its own set of security controls. This helps to limit the spread of potential threats by containing them within specific network segments.

In addition, organizations can leverage network security groups to apply firewall rules to specific subnets or virtual machines, further enhancing the security of your network.

Another best practice is implementing Azure Firewall to control outbound traffic from virtual networks. Azure Firewall allows you to create and enforce outbound security rules based on the source IP address, destination IP address, and port number, helping to prevent unauthorized access to your resources.

By integrating Azure networking services with Azure containerized applications, organizations can leverage the strengths of both components to build a reliable, secure, and scalable cloud environment. Azure networking services provide the backbone for communication between resources in the cloud, while Azure containerized applications offer a flexible and cost-effective way to deploy and manage applications.

Now, let's review recommendations when working with Azure containers.

Azure containers

Azure containers provide a flexible and cost-effective way to deploy and manage applications in the cloud. They must be configured and managed to support the organization's needs.

You can use Azure Container Registry to store and manage container images, leverage Azure Kubernetes Service to manage and orchestrate containers, and utilize Azure Security Center to monitor and protect containers from security threats.

When using Azure Container Instances, it is essential to note that they are stateless by default. Any state stored within the container will be lost if it is restarted, crashes, or stops. Therefore, to maintain a state beyond the container's lifetime, it is recommended to mount a volume from an external store.

One option for achieving this is to mount an Azure file share created with Azure Files, as it provides fully managed file shares hosted in Azure Storage that can be accessed via the **Server Message Block (SMB)** protocol. By using an Azure file share with Azure Container Instances, you can benefit from file-sharing capabilities similar to those available when using an Azure file share with Azure virtual machines.

Another best practice when working with Azure Containers is to leverage a liveness probe to ensure your container runs correctly and trigger an automatic restart if necessary.

A liveness probe is a diagnostic tool used in container orchestration systems such as Kubernetes and Azure Container Instances to determine whether a container is still running and can respond to requests. A liveness probe can check for a wide variety of issues, such as deadlocks, resource starvation, and network connectivity problems. Suppose a liveness probe determines that a container is not responsive. In that case, the container can be restarted or replaced automatically by the orchestration system, helping to maintain the availability of the application.

To implement a liveness probe in Azure Container Instances, you can use a simple HTTP endpoint that returns a status code indicating whether the container is healthy or not. The container will be restarted if the endpoint returns a non-successful status code. You can also configure the liveness probe to check the container's filesystem or execute a custom command to determine the container's health. Using a liveness probe ensures that your containerized application is always available and responsive to requests.

As with any container-based solution, security risks need to be considered. The Azure Container Instances Security Baseline guides how to secure ACI deployments by following best practices and implementing recommended security controls.

The baseline covers a range of security topics, including network security, access management, logging and monitoring, and data protection. For example, it recommends restricting network access to ACI by using a virtual network with appropriate network security groups.

It also recommends using RBAC to manage access to Azure Container Instances resources and implementing MFA for administrative access. Additionally, the baseline provides guidance on configuring Azure Container Instances logging and monitoring to detect and respond to security events and encrypt sensitive data in transit and at rest. By implementing the Azure Container Instances Security Baseline, organizations can help to ensure that their container-based workloads in Azure are secure and protected against potential threats.

Summary

This chapter summarized the top best practices you could implement in your organization. It started with the importance of Azure governance in ensuring a secure, compliant, and efficient cloud environment. It highlighted the need for a clear organizational hierarchy, defining organizational roles and responsibilities, and using Azure initiatives and policies to manage and govern Azure resources. We suggested leveraging Azure landing zones with the Bicep language and Azure Blueprints and continuously monitoring and auditing the Azure environment. We recommended using tools such as the Cloud Adoption Framework Governance Benchmark Tool and IaC for Azure, such as ARM templates or the Bicep language, to automate the deployment of resources and ensure compliance with Azure governance policies.

We also highlighted the importance of creating a dedicated team to manage and maintain Azure's monitoring solution. We emphasized the significance of clear communication channels between the monitoring team and other organizational stakeholders, such as developers and operations teams, to ensure everyone knows the monitoring goals and objectives and can work together to achieve them. We provided some best practices for planning, designing, and setting up effective alerts in Azure Monitor, such as defining clear alert criteria, considering the Dynamic Thresholds feature, and using Azure Monitor suppression rules to prevent duplicate or unnecessary alerts.

Moving on, we shared best practices for managing access to resources using Azure AD and Microsoft Sentinel. By implementing RBAC and MFA, organizations such as SpringToys can ensure that authorized users only access sensitive resources, reducing the risk of security incidents and complying with regulations. Regularly reviewing and revoking access to resources is another crucial practice for enforcing security policies effectively. Integrating Azure AD with Microsoft Sentinel can provide a comprehensive security solution, allowing organizations to detect and respond to potential threats. Additionally, integrating Microsoft Sentinel with Azure networking services can enhance the cloud environment's overall security posture and reliability. Regularly reviewing and updating configurations and policies in Microsoft Sentinel is also essential to maintain alignment with the organization's changing needs and evolving security landscape.

We also emphasized configuring and regularly reviewing Azure networking services to meet an organization's security, performance, and compliance requirements. Network security is a crucial aspect of cloud deployment, and Azure provides many features, such as virtual networks, network security groups, and Azure Firewall, to help secure workloads. Network segmentation, security groups, and Azure Firewall are highlighted as best practices to secure an organization's Azure network. Implementing these best practices can limit the spread of potential threats, apply firewall rules to specific subnets or virtual machines, and control outbound traffic to prevent unauthorized access to resources.

Lastly, we reviewed how Azure containers provide a cost-effective and flexible way to deploy and manage applications in the cloud. Azure offers various tools such as Azure Container Registry, Azure Kubernetes Service, and Azure Security Center to store, manage, and secure container images. However, it's essential to consider the stateless nature of Azure Container Instances, which requires external storage to maintain any state beyond the container's lifetime. Leveraging a liveness probe was also recommended to ensure container availability and request responsiveness. To mitigate security risks, organizations can follow the Azure Container Instances Security Baseline guidelines, which cover topics such as network security, access management, logging and monitoring, and data protection. By implementing these best practices, organizations can ensure that their container-based workloads in Azure are secure and protected against potential threats.

Congratulations on reaching the final pages of this book! We are thrilled to extend our heartfelt appreciation for your dedication and perseverance in delving into the intricate world of Azure and mastering its architectural concepts.

As you continue on your path of success, we eagerly await your thoughts and reflections on your learning journey with this book. We invite you to provide your feedback and comments on the Amazon web page for the book.

Once again, congratulations on this remarkable milestone! We wish you continued success in all your future endeavors and eagerly await your thoughts on your learning journey with this book.

Index

A

D

E

F

G

H

N

O

P

R

S

T

U

V

W

Y

Z

Other Books You May Enjoy

If you enjoyed this book, you may be interested in these other books by Packt:

<packt>
FinOps Handbook for Microsoft Azure
Empowering teams to optimize their Azure cloud spend with FinOps best practices
MAULIK SONI

FinOps Handbook for Microsoft Azure

Maulik Soni

ISBN: 978-1-80181-016-6

- Get the grip of all the activities of FinOps phases for Microsoft Azure
- Understand architectural patterns for interruptible workload on Spot VMs
- Optimize savings with Reservations, Savings Plans, Spot VMs
- Analyze waste with customizable pre-built workbooks
- Write an effective financial business case for savings
- Apply your learning to three real-world case studies
- Forecast cloud spend, set budgets, and track accurately

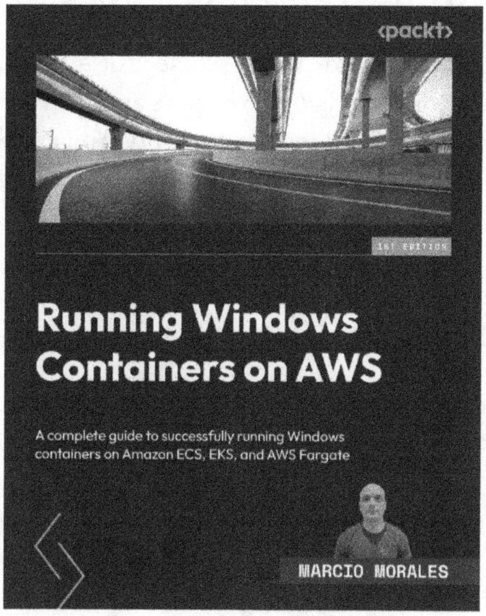

<packt>
1ST EDITION
Running Windows Containers on AWS
A complete guide to successfully running Windows containers on Amazon ECS, EKS, and AWS Fargate
MARCIO MORALES

Running Windows Containers on AWS

Marcio Morales

ISBN: 978-1-80461-413-6

- Get acquainted with Windows container basics
- Run and manage Windows containers on Amazon ECS, EKS, and AWS Fargate
- Effectively monitor and centralize logs from Windows containers
- Properly maintain Windows hosts and keep container images up to date
- Manage ephemeral Windows hosts to reduce operational overhead
- Work with the container image cache to speed up the container's boot time

Packt is searching for authors like you

If you're interested in becoming an author for Packt, please visit authors.packtpub.com and apply today. We have worked with thousands of developers and tech professionals, just like you, to help them share their insight with the global tech community. You can make a general application, apply for a specific hot topic that we are recruiting an author for, or submit your own idea.

Share Your Thoughts

Now you've finished *Azure Architecture Explained*, we'd love to hear your thoughts! Scan the QR code below to go straight to the Amazon review page for this book and share your feedback or leave a review on the site that you purchased it from.

https://packt.link/r/1837634815

Your review is important to us and the tech community and will help us make sure we're delivering excellent quality content.

Download a free PDF copy of this book

Thanks for purchasing this book!

Do you like to read on the go but are unable to carry your print books everywhere?

Is your eBook purchase not compatible with the device of your choice?

Don't worry, now with every Packt book you get a DRM-free PDF version of that book at no cost.

Read anywhere, any place, on any device. Search, copy, and paste code from your favorite technical books directly into your application.

The perks don't stop there, you can get exclusive access to discounts, newsletters, and great free content in your inbox daily

Follow these simple steps to get the benefits:

1. Scan the QR code or visit the link below

`https://packt.link/free-ebook/9781837634811`

2. Submit your proof of purchase
3. That's it! We'll send your free PDF and other benefits to your email directly

www.ingramcontent.com/pod-product-compliance
Lightning Source LLC
Chambersburg PA
CBHW060646060326
40690CB00020B/4539